Birkhäuser

Control Engineering

More information about this series at http://www.springer.com/series/4988

Harry G. Kwatny · Karen Miu-Miller

Power System Dynamics and Control

Birkhäuser

Harry G. Kwatny
Department of Mechanical Engineering
and Mechanics
Drexel University
Philadelphia, PA
USA

Karen Miu-Miller
Department of Electrical and Computer
Engineering
Drexel University
Philadelphia, PA
USA

ISSN 2373-7719 ISSN 2373-7727 (electronic)
Control Engineering
ISBN 978-0-8176-4673-8 ISBN 978-0-8176-4674-5 (eBook)
DOI 10.1007/978-0-8176-4674-5

Library of Congress Control Number: 2016940325

Mathematics Subject Classification (2010): 78A25, 83C50, 70H03, 70G10, 34H05, 34H20, 37C75, 37G10, 49L20, 93C05, 93C10, 93C30

Printed on acid-free paper

This book is published under the trade name Birkhäuser
The registered company is Springer Science+Business Media LLC New York
(www.birkhauser-science.com)

To Mim,

The engineer is concerned to travel from the abstract to the concrete. He begins with an idea and ends with an object. He journeys from theory to practice. The scientist's job is the precise opposite. He explores nature with his telescopes or microscopes, or more sophisticated techniques, and feeds into a computer what he finds or sees in an attempt to define mathematically its significance and relationships. He travels from the real to the symbolic, from the concrete to the abstract. The scientist and the engineer are the mirror image of each other.

—Gordon Lindsay Glegg, "The Development of Design," 1981

Preface

"... I repeat that you must lay aside all prejudice on both sides, and neither believe nor reject any thing because any other person, or description of persons have rejected or believed it. Your own reason is the only oracle given you ..."

—Thomas Jefferson, letter to his nephew, Peter Carr, August 10, 1787.

Electric power systems are a vital part of modern life. We are reminded of this whenever there is a significant blackout in any of the world's developed economies. Even in countries that have a minimal reliance on electricity, we see the powerful impact of supply disruptions on the quality of daily life. The development of new technologies, including new forms of electric power generation and storage and new mechanisms to regulate the flow of power, is having an enormous impact on power systems large and small. New and more widespread applications of electric power such as vehicle propulsion and distributed and remote generation continue to expand.

Society has become increasingly reliant on electric power and consequently more vulnerable to service breakdowns. Recent events worldwide have brought with them the startling realization that civil infrastructure systems are vulnerable to assault by small organized groups with malicious intent. Whereas power systems have traditionally been designed with a focus on protecting them from routine component failures and atypical user demand, we now also confront the fact that deliberat attacks intended to cause maximum disruption are a real possibility.

In response to this changing environment, new concepts and tools have emerged that address many of the issues facing power system operation today. This book is aimed at introducing these ideas to practicing power system engineers, control system engineers interested in power systems, and graduate students in these areas.

This book is intended to provide sufficient information about power system modeling and behavior, so that a control engineer without a background in power systems can think coherently about power system control. But it is not intended to duplicate the material that would be found in a traditional power system course. Similarly, the control system material is intended to provide a power system engineer with sufficient information about new and emerging control-theoretic ideas to encourage their application to power systems. But the material covered is far from standard fare in a control system plan of study.

This book is focused on two main themes: the nonlinear dynamics of power systems, and the discrete event mechanisms that are a dominating factor in power system operations. Stability, voltage collapse, power transfer limits, power flow oscillations, and other important aspects of power system behavior have been elucidated through the application of advances in dynamical systems and nonlinear control theory. The interaction of discrete protection systems and control actions such as load shedding with the nonlinear continuous dynamics of the system are central to the behavior of power systems, especially during emergencies. New methods of modeling, analysis, and design of such *hybrid* systems continue to be a central theme of present-day control systems research. In this work, we examine these ideas and consider how they can be applied to improve our understanding of power system behavior and how they can be used to design better control systems.

This book is supplemented by a software (Mathematica) package that will enable the reader to work out nontrivial examples and problems. Also available is a set of tutorial Mathematica notebooks that provide detailed solutions of the worked examples in the text. Besides Mathematica, simulations are carried out using Simulink with Stateflow. These can all be obtained at the Web site http://www.pages.drexel.edu/~hgk22/.

The authors are fortunate to have had the opportunity to participate in research aimed at improving power system operations. Support over many years from the National Science Foundation, Department of Energy, Middle Atlantic Power Research Council, Electric Power Research Institute, and the Office of Naval Research is gratefully acknowledged. Many individuals have influenced our view of power systems. We are indebted to all of them.

Contents

Chapter 1
Introduction

"There is scarcely a subject that cannot be mathematically treated and the effects calculated or the results determined beforehand from the available theoretical and practical data."
—Nikola Tesla, "The Electrical Experimenter," Volume VI, No. 70, February, 1919

1.1 Goals and Motivation

Electric power systems continue to evolve as new technologies for generation, storage, delivery, and consumption take form. Accordingly, system management practices are continually reexamined in terms of providing an efficient and reliable supply. But most importantly, recent recognition that "severe weather is the leading cause of power outages in the United States"[1] highlights the importance of an accelerated focus on improving the *resilience* of the power grid to withstand weather-related disturbances and to reduce the time it takes to restore service following an outage. The report [160], along with an earlier report: A Policy Framework for the 21st Century Grid: Enabling Our Secure Energy Future [175], has fueled an already ongoing discussion about grid modernization.

Modern communication and computational technologies that enable innovative *smart grid* [153] capabilities and the increasing penetration of distributed generation [135] bring opportunity and complexity to the emerging power network. The application of computational tools for real-time assessment and management of the network requires computationally efficient models and a sound understanding of how power systems work along with a clear, quantifiable notion of performance. Real-time decision-making tools, including optimization that involves dynamical constraints as well as logical constraints, are necessary, as are estimation tools for situational assessment.

[1] Excerpt from the report [160] issued by the Executive Office of the President.

H.G. Kwatny and K. Miu-Miller, *Power System Dynamics and Control*,
Control Engineering, DOI 10.1007/978-0-8176-4674-5_1

The goal of this book is to lay out a consistent modeling and analysis framework that provides tools for building efficient power system models, provide the essential concepts for the analysis of static and dynamic network stability, review the structure and design of basic voltage and load-frequency regulators, and provide an introduction to power system optimal control with reliability constraints.

1.2 Content

A key premise of this book is that the design of successful control systems requires a deep understanding of the processes to be controlled. Accordingly, the technical discussion begins with the physical foundations of electricity and magnetism. Chapter 2 provides a concise review of the basics of electricity and magnetism. It is fitting that James Clerk Maxwell plays a major role in this discussion, as his extraordinary contributions include the first major work on control theory [143].

The material of Chapter 2 forms a basis for the discussion on electric circuits and devices in Chapter 3. The approach to network modeling is based on a form of Lagrange equations referred to as the *generalized* Lagrange equations. Lagrange methods are particularly useful when assembling models of devices that include both electrical and mechanical elements such as electrical machines. But beyond that, these techniques are linked to energy concepts that provide tools for stability analysis. Other important aspects associated with symmetry and first integrals also have implications with respect to power networks.

Chapter 4 is focused on AC power systems. The basic concepts are reviewed, and models for the basic components, including transformers, transmission lines, and machines, are derived. The load flow equations are assembled and explained, and classical simplified models for balanced networks are constructed.

Chapters 5 and 6 address power system dynamics. Chapter 5 is focused on power systems as ordinary differential equations (ODEs). The basic properties of ODEs are reviewed, and then, stability analysis via Lyapunov techniques is considered. Chapter 6 treats the more complex, differential-algebraic equation (DAE) model of a power network. The focus turns to bifurcation analysis and the behavior of a networks as it approaches voltage instability.

Classical problems of power system control are the subject of Chapter 7. Two classic control problems are addressed in this chapter—voltage regulation and load frequency control. The classic voltage control system uses an excitation system to adjust the field voltage in order to regulate the generator terminal bus voltage. The Load frequency control (AGC) problem within a single area is also treated in the classical way. Load frequency control has two goals—regulate the electrical frequency supplied by each generator to synchronous frequency and insure that the total real power supplied by the generators is distributed among them in specified proportions. Automatic generation control (AGC) is intended to regulate frequency as well as power interchanges between multiple interconnected control areas. The AGC problem is also considered in this chapter. Economic dispatch, the allocation of generation among plants within an area in order to achieve minimal fuel consumption, is also discussed from the point of view of coordinating dispatch with AGC.

Chapter 8 addresses a class of control problems that involve operation in highly nonlinear regimes where failure events cause abrupt changes in the controlled system which may require a corresponding discrete change in control strategy. This material deals with a class of control problems which is relatively new and particularly relevant to modern power system concerns about efficiency, reliability, and resilience.

Chapter 2
Basics of Electricity and Magnetism

"My direct path to the special theory of relativity was mainly determined by the conviction that the electromotive force induced in a conductor moving in a magnetic field is nothing more than an electric field."
—Albert Einstein, message to the centennial of Albert Michelson's birth, December 19, 1952.

2.1 Introduction

This chapter provides a succinct review of the essential physics of electricity and magnetism that forms the basis for understanding how electric power systems work. Later chapters will use this foundational material to build models of power system components and systems. Electric fields, magnetic fields, and Maxwell's equations are the topics of the three sections of this chapter. Examples are given that illustrate the basic characteristics of core electrical components and electromechanical devices.

2.2 The Electric Field

Coulomb showed that two like point charges repel each other.[1] In fact, the force between two stationary particles of charge q_1 and q_2 a distance r apart in free space is given by the formula

$$F = k\frac{q_1 q_2}{r^2} \tag{2.1}$$

This is *Coulomb's law*. A positive F is repulsion and a negative F is attraction. In SI units, the unit of charge is the coulomb, the unit of distance is the meter, and the

[1] Material in this section can be found more fully discussed in any book on theoretical physics, Example [162].

H.G. Kwatny and K. Miu-Miller, *Power System Dynamics and Control*, Control Engineering, DOI 10.1007/978-0-8176-4674-5_2

unit of force is the newton. In this case, $k = 1/4\pi\varepsilon_0$, where ε_0 is the permittivity of free space, $\varepsilon_0 = 8.85 \times 10^{-12}\ C^2/N \cdot m^2$.

A region in space contains an *electrical field* if a charge fixed in it experiences a force. The *electric intensity* **E** at a point in the region is the force exerted on a unit positive charge. Thus, the force **F** on a particle of charge q at a point of electrical intensity **E** is

$$\mathbf{F} = q\mathbf{E}$$

Since the force between two stationary charges depends only on the distance between them, it follows that the electrical intensity is derivable from a potential function; i.e., it is the gradient of a scalar potential function $\varphi(x, y, z)$. The work done in moving a unit charge against the field by an amount $d\mathbf{r}$ is the increase in potential $d\varphi$, $d\varphi = \mathbf{E} \cdot d\mathbf{r}$, or

$$\mathbf{E} = -\nabla\varphi$$

Example 2.1 We will compute the potential V produced by a point charge q. At any distance r from q, from (2.1), we have

$$\mathbf{E} = k\frac{q}{r^2}\frac{\mathbf{r}}{r}$$

so that

$$dV = -k\frac{q}{r^3}\mathbf{r} \cdot d\mathbf{r} = -k\frac{q\,dr}{r^2}$$

Integrating, we obtain

$$V(r) = V_0 + k\frac{q}{r}$$

If we impose the boundary condition $V(0) = 0$, then $V_0 = 0$.

We will be interested in material media that can be classified as *conductors* or *dielectrics*. Conductors contain free electrons which are under the influence of an electric field can flow freely through conductors. So, conductors admit the flow of current. A dielectric is an electrical insulator in that it is highly resistant to current flow. An electrical field applied to a dielectric does cause motion of charges within it. The resultant motion or current is composed of two parts, a negligible conduction current and a displacement current. The neutrally charged atoms or molecules that make up the dielectric typically have the center of positive charge and the center of negative charge displaced. Such an arrangement constitutes an *electrical dipole*. Even if the charge centers are not displaced, the application of an electric field generally induces a displacement. If the dipole consists of a charge $-q$ and a charge q separated by a distance l, then we associate it with a *dipole moment* **p**, a vector of magnitude ql (coulomb meter^2) and direction pointing from $-q$ to q. An applied electric field imposes a force and a torque on the dipole. If the imposed field is constant over the domain of the dipole (which is ordinarily the case at the microscopic scale), then the net force acting on a dipole is negligible and the torque is given by

$$\tau = \mathbf{p} \times \mathbf{E}$$

At the microscopic level, the stretching and twisting of the dipoles that occurs under the influence of the applied electric field alters the potential function defining the electric field within the dielectric, thereby modifying the field. The change in the field is denoted **P**, called the *polarization* and the modified field is denoted **D**, called the *(displacement) electric flux density*, so we have

$$\mathbf{D} = \mathbf{E} + \mathbf{P} \tag{2.2}$$

In isotropic media, the polarization is proportional to the electric field intensity, $\mathbf{P} = \chi \mathbf{E}$ and as a consequence $\mathbf{D} = \varepsilon \mathbf{E}$. χ is known as the *electric susceptibility* and ε as the *permittivity* of the dielectric. In free space $\mathbf{P} = \mathbf{0}$.

2.3 The Magnetic Field

The movement of electrical charge gives rise to a force field known as a *magnetic field*. The magnetic field is characterized by a vector field **B** known as the *magnetic flux density* which has SI units weber/meter^2, equivalently, volt-second/meter^2. Such fields arise on a macroscopic level, as when current flows through a wire, or on an atomic scale, as electron spin in an atom. Consider that a current **I** flows along a differential element $d\mathbf{l}$, then the differential magnetic field produced is given by the Biot–Savart law

$$d\mathbf{B} = \frac{\mu_0}{4\pi} \frac{I d\mathbf{l} \times (\mathbf{r}/r)}{r^2} \tag{2.3}$$

where μ_0 is the permeability of free space, I is the current, $\mathbf{r}$ is the displacement vector from the current element to the field point.

Example 2.2 Current in a Thin Wire. We can use the Biot–Savart law to compute the field produced by a constant current i flowing in a long thin straight wire as shown in Figure 2.1.

In accordance with Figure 2.1, the Biot–Savart law (2.3) can be written

$$\mathbf{B} = \frac{\mu_0 i}{4\pi} \oint \frac{d\mathbf{z} \times \mathbf{r}/r}{r^2}$$

We can express dz and r in terms of ρ and α: $dz = \rho \sec^2 \alpha$, $r = \rho \sec \alpha$, to obtain

$$\mathbf{B} = \frac{\mu_0 i}{4\pi} \int_{-\pi/2}^{\pi/2} \frac{\cos \alpha \, d\alpha}{\rho} \mathbf{u}_\varphi = \frac{\mu_0}{2\pi} \frac{i}{\rho} \mathbf{u}_\varphi$$

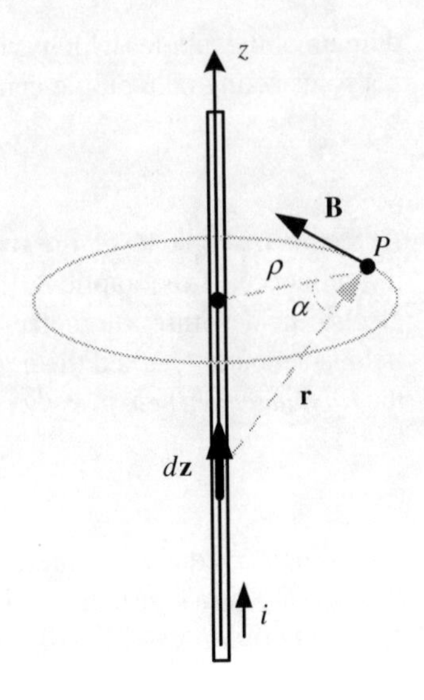

Fig. 2.1 An infinite wire carrying constant current

Some materials have an atomic structure in which the electron spins are aligned, thereby giving rise to a *permanent magnet.* The bar magnet is a familiar example of a magnetic dipole. Magnetic dipoles on atomic or molecular scale are the basic building blocks of all magnetic materials. We associate with a magnetic dipole its *magnetic dipole moment* **m**, with units ampere-meter^2. When a magnetic dipole is placed in a magnetic field, it experiences a moment

$$\tau = \mathbf{m} \times \mathbf{B}$$

When a magnetic field is applied, the atomic scale magnetic dipoles in the material tend to align with it. Materials can be classified in accordance with degree of alignment produced by the field. Diamagnetic and paramagnetic materials have relatively small alignment, whereas ferromagnetic materials have virtually complete alignment. The dipole alignment in a material gives rise to a macroscopic dipole moment per unit volume, **M**, called the *magnetization* of the material. So, for example, if each atom in a material has a dipole moment **m** and there are N atoms per unit volume, then $\mathbf{M} = N\mathbf{m}$. $\mathbf{M} = 0$ in free space.

The *magnetic field intensity*, **H**, is defined by the relation

$$\frac{1}{\mu_0}\mathbf{B} = \mathbf{H} + \mathbf{M} \tag{2.4}$$

Notice the similarity of (2.4) with (2.2). For linear materials, the relationship between **H** and **M** is

$$\mathbf{M} = \chi_m \mathbf{H},$$

where χ_m is the *magnetic susceptibility* of the material. For paramagnetic materials, χ_m is positive and for diamagnetic materials it is negative. Ferromagnetic materials are generally not linear. For linear materials, we have

$$\mathbf{B} = \mu_0 (\mathbf{H} + \mathbf{M}) = \mu_0 (1 + \chi_m) \mathbf{H} = \mu \mathbf{H},$$

where $\mu = \mu_0 (1 + \chi_m)$ is the *permeability* of the material. The ratio μ/μ_0, $\mu_r = (1 + \chi_m)$, is called the *relative permeability*.

2.4 Maxwell's Equations

We will summarize some basic concepts about magnetic and electric fields.

Table 2.1 defines the symbols used in the following discussion.

Four basic equations, called Maxwell's equations, describe the behavior of electromagnetic fields. These include the following:

1. Gauss law describes how charge produces an electrical field,

$$\nabla \cdot \mathbf{D} = \rho \text{ or } \int_S \mathbf{D} \cdot d\mathbf{s} = \int_V \rho dv$$

This implies that the integral of the electrical flux density over a surface S that encloses a volume V must equal the total charge contained in V,

Table 2.1 Electromagnetic Fields Nomenclature

Symbol	Quantity	Units
$\mathbf{E}$	electric field intensity	volt per meter
$\mathbf{D}$	electric flux density	coulomb per meter^2
$\mathbf{H}$	magnetic field intensity	ampere per meter
$\mathbf{B}$	magnetic flux density	weber per meter^2
Φ	magnetic flux	weber
ρ	electric charge density	coulomb per meter^3
$\mathbf{J}$	current density	ampere per meter^2
$d\mathbf{s}$	differential vector element of surface area with direction perpendicular to surface S	meter^2
dv	differential element of volume enclosed by surface S	meter^3
$d\mathbf{l}$	differential vector element of path length tangential to contour C enclosing surface S	meter

$$\int_S \mathbf{D} \cdot d\mathbf{s} = Q_{enclosed}$$

2. Gauss law for magnetism asserts the absence of magnetic sources

$$\nabla \cdot \mathbf{B} = 0 \text{ or } \int_S \mathbf{B} \cdot d\mathbf{s} = 0,$$

where the surface S again encloses a volume V. The magnetic flux density **B** is a vector field defined in three-dimensional space. The integral curves of **B** are the "magnetic flux lines" or "magnetic field lines." The integral of magnetic flux density over any closed surface must be zero implies that these lines are closed loops.
The magnetic flux through area S bounded by a closed curve C, Φ, is defined as

$$\Phi = \int_S \mathbf{B} \cdot d\mathbf{s}$$

3. The Maxwell–Faraday equation describes how changing magnetic fields produce electrical fields

$$\nabla \times \mathbf{E} = -\frac{\partial \mathbf{B}}{\partial t} \text{ or } \oint_C \mathbf{E} \cdot d\mathbf{l} = -\frac{d}{dt} \int_S \mathbf{B} \cdot d\mathbf{s},$$

where S is a surface bounded by the closed curve C. The equation shows how an electric field is produced by varying the magnetic flux passing through a given cross-sectional area. As will be seen, this is the fundamental principle underlying the operation of electric motors and generators.
4. The Ampère–Maxwell law describes how the magnetic fields are produced by currents and changing electrical fields

$$\nabla \times \mathbf{H} = \mathbf{J} + \frac{\partial \mathbf{D}}{\partial t} \text{ or } \oint_C \mathbf{H} \cdot d\mathbf{l} = \int_S \mathbf{J} \cdot d\mathbf{s} + \frac{d}{dt} \int_S \mathbf{D} \cdot d\mathbf{s},$$

where C denotes the closed edge (or boundary) of an open surface S. Define the encircled current

$$I_{encircled} = \int_S \mathbf{J} \cdot d\mathbf{s}$$

If **D** is very slowly varying, then the Ampère–Maxwell law reduces to Ampère's law

$$\oint_C \mathbf{H} \cdot d\mathbf{l} = I_{encircled}$$

Another important result is the *Lorentz force equation* which describes the force $\mathbf{F}$ acting on a particle of charge q moving through an electromagnetic field with velocity $\mathbf{v}$

$$\mathbf{F} = q\,(\mathbf{E} + \mathbf{v} \times \mathbf{B})$$

In addition, *Ohm's Law* states that the current density in a conductor is proportional to the electric field:

$$\mathbf{J} = \sigma \mathbf{E}$$

where σ is the *conductivity* with units ohms per meter.

Remark 2.3 (*Scalar and Vector Potentials*) Gauss law for magnetism states that the divergence of the magnetic field vanishes, thereby implying that $\mathbf{B}$ can be expressed

$$\mathbf{B} = \nabla \times \mathbf{A}, \tag{2.5}$$

where $\mathbf{A}$ is some magnetic vector potential. Then, Faraday's law can be written

$$\nabla \times \left[\mathbf{E} + \frac{\partial \mathbf{A}}{\partial t}\right] = 0$$

But the fact that the curl of a vector vanishes implies that the vector can be expressed as the gradient of a scalar potential, φ. Hence,

$$\mathbf{E} = -\nabla\varphi - \frac{\partial \mathbf{A}}{\partial t} \tag{2.6}$$

Further substitutions in Maxwell's equations and some algebra lead to partial differential equations for $\mathbf{A}$ and φ [167]:

$$\mu\varepsilon \frac{\partial^2 \mathbf{A}}{\partial t^2} - \nabla^2 \mathbf{A} = \mu \mathbf{J}$$

$$\mu\varepsilon \frac{\partial^2 \varphi}{\partial t^2} - \nabla^2 \varphi = \frac{\rho}{\varepsilon}$$

The implication of this is that the scalar potential φ depends on the charge distribution, whereas the vector potential $\mathbf{A}$ depends on the current density.

Remark 2.4 (*Electromotive Force*) The *electromotive force* (EMF), E, produced by some generating mechanism is the energy per unit charge, i.e., the voltage change, made available by the generating mechanism. The energy required to move a unit charge along a path from point a to point b through an electric field $\mathbf{E}$ is

$$E = \int_a^b \mathbf{F} \cdot d\mathbf{l} = \int_a^b (\mathbf{E} + \mathbf{v} \times \mathbf{B}) \cdot d\mathbf{l}$$

Remark 2.5 (*Magnetomotive Force*) The *magnetomotive force* (MMF), F, plays a role in magnetic circuits similar to that of E in electrical circuits. F is defined by

$$F = \int_a^b \mathbf{H} \cdot d\mathbf{l}$$

In a magnetic circuit comprised of a loop of uniform magnetic material of length l and cross-sectional area A, it is useful to define the *reluctance*, R:

$$R = \frac{l}{\mu_0 \mu_r A}$$

Then

$$F = \oint \mathbf{H} \cdot d\mathbf{l} = R\,\Phi$$

This formula is similar to Ohm's law governing the flow of current through a resistor.

Remark 2.6 (*Continuity of Charge*) Note that the taking the divergence of the Ampère–Maxwell law yields

$$\nabla \cdot \mathbf{J} = -\frac{\partial \nabla \cdot \mathbf{D}}{\partial t}$$

and using Gauss electric field law gives

$$\nabla \cdot \mathbf{J} = -\frac{\partial \rho}{\partial t} \text{ or } \int_S \mathbf{J} \cdot d\mathbf{s} = -\int_V \rho dV$$

This of course asserts the principle of continuity (or conservation) of charge.

Example 2.7 Capacitor. A *capacitor* is a device that stores charge. A typical capacitor consists of two conductors separated by a dielectric. The simplest example is the parallel plate capacitor shown in Figure 2.2.

In its uncharged state, both plates have zero charge. The capacitor can be charged using a battery or other means to a charge level Q, in which case one plate becomes positively charged with charge $+Q$ and the other negatively charged with charge $-Q$. The potential difference, ΔV, across the two plates can be obtained by integrating the electric field along a path through the dielectric from the positively to the negatively charged plate. Then, *capacitance*, C, of the device is the ratio of the charge to the potential difference, $C = Q/\Delta V$. The SI unit of capacitance is the farad (F). Thus, one farad is one coulomb per volt.

To compute the capacitance of the capacitor in Figure 2.2, first compute the electric field in the dielectric between the plates. Apply Gauss law

$$\int_S \varepsilon \mathbf{E} \cdot d\mathbf{s} = \int_V \rho dv$$

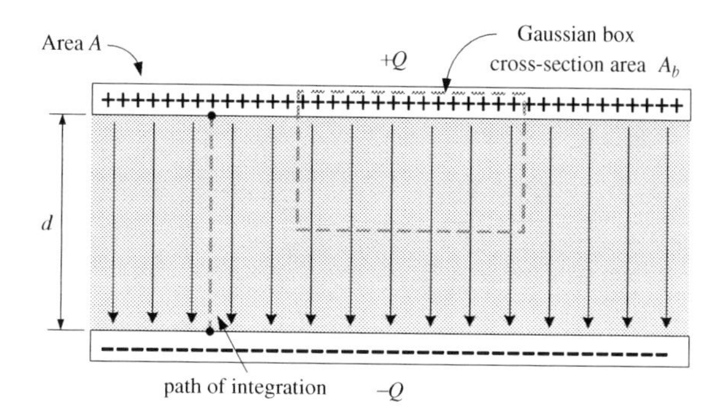

Fig. 2.2 A two-plate capacitor with very large plate area

to the box shown in the figure to find

$$\varepsilon E A_b = \sigma A_b,$$

where σ is the charge per unit area, Q/A, and ε is the permittivity of the dielectric. Thus, $E = \sigma/\varepsilon$. Now integrate from the positive plate to the negative plate along the integration path shown to get the potential difference:

$$\Delta V = -\int_{+}^{-} \mathbf{E} \cdot d\mathbf{l} = E d$$

Thus, the capacitance is

$$C = \frac{Q}{\Delta V} = \frac{\varepsilon A}{d}$$

Example 2.8 Wire Revisited. From Example 2.2, we know that

$$\mathbf{H} = \frac{i}{2\pi\rho}\mathbf{u}_{\varphi}$$

Let us verify Ampère's law. Choose for C a circular path of radius ρ in a plane with z constant.

$$\oint_C \mathbf{H} \cdot d\mathbf{l} = \int_0^{2\pi} \frac{i}{2\pi\rho}\rho d\varphi = i$$

Example 2.9 Infinite Solenoid. Consider an infinite solenoid composed of a tightly wound, thin wire coil with a core as shown in Figure 2.3. The solenoid has n turns per unit length, cross-sectional area, A, length, l and a constant current i passes through it. The core has permeability $\mu = \mu_0\mu_r$.

By symmetry, the induced magnetic field is horizontal, and in view of the winding direction and current flow the field vectors point to the right. We can use Ampère's

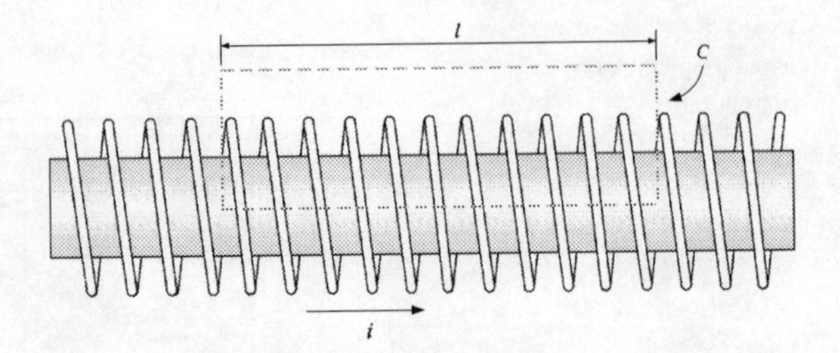

Fig. 2.3 Solenoid

law to compute its magnitude. Choose a rectangular contour with horizontal lines outside the coil, one on each side — above and below the coil. Since the net encircled current is zero, the field outside of the coil is zero. To determine the field inside the coil, choose a contour C as shown. Application of Ampère's law yields $Hl = nil$. Consequently, we have

$$\mathbf{H} = \begin{cases} 0 & \text{outside the coil} \\ ni\mathbf{u}_z & \text{inside the coil} \end{cases}$$

Inside the coil, the magnetic flux density is $\mathbf{B} = \mu_0\mu_r ni\,\mathbf{u}_z$ and the magnetic flux through a cross section is $\Phi = \mu_0\mu_r i\,A$. The number of loops in a section of length l i nl so the effective area through which $\mathbf{B}$ passes is nlA. Consequently, the effective flux within the coil section is $\lambda = \mu_0\mu_r nli\,A$. λ is called the *flux linkage*.

Example 2.10 Inductive Loop. A single, perfectly conducting wire loop encircles a core of permeable magnetic material in Figure 2.4.

As in the previous example, application of Ampère's law enables computation of the magnetic flux density in the core, $|\mathbf{B}| = \mu i\,(t)$. It follows that the induced back EMF is

$$E = -\frac{d\lambda}{dt}, \quad \lambda = \mu A i\,(t)$$

and so the applied voltage is related to the current in the wire loop by

$$v\,(t) = \mu A \frac{d\,i\,(t)}{dt}$$

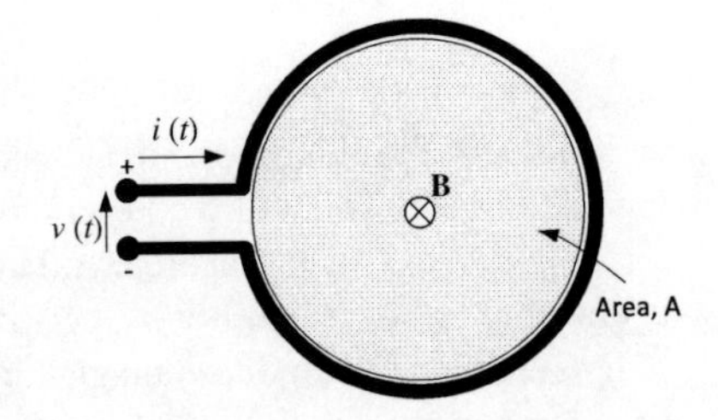

Fig. 2.4 A single-wire loop surrounds a magnetic core

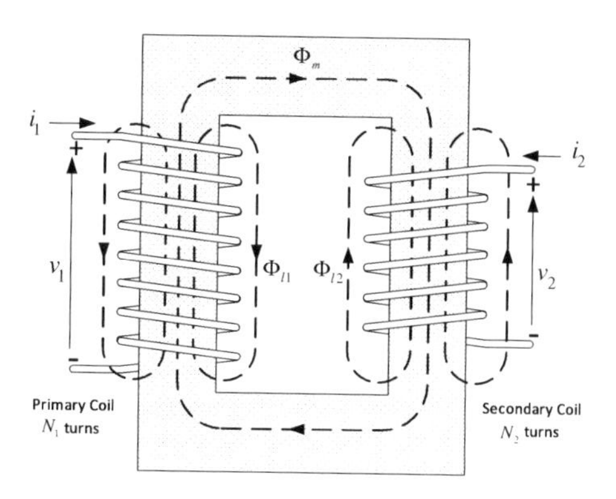

Fig. 2.5 A transformer formed of two coupled coils with a common core

For a tightly wound coil of N loops, the voltage–current relationship is

$$v(t) = \frac{d\lambda(t)}{dt} = \mu NA\frac{di(t)}{dt} = L\frac{di(t)}{dt},$$

where L is the *inductance*.

Example 2.11 Transformer. The transformer in Figure 2.5 has a primary coil with N_1 turns and a secondary coil with N_2 turns.

Consider the ideal case, in which the transformer has the following characteristics:

1. no losses
2. zero leakage flux, i.e., $\Phi_{l,1} = \Phi_{l,2} = 0$
3. zero reluctance.

Faraday's law yields

$$v_1 = N_1\frac{d\Phi_m}{dt}, \quad v_2 = N_2\frac{d\Phi_m}{dt}$$

which implies

$$\frac{v_2}{v_1} = \frac{N_2}{N_1}$$

In addition, zero reluctance implies that the magnetomotive force around a closed loop in the core sums to zero, so that

$$N_1 i_i + N_2 i_2 = 0$$

Thus,

$$i_2 = -\frac{N_1}{N_2} i_1$$

Notice that $v_2 i_2 = -v_1 i_1$ which implies that the instantaneous power entering on the left is the same as that exiting on the right, as expected for this lossless transformer.

Chapter 3
Electric Circuits and Devices

> *"Science is built up of facts, as a house is with stones. But a collection of facts is no more a science than a heap of stones is a house."*
>
> —Henri Poincaré

3.1 Introduction

In this chapter, nonlinear circuits that include resistors, inductors, capacitors, and memristors (RLCM) are discussed. Analysis of the individual elements is followed by examination of basic circuit mathematical models and theorems. RLCM circuit dynamic models are constructed using a Lagrange formulation based on the *generalized Euler–Lagrange* equations of Noble and Sewell [157]. The discussion is based on the work reported in [117] which extended earlier work of [138, 54, 180, 150]. Alternative forms of the network equations, such as Brayton–Moser equations and the generalized Hamilton equations, are also discussed.

The use of Lagrange equations has several benefits including the connection with energy functions and other first integrals associated with stability analysis (to be discussed in Chapter 5). The ability to model electromechanical devices within a single coherent framework is another. The chapter ends with a discussion of such devices including a basic motor generator.

3.2 Circuits and Circuit Elements

A two-terminal circuit element is a conducting path between two points a and b called terminals. A two-terminal element is also called a one-port element. Various types of two-terminal elements exist, each type associated with a specific relationship

H.G. Kwatny and K. Miu-Miller, *Power System Dynamics and Control*, Control Engineering, DOI 10.1007/978-0-8176-4674-5_3

between the voltage across the terminals and the current through the element. We will consider several lumped parameter elements below. A simple electrical circuit is a closed conducting path composed of a finite number of serially connected one-port elements. An electrical network is an arbitrary interconnection of circuit elements. Usually, a network contains several circuits.

In electromagnetic circuits, there are four basic variables: *current* i, *voltage* v, *charge* q, and *flux linkage* λ. Among these variables, there are two fundamental (kinematic) relationships

$$\frac{dq\,(t)}{dt} = i\,(t)\,, \quad \frac{d\lambda\,(t)}{dt} = v\,(t) \tag{3.1}$$

From the four variables, it is possible to identify six distinct pairs, and consequently, there can be six pairwise relationships of which two are already known (3.1). Thus, four remain. These will be called *constitutive relations* and will be used to define the basic circuit elements.

Two-terminal or *one-port* circuit elements are characterized by two port variables: the voltage v across the terminal pair and the current i flowing through the element. Each element will have an appropriate causality in which one of the port variables is considered the input and the other the output of the device. Four different elements can be defined in terms of the four constitutive relationships:

1. the *resistor*, defined by a relationship between v and i,
2. the *inductor*, defined by a relationship between λ and i,
3. the *capacitor*, defined by a relationship between v and q,
4. the *memristor*, defined by a relationship between λ and q.

The first three, of course, are the classical circuit elements. The existence of the fourth was postulated by Leon Chua in [53] and recently confirmed in [179]. Actually, the missing element was also anticipated by Paynter [165] who recognized the missing constitutive relation as indicated by the dashed line in Figure 3.1.

A complete model of a circuit element is one in which specification of one of the port variables (v or i) enables the determination of the other. A complete model

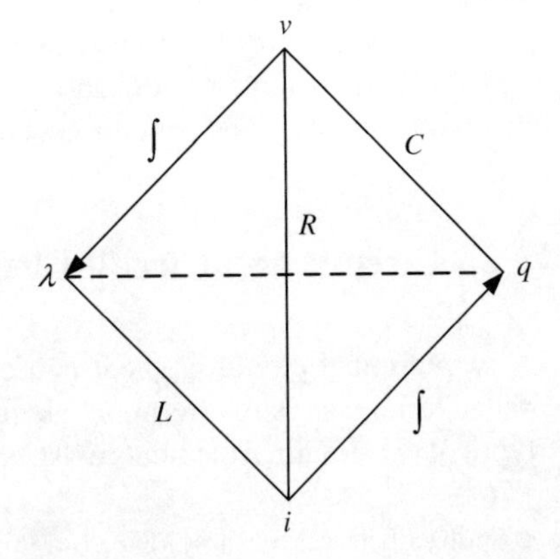

Fig. 3.1 The state tetrahedron adapted from [165]. The lines denoted R, L, and C connect the variables appearing in the respective constitutive relations. The dashed line denotes the missing relationship

of each of the four circuit elements requires combining the appropriate constitutive relation with, if necessary, one or more of the kinematic relations.

1. the *resistor*, $\phi_R(v,i)=0$,
2. the *inductor*, $\phi_L(\lambda,i)=0$, $\dot{\lambda}=v$,
3. the *capacitor*, $\phi_C(v,q)=0$, $\dot{q}=i$,
4. the *memristor*, $\phi_M(\lambda,q)=0$, $\dot{\lambda}=v$, $\dot{q}=i$.

Remark 3.1 In the above definitions, the implicit relations are intended to define a one-dimensional regular submanifold[1] of R^2. As an example, in the case of the resistor, the manifold is implicitly defined by $M_R=\left\{(v,i)\in R^2\,|\phi_R(v,i)=0\right\}$. In some instances, it might be more appropriate to define the manifold parametrically, $M_R=\left\{(v,i)\in R^2\,|v=\varphi_1(s)\,,i=\varphi_2(s)\,,s\in[a,b]\subset R\right\}$.

An element is linear if its constitutive relationship is linear. In which case, for the resistor, we have $\phi_R=v-Ri$, for the inductor, $\phi_L=\lambda-Li$, and for the capacitor, $\phi_C=v-Cq$. Here, R,L,C represent the usual resistance, inductance, and capacitance, respectively. For the memristor, notice that upon differentiation with respect to t yields

$$\frac{\partial\phi_M}{\partial\lambda}\dot{\lambda}+\frac{\partial\phi_M}{\partial q}\dot{q}=0 \tag{3.2}$$

or

$$\frac{\partial\phi_M}{\partial\lambda}v+\frac{\partial\phi_M}{\partial q}i=0 \tag{3.3}$$

Consequently, if ϕ_M is linear, i.e., $\phi_M=A\lambda-Bq$, (3.51) becomes

$$Av-Bi=0$$

so that the memristor is simply a resistor with resistance $R=B/A$. This point is noted in [53].

Example 3.2 (*Linear capacitor*) The capacitance of a parallel plate capacitor separated by distance a and with dielectric permittivity ε is

$$\frac{\varepsilon}{4\pi d}\text{per unit area}$$

The capacitance of a cylindrical capacitor of radii a,b with dielectric permittivity ε is

$$\frac{\varepsilon}{2\log(b/a)}\text{per unit length}$$

[1] Throughout the book, basic terminology of differential geometry will be used. Basic references are [186] and [28]

Example 3.3 (Linear Inductor) Consider the solenoid of Example 2.9. The self-inductance is

$$L = \frac{\mu_0 \mu_r n i A}{i} = \mu_0 \mu_r n A$$

where L has units of henry per unit length.

In order to address nonlinear elements, we need some additional terminology.

Definition 3.4 (*Controlled Elements*)

(a) A resistor is *current-controlled* (*voltage-controlled*) if $\phi_R(v, i) = 0$ is satisfied by a single-valued function of the current $v = \varphi(i)$ (of the voltage $i = \varphi(v)$).
(b) An inductor is *flux-controlled* (*current-controlled*) if $\phi_L(\lambda, i) = 0$ is satisfied by a single-valued function of the flux, $i = \varphi(\lambda)$ (current, $\lambda = \varphi(i)$).
(c) A capacitor is *charge-controlled* (*voltage-controlled*) if $\phi_C(v, q) = 0$ is satisfied by a single-valued function of charge, $v = \varphi(q)$ (voltage, $q = \varphi(v)$).
(d) A memristor is *flux-controlled* (*charge-controlled*) if $\phi_M(\lambda, q) = 0$ is satisfied by a single-valued function of flux, $q = \varphi(\lambda)$ (charge, $\lambda = \varphi(q)$).

It is well known that the inductor and capacitor are energy storage devices. This fact is easily established. For example, consider a flux-controlled inductor. The energy supplied to the device over a time interval (t_0, t_1) is

$$\mathcal{E} = \int_{t_0}^{t_1} iv\,dt = \int_0^{\lambda} \varphi(\lambda)\, d\lambda$$

Now, if we assume that the graph of $\varphi(\lambda)$ is confined to the first and third quadrants, i.e., $\lambda\varphi(\lambda) \geq 0$, then $\mathcal{E} \geq 0$. Furthermore, it is clear that any energy increase obtained by changing the flux linkage from λ_0 to λ_1 is recovered when the flux linkage is reduced to λ_0. A similar conclusion is obtained for the capacitor. On the other hand, for a current-controlled resistor, compute the power supplied to it

$$\mathcal{P} = iv = i\varphi(i)$$

Thus, if the graph of $\varphi(i)$ is confined to the first and third quadrants, i.e., $i\varphi(i) \geq 0$, it follows that $\mathcal{P} \geq 0$. Thus, power is always injected and cannot be retrieved, so the resistor is purely dissipative.

Consider a charge-controlled memristor. Then,

$$\lambda = \varphi(q) \Rightarrow v = M(q)\, i, \quad M(q) = \frac{\partial \varphi(q)}{\partial q}$$

Consequently, if $M(q) \geq 0$, $\forall q$, it follows that

$$\mathcal{P} = iv = M(q)\, i^2 \geq 0$$

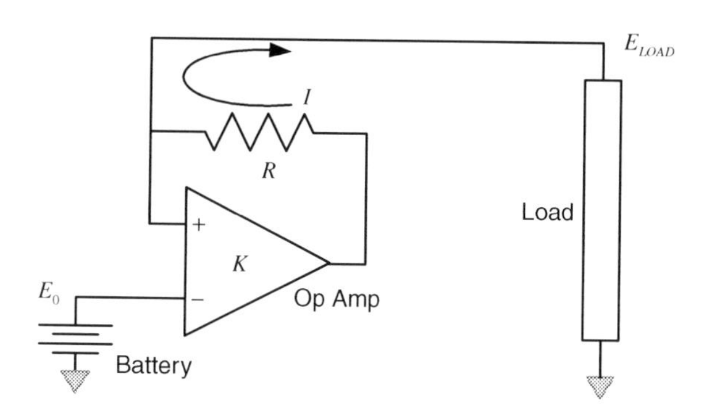

Fig. 3.2 A current source using a battery and operational amplifier

The memristor is purely dissipative. Recall that a one-port device is *passive* if the energy that can be extracted from it with any initial state is x, i.e., the *available energy* $\mathcal{E}_A(x)$ is bounded [201]. Chua proved in [53] that a charge-controlled memristor is passive if and only if $M(q) \geq 0$. Similar conclusions can be made for a flux-controlled memristor for which the constitutive relation is

$$q = \varphi(\lambda) \Rightarrow i = W(\lambda)\, v, \quad W(\lambda) = \frac{\partial \varphi(\lambda)}{\partial \lambda}$$

Sources provide the driving inputs in a circuit. We will be concerned with two types of sources. An *ideal voltage source* is a two-terminal element in which the voltage across it is independent of the current through it. A battery is a common voltage source. An *ideal current source* is a two-terminal element in which the current through it is independent of the voltage across it.

Example 3.5 Current sources can be constructed in many ways. A simple one can be constructed using a battery and an operational amplifier as shown in Figure 3.2. The current supplied to the load is given by

$$I = \left.\frac{K\left(E_{LOAD} - E_0\right) - E_{LOAD}}{R}\right|_{K=1} = \frac{-E_0}{R}$$

Notice that the current is constant and independent of the load.

3.3 Network Modeling

The classical formulation of electric network models is based on Kirchhoff's two-circuit laws. Given a network, consider an arbitrary node connecting n branches. Kirchhoff's current law (KCL) states that the sum of the currents into the node is zero, i.e.,

$$\sum_{j=1}^{n} I_j = 0 \tag{3.4}$$

Kirchhoff's current law can be obtained from Maxwell equations. Suppose S is a closed surface. Then, the Ampère–Maxwell relation becomes

$$0 = \int_S \mathbf{J} \cdot d\mathbf{s} + \frac{\partial}{\partial t} \int_S \mathbf{D} \cdot ds$$

From Gauss law, this becomes

$$0 = \int_S \mathbf{J} \cdot d\mathbf{s} + \frac{\partial}{\partial t} \int_V \rho dv$$

Assuming that the node is isolated, S can be chosen sufficiently small so that there is no charge within S. Consequently,

$$0 = \int_S \mathbf{J} \cdot d\mathbf{s}$$

This, of course, leads directly to (3.4).

Now consider a circuit within the network containing m elements. Kirchhoff's voltage law (KVL) states that the sum of the voltage drops across the elements around the circuit is zero, i.e.,

$$\sum_{j=1}^{m} E_j = 0 \tag{3.5}$$

Kirchhoff's voltage law can be derived from Faraday's law under the assumption that the circuit does not enclose a fluctuating magnetic field. In this case, Faraday's law states:

$$\oint_C \mathbf{E} \cdot d\mathbf{l} = 0$$

which obviously leads to (3.5). Ordinarily, we assume that any varying magnetic field influencing the circuit is concentrated within any inductors so that the only effect is to produce the voltage drop across the inductor, thereby properly accounting for it. When this is not the case, we need to include in (3.5) the back-electromotive force induced by the field.

Example 3.6 Consider the RLC circuit shown in Figure 3.3. Notice that the network has four branches and three nodes. The KCL applied to the three nodes leads to the equations

$$I_E = -I_L$$
$$I_L = I_C + I_R$$
$$I_E = -I_C - I_R$$

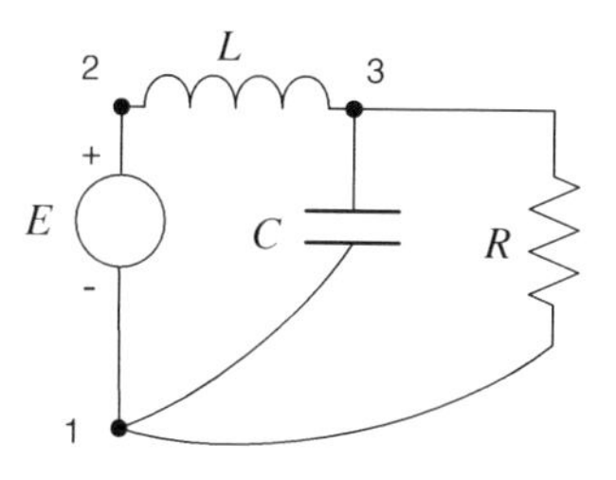

Fig. 3.3 A simple RLC circuit

Three circuits can be identified to which application of the KVL produces

$$E = E_L + E_C$$
$$E = E_L + E_R$$
$$E_R = E_C$$

These six equations are clearly redundant. Recall the element equations

$$\frac{1}{C}\frac{dE_C}{dt} = I_C$$

$$L\frac{dI_L}{dt} = E_L$$

$$E_R = R I_R$$

Now, we need only the second KCL equation and the first and third KVL equations to obtain the closed system of differential equations that characterize the network

$$\frac{1}{C}\frac{dE_C}{dt} = I_L - \frac{1}{R}E_C$$

$$L\frac{dI_L}{dt} = E - E_C$$

Kirchhoff's laws along with the element constitutive relationships are used to formulate the network equations. Consider a network consisting of m branches with each branch identified with exactly one element. Clearly, the network is completely solved if each branch voltage and current are known. In the above example, this is accomplished with a subset of Kirchhoff equations. This is the usual case. So our goal is to determine a systematic method to identify a necessary and sufficient set of equations. To do this, we follow a graph theoretic approach.

A *graph*, like a network, is composed of nodes and branches. A *directed graph* is one in which each branch is assigned a direction. It is connected if there is a continuous path between every pair of nodes. A network can be associated with a connected directed graph in which the branch directions denote positive current flow.

Positive voltage is the opposite so that positive branch current flows from the node with higher voltage to the node with lower voltage.

A *loop* is a set of branches that form a closed path. A *tree* is a maximal set of branches that do not contain any loops. For any tree, the remaining set of branches is called a *co-tree*. The member branches of a co-tree are called *links*. Insertion of a link in a tree forms a loop. A graph typically contains several trees. If a graph contains n nodes, then each tree contains exactly $n - 1$ branches—the number of branches needed to connect n nodes. If the graph contains m branches, then there are exactly $m - n + 1$ links.

Consider a network with n nodes and m branches. Select a tree and remove all of the links. The resulting network has no loops, so all of the branch currents are zero. By inserting any link, the flow of current is enabled. In fact, each link can be used to create a distinct flow. The set of loops created by inserting one link at a time is called a *tieset*. We can state the following:

Proposition 3.7 *A network having n nodes and m branches has $m - n + 1$ independent current variables. The link currents of any co-tree provide a complete set of current variables. The tieset loops provide $m - n + 1$ independent KVL equations.*

Consider a graph with n nodes and a tree of the graph with $n - 1$ branches. The voltage across each branch of the tree can be arbitrarily specified, so they form a set of independent voltage variables. Insertion of a link forms a loop, and with the tree branch voltages specified, KVL determines the link voltage. Specifically, if the tree branch voltages are zero, then all branch voltages are zero.

Given a graph, a *cutset* is a set of branches that when cut divides the graph into two disconnected pieces. Given a graph and a tree, a *basic cutset* is a cutset that contains only one tree branch. In a graph with n nodes, each tree has $n - 1$ branches, so there are $n - 1$ basic cutsets.

Consider one of the two disconnected subnetworks associated with a basic cutset. We can collapse all the nodes into one and apply the Ampère–Maxwell law, as above, to obtain the generalized form of KCL, (3.4): The sum of the currents into the subnetwork over the cutset branches is zero.

$$\sum_{cutset} I_j = 0 \tag{3.6}$$

Applying this formula to the $n - 1$ basic cutsets, we obtain $n - 1$ independent current equations. They are independent because they each contain a unique tree branch.

Example 3.8 For example, consider the circuit graph in Figure 3.4.

One of the four basic cutsets consists of the branches $\{b, f, g\}$ including the single tree branch b. Consider the lower left subnetwork containing nodes $\{1, 2, 4\}$. These are collapsed as shown. With respect to the indicated directions of positive current, the resulting KCL is

$$i_b = i_f + i_g$$

These results can be summarized in the following statement.

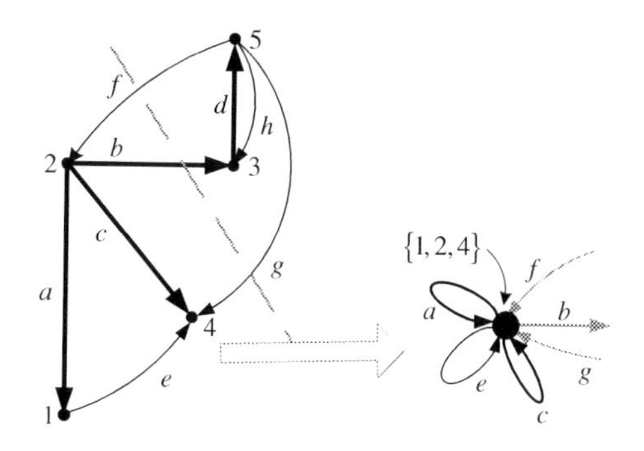

Fig. 3.4 This simple graph has 5 nodes, and hence, each tree has 4 branches

Proposition 3.9 *An n-node network has $n-1$ independent voltage variables. The $n-1$ branches of any tree provide a complete set of voltage variables. The basic cutsets provide $n-1$ independent KCL equations.*

Example 3.10 Consider the circuit in Figure 3.3. Its graph is shown in Figure 3.5 along with two different choices of tree and co-tree. Based on the tree on the right, the tieset includes two loops: $\{e, l, c\}$ and $\{c, r\}$. Thus, we can write two KVL equations:

$$
\begin{aligned}
E &= E_L + E_C \\
E_C &= E_R
\end{aligned}
$$

There are two basic cutsets: $\{e, l\}$ and $\{l, c, r\}$. Thus, there are two KCL equations:

$$
\begin{aligned}
I &= -I_L \\
I_L &= I_C + I_R
\end{aligned}
$$

These equations are a subset of the KVL and KCL equations derived previously in Example 3.6. These and the element constitutive equations form a complete set in that they completely define all voltages and currents in the network.

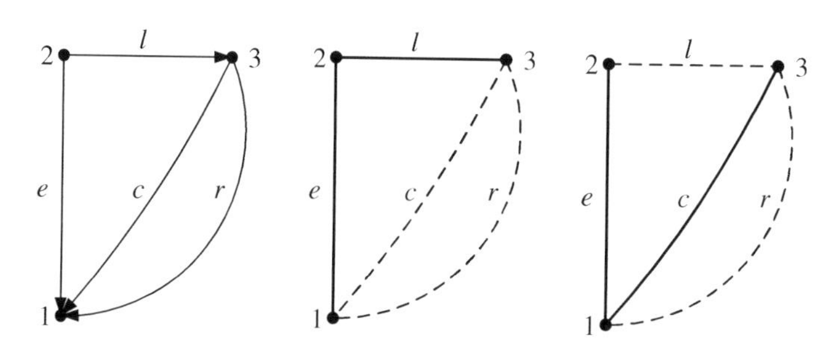

Fig. 3.5 The graph corresponding to the circuit in Figure 3.3 is shown on the left. Two different trees and the corresponding co-trees are also shown

Let us summarize the process of assembling the equations for a given network. Consider a network with n nodes and m branches. Solution of the network requires the determination of the m branch voltages and m branch currents. Construct a directed graph for the network and identify a tree.

1. Identify the tieset with its $m - n + 1$ loops and construct the $m - n + 1$ KVL equations, relating the m branch voltages.
2. Identify the $n - 1$ basic cutsets and construct the $n - 1$ KCL equations relating the m branch currents.
3. Assemble the m circuit element constitutive relations relating branch voltage and current.
4. Altogether, there are $2m$ equations relating the $2m$ variables.

3.4 The Incidence Matrix and Tellegen's Theorem

Consider a network with n nodes and m branches. First, we define the network incidence matrix.

Definition 3.11 The branch-to-node incidence matrix of a network with m branches and n nodes is the matrix $A \in R^{m \times n}$ with elements

$$a_{ij} = \begin{cases} 1, & \text{if current } i \text{ leaves node } j \\ -1, & \text{if current } i \text{ enters node } j \\ 0, & \text{if branch } i \text{ is not incident on node } j \end{cases}$$

The incidence matrix allows succinct determination of the KCL and KVL equations. Let v be the n-vector of node voltages and I the m-vector of branch currents. Then, KCL can be expressed

$$A^T I = 0$$

The m vector of branch voltages, V, can be expressed in terms of the node voltages

$$V = Av$$

which can be interpreted as the KVL.

Example 3.12 Consider the example in Figure 3.6. The incidence matrix is

$$A^T = \begin{bmatrix} -1 & 0 & -1 & -1 \\ 1 & 1 & 0 & 0 \\ 0 & -1 & 1 & 1 \end{bmatrix}$$

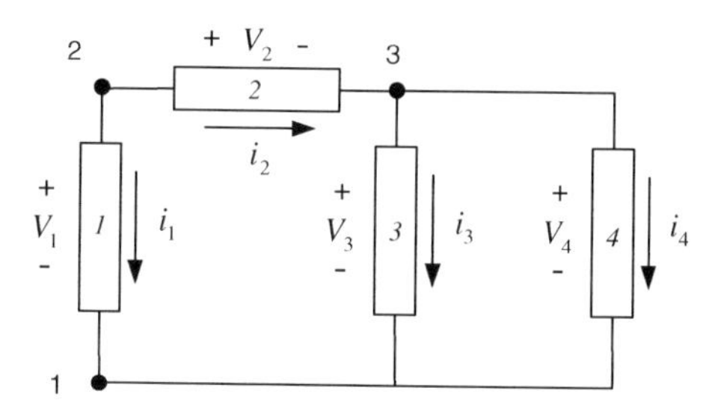

Fig. 3.6 A simple network with 3 nodes and 4 branches

producing the node equations

$$
\begin{aligned}
-I_1 - I_3 - I_4 &= 0 \\
I_1 + I_2 &= 0 \\
-I_2 + I_3 + I_4 &= 0
\end{aligned}
$$

The two independent loop equations are

$$
\begin{aligned}
V_1 &= V_2 + V_3 \\
V_1 &= V_2 + V_4
\end{aligned}
$$

The branch voltages are given by

$$
\begin{bmatrix} V_1 \\ V_2 \\ V_3 \\ V_4 \end{bmatrix} = A^T v = \begin{bmatrix} v_2 - v_1 \\ v_2 - v_3 \\ v_3 - v_1 \\ v_3 - v_1 \end{bmatrix}
$$

which satisfy the loop equation.

Theorem 3.13 (Tellegen's Theorem) *Given a network with m branches and n nodes. Let the branch potential differences $V_1, \ldots, V_m$ satisfy KVL and the branch currents $I_1, \ldots, I_m$ satisfy KCL. Then,*

$$
\sum_{j=1}^{m} V_j I_j = 0
$$

Proof KVL provides

$$
V^T I = (Av)^T I = v^T A^T I
$$

But $A^T I = 0$ by KCL, so

$$
\sum_{j=1}^{m} V_j I_j = V^T I = 0
$$

□

Remark 3.14 (Conservation of Energy) Tellegen's theorem states that the sum of the instantaneous power flows over all branches of a circuit is zero. Thus, it expresses a concept of conservation of power. If the branches are divided into two subsets, one consisting of independent sources and the other consisting of the remaining branches referred to as components, then Tellegen's theorem can be rephrased as "the sum of the powers delivered by the independent sources is equal to the sum of the powers absorbed by the components." This statement is often interpreted as a statement of conservation of energy.

3.5 Generalized Lagrange Equations

3.5.1 Introduction

The use of Hamilton's principle and Lagrange equations, example [57], has a long history. The traditional method chooses either capacitor charges *or* inductor flux linkages as the independent coordinates. Thus, it is limited in the class of circuits that can be addressed. This problem has been addressed in several publications including [54, 139, 117]. The following discussion employs the method introduced in [117]. It is based on the *generalized Lagrange equations* [157], which can be viewed as a variant of *Poincaré equations* [10, 45,46] also referred to as *Lagrange equations in quasi-coordinates* [148, 155] or *pseudo-coordinates* [75], or the *Euler–Poincaré equations* [172].

Poincaré equations preserve the underlying theoretical structure and elegance of the Lagrange formulation. In the case of circuits, Poincaré equations take the simplified form

$$\dot{q} = V\ p \tag{3.7}$$

$$\frac{d}{dt}\frac{\partial L\,(p,q)}{\partial p} - \frac{\partial L\,(p,q)}{\partial q}V = Q^T V \tag{3.8}$$

where p is a vector of quasi-velocities and q is the generalized coordinate vector. V is the velocity transformation matrix. It relates the coordinate velocities to the quasi-velocities. Q is a vector of externally applied generalized forces and other nonconservative forces. The function $L(p, q)$ is the Lagrangian function. In the general case, the velocity transformation matrix V is dependent on the coordinates q.

3.5.1.1 Energy Functions and the Classical Lagrange Equations

In Section 3.2, it was shown that capacitors and inductors are energy storage elements. The electrical energy stored in a charge-controlled capacitor is easily computed, given the constitutive relation $v = v(q)$, by integrating the work done over time of the power delivered

$$T_C(q) = \int v\,i\,dt = \int_0^q v(q)\,dq$$

Similarly, if the capacitor is voltage-controlled, its co-energy can be computed as Legendre transformation of the energy, given the constitutive relation $q = q(v)$

$$T_C^*(v) = \left[q\,v - T_C(q)\right]_{q \to \varphi(v)} = \int_0^q q(v)\,dv$$

In the same way, the inductor energy, $U_L(\lambda)$, and co-energy, $U_L^*(i)$ can be obtained

$$U_L(\lambda) = \int_0^\lambda i(\lambda)\,d\lambda$$

$$U_L^*(i) = \int_0^i \lambda(i)\,di$$

These definitions were given in [44]. A companion paper [151] provides the following definitions of resistor content, $G_R(i)$, and co-content, $G_R^*(v)$

$$G_R(i) = \int_0^i v(i)\,di$$

$$G_R^*(v) = \int_0^i i(v)\,dv$$

as well as voltage source content, $G_E(i, t)$, and current source co-content, $G_J^*(v, t)$

$$G_E(i, t) = \int_0^i v(t)\,di$$

$$G_J^*(v, t) = \int_0^v i(t)\,dv$$

Hamilton's principle of stationary action [85, 159] was originally stated for classical mechanical systems but has been extended to other systems including electric circuits. In its basic form, this principle states the natural evolution of a system described by a set of configuration coordinates, $q(t)$ from a fixed initial configuration at time t_1 to another configuration at time $t_2 > t_1$ a stationary point of the action integral

$$S = \int_{t_1}^{t_2} L(\dot{q}(t), q(t), t)\,dt$$

where L is the Lagrangian function. This means that the arbitrary small variation of I from an admissible trajectory $q(t)$ is zero

$$\delta S = \int_{t_1}^{t_2} [L(\dot{q}(t) + \dot{\varepsilon}(t), q(t) + \varepsilon(t), t) - L(\dot{q}(t), q(t), t)]\, dt = 0$$

where $\varepsilon(t)$ is an arbitrary small function of t. From this statement, it is possible to derive the basic form of Lagrange equations

$$\frac{d}{dt}\left(\frac{\partial L}{\partial \dot{q}}\right) - \frac{\partial L}{\partial q} = 0 \tag{3.9}$$

Also, the variational indicator can be expanded to account for nonconservative elements, example resistors and sources leading to Lagrange equations in the form:

$$\frac{d}{dt}\left(\frac{\partial L}{\partial \dot{q}}\right) - \frac{\partial L}{\partial q} = Q \tag{3.10}$$

where the generalized forces Q are obtained from the variation of the total work expression for all nonconservative elements

$$\delta W = \sum_{q} Q_i \, \delta q_i$$

The ordinary application to circuits selects the coordinates to be either an independent set of capacitor charges or an independent set of inductor flux linkages. For the capacitor charge formulation, the Lagrangian is

$$L(\dot{q}, q) = U^*(i) - T(q)$$

where U^* is the sum of all inductor co-energies and T is the sum of all capacitor energies. In the case of inductor flux linkage formulation, the Lagrangian is

$$L(\dot{\lambda}, \lambda) = T^*(v) - U(\lambda)$$

where T^* is the sum of all capacitor co-energies and U is the sum of all inductor energies.

Example 3.15 Simple Circuit Lagrange Equations. Consider the network shown in Figure 3.7. Assume that all the inductors, capacitors, and resistor are linear. Choose the capacitor charges, q_1, q_2 to be the coordinates. Then, the velocities are

$$i_1 = \frac{dq_1}{dt}, \quad i_2 = \frac{dq_2}{dt}$$

Fig. 3.7 Network with 5 nodes and 6 branches

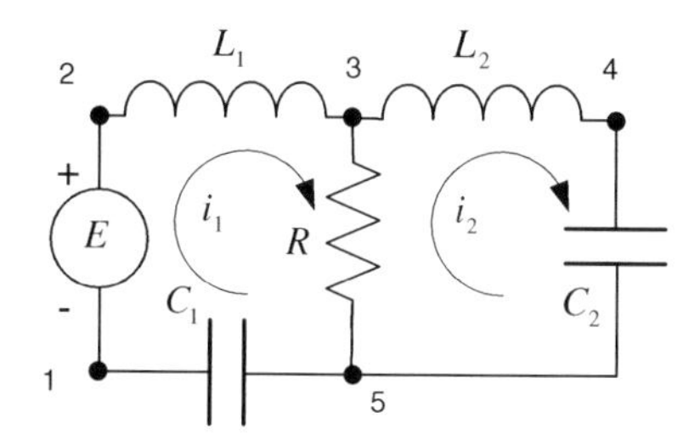

Consequently, the Lagrangian is

$$L = \tfrac{1}{2}L_1\dot{q}_1^2 + \tfrac{1}{2}L_2\dot{q}_2^2 - \frac{q_1^2}{2C_1} - \frac{q_2^2}{2C_2}$$

The work variation is

$$\delta W = E\,\delta q_1 - R\,(\dot{q}_2 - \dot{q}_1)\,(\delta q_2 - \delta q_1)$$

Thus,

$$Q_1 = E + R\,(\dot{q}_2 - \dot{q}_1)\,,\, Q_2 = -R\,(\dot{q}_2 - \dot{q}_1)$$

Lagrange equations are

$$L_1\ddot{q}_1 + R\,(\dot{q}_1 - \dot{q}_2) + \frac{q_1}{C_1} = E$$

$$L_2\ddot{q}_2 + R\,(\dot{q}_2 - \dot{q}_1) + \frac{q_2}{C_2} = 0$$

3.5.2 *State Variables*

3.5.2.1 Selection of State Variables

Given a network $\mathcal{N}$ defines a *normal tree* (as in [139]):

Definition 3.16 A normal tree is a tree containing a maximum number of capacitors and a minimum number of inductors.

Consider $\mathcal{N}$ and choose a normal tree $\mathcal{T}$ and let $\mathcal{L}$ be its co-tree. Define the *dynamic transformation matrix* as follows:

Definition 3.17 Consider a network with n nodes and m branches. The chord to tree branch dynamic transformation matrix is the matrix $D \in R^{(n-1)\times(n-m+1)}$ with elements

$$d_{jk} = \begin{cases} 1, & \text{tree branch } j \text{ and chord } k \text{ currents have the same direction} \\ -1, & \text{tree branch } j \text{ and chord } k \text{ currents have opposite directions} \\ 0, & \text{tree branch } j \text{ does not lie in a loop formed by insertion of chord } k \end{cases}$$

Consequently, Kirchhoff's laws imply

$$i_t = Di_c \tag{3.11}$$

$$v_c = -D^T v_t \tag{3.12}$$

where i_t, v_t are the tree branch currents and voltages, respectively, and i_c, v_c are the chord currents and voltages.

Example 3.18 Dynamical Transformation Matrix

Consider the circuit in Figure 3.6. Its graph is shown in Figure 3.8. Choose the tree to be composed of two branches, 1 and 2. The co-tree is then composed of branches 3 and 4. According to Definition 3.17, the dynamic transformation matrix is

$$D = \begin{bmatrix} -1 & -1 \\ 1 & 1 \end{bmatrix}$$

Notice that Equation (3.11) simply states

$$\begin{aligned} i_1 &= -i_3 - i_4 \\ i_2 &= i_3 + i_4 \end{aligned}$$

Note that

$$-D^T = \begin{bmatrix} 1 & -1 \\ 1 & -1 \end{bmatrix}$$

so that Equation (3.12) becomes

$$\begin{aligned} V_3 &= V_1 - V_2 \\ V_4 &= V_1 - V_2 \end{aligned}$$

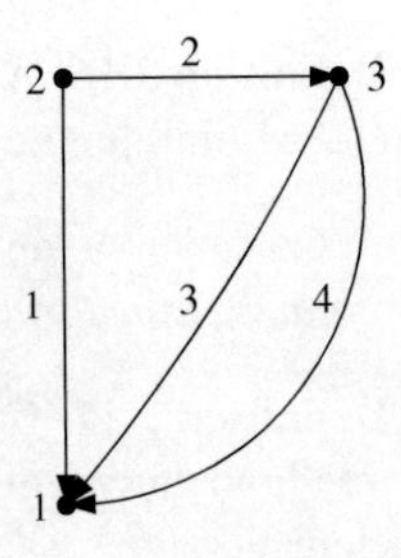

Fig. 3.8 Network directed graph with 3 nodes and 4 branches

Definition 3.19 Independent Sets of Capacitors and Inductors. If a set of capacitors form a loop, then their voltages (and charges) are dependent. Thus, the set of capacitors is said to be *dependent*. Otherwise, the set is called *independent*. If a set of inductors form a cutset, then their currents are dependent and the set is said to be *dependent*. Otherwise, the set is called *independent*.

In general, the elements of network can be divided into three categories: 1) *state*—independent elements that store energy, but do not inject or dissipate energy from the network, including ideal capacitors and inductors, which will be used to define the dynamic state, 2) *nonstate* elements—elements that do change the network total energy including resistors and sources, and 3) excess elements [139]. Accordingly, it is useful to partition Equations (3.11) and (3.12):

$$\begin{bmatrix} i_{ts} \\ i_{tn} \\ i_{te} \end{bmatrix} = \begin{bmatrix} D_{ss} & D_{sn} & D_{se} \\ D_{ns} & D_{nn} & D_{ne} \\ D_{es} & D_{en} & D_{ee} \end{bmatrix} \begin{bmatrix} i_{cs} \\ i_{cn} \\ i_{ce} \end{bmatrix} \tag{3.13}$$

$$\begin{bmatrix} v_{cs} \\ v_{cn} \\ v_{ce} \end{bmatrix} = \begin{bmatrix} -D_{ss}^T & -D_{ns}^T & -D_{es}^T \\ -D_{sn}^T & -D_{nn}^T & -D_{en}^T \\ -D_{se}^T & -D_{ne}^T & -D_{ee}^T \end{bmatrix} \begin{bmatrix} v_{ts} \\ v_{tn} \\ v_{te} \end{bmatrix} \tag{3.14}$$

Given a network $\mathcal{N}$ chooses a tree with a maximum number of independent capacitors and a co-tree with a maximal number of independent inductors [34, 139]. The only situation preventing all of the network capacitors from being included in the tree is the presence of capacitor-only loops. In this case, the loop is divided into a set of independent capacitors and a set of dependent capacitors. The independent set is included in the tree and the dependent set in the co-tree. Similarly, the only situation preventing the inclusion of all inductors in the co-tree is the existence of an inductor-only cutset. In this situation, the cutset is divided into an independent set of inductors which is included in the co-tree and the set of excess inductors included in the tree. The independent capacitors and independent inductors comprise the state elements of the network.

By treating each capacitor-only loop and each inductor-only cutset in this way, MacFarlane [139] defines a *state tree* for the network. For any given network, following this procedure a state tree and its co-tree are identified to have the following properties:

- ST1 The state tree contains a maximal independent set of capacitors and minimal set of dependent inductors.
- ST2 The co-tree contains a maximal independent set of inductors and a minimal dependent set of capacitors.
- ST3 The only type of loop that can be formed by the insertion of chord capacitor into a state tree is a capacitor-only loop.

ST4 No loop formed by the insertion of a chord resistor into the state tree can contain an inductor.

Note that item 4 follows from the fact that the inductor would not have belonged to an inductor-only cutset and therefore should not have been included in the state tree.

It follows from (ST3) and (ST4) that the dynamic transformation matrix takes the form:

$$D = \begin{bmatrix} D_{ss} & D_{sn} & D_{se} \\ D_{ns} & D_{nn} & 0 \\ D_{es} & 0 & 0 \end{bmatrix} \tag{3.15}$$

Now, the tree $\mathcal{T}$ and co-tree $\mathcal{L}$ are each divided into two parts, respectively, $\mathcal{T}_1, \mathcal{T}_2$, and $\mathcal{L}_1, \mathcal{L}_2$ according to the following criteria:

A1 All independent voltage sources belong to $\mathcal{T}_1$, and all independent current sources belong to $\mathcal{L}_2$.
A2 All resistors are divided between $\mathcal{T}_1$ and $\mathcal{L}_2$ such that all current-controlled resistors belong to $\mathcal{T}_1$ and all voltage-controlled resistors belong to $\mathcal{L}_2$.
A3 All inductors in the tree are current-controlled and belong to $\mathcal{T}_1$. All capacitors in the co-tree are voltage-controlled and belong to $\mathcal{L}_2$.
A4 Elements in $\mathcal{L}_2$ do not make fundamental loops with elements in $\mathcal{T}_1$. Elements in $\mathcal{T}_2$ do not belong to cutsets with elements in $\mathcal{L}_1$.
A5 $\mathcal{T}_1$ does not contain any voltage-controlled capacitors, and $\mathcal{T}_2$ does not contain any charge-controlled capacitors. Similarly, $\mathcal{L}_1$ does not contain any flux-controlled inductors, and $\mathcal{L}_2$ does not contain any current-controlled inductors.

These conditions imply that all nonstate elements and excess elements belong to $\mathcal{T}_1$ or $\mathcal{L}_2$. Since elements in $\mathcal{L}_2$ do not make fundamental loops with elements in $\mathcal{T}_1$, it follows that under the above conditions, D has the form:

$$\begin{bmatrix} i_{C\mathcal{T}_1} \\ i_{C\mathcal{T}_2} \\ \hline i_{E\mathcal{T}_1} \\ i_{R\mathcal{T}_1} \\ \hline i_{L\mathcal{T}_1} \end{bmatrix} = \left[\begin{array}{cc|cc|c} D_{ss11} & 0 & 0 & 0 & 0 \\ D_{ss21} & D_{ss22} & D_{sn21} & D_{sn22} & D_{se22} \\ \hline D_{ns11} & 0 & 0 & 0 & 0 \\ D_{ns21} & 0 & 0 & 0 & 0 \\ \hline D_{es1} & 0 & 0 & 0 & 0 \end{array}\right] \begin{bmatrix} i_{L\mathcal{L}_1} \\ i_{L\mathcal{L}_2} \\ \hline i_{J\mathcal{L}_2} \\ i_{R\mathcal{L}_2} \\ \hline i_{C\mathcal{L}_2} \end{bmatrix} \tag{3.16}$$

$$\begin{bmatrix} v_{L\mathcal{L}_1} \\ v_{L\mathcal{L}_2} \\ \hline v_{J\mathcal{L}_2} \\ v_{R\mathcal{L}_2} \\ \hline v_{C\mathcal{L}_2} \end{bmatrix} = -\left[\begin{array}{cc|cc|c} D_{ss11}^T & D_{ss21}^T & D_{ns11}^T & D_{ns21}^T & D_{es1}^T \\ 0 & D_{ss22}^T & 0 & 0 & 0 \\ \hline 0 & D_{sn21}^T & 0 & 0 & 0 \\ 0 & D_{sn22}^T & 0 & 0 & 0 \\ \hline 0 & D_{se2}^T & 0 & 0 & 0 \end{array}\right] \begin{bmatrix} v_{C\mathcal{T}_1} \\ v_{C\mathcal{T}_2} \\ \hline v_{E\mathcal{T}_1} \\ v_{R\mathcal{T}_1} \\ \hline v_{L\mathcal{T}_1} \end{bmatrix} \tag{3.17}$$

The perspective to be taken here is to view the capacitor charges in $\mathcal{T}_1$ and the inductor fluxes in $\mathcal{L}_2$ as generalized coordinates. The capacitor voltages in $\mathcal{T}_2$ and the inductor

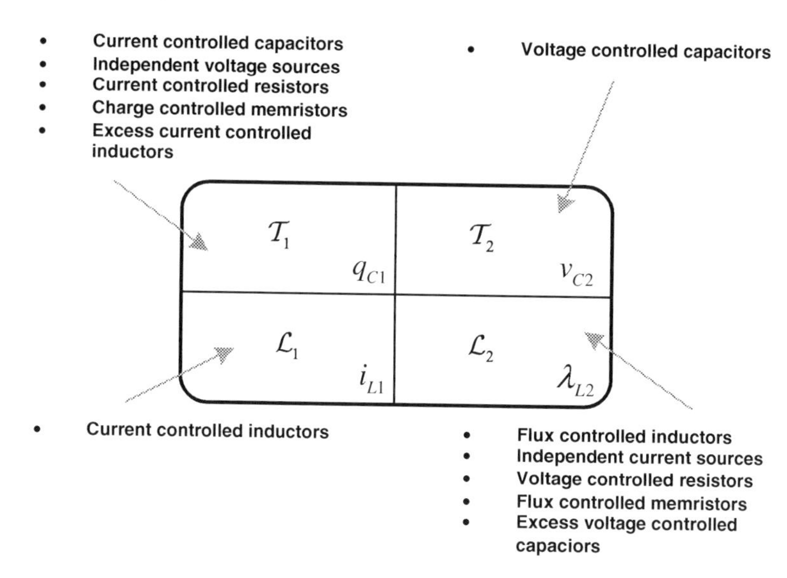

Fig. 3.9 Tree and co-tree division of circuit elements and state coordinates

currents in $\mathcal{L}_1$ are generalized velocities. The distribution of circuit elements within the tree and co-tree is summarized in Figure 3.9.

Remark 3.20 A Special Case. The notion of generalized coordinates and generalized velocities as used here can be clearly illustrated by considering a simple circuit composed of linear inductors and capacitors without any excess elements or non-conservative (dissipative or converter) elements. Then, in view of (3.15), (3.11) and (3.12) are reduced to

$$i_t = D_{ss} i_c, \quad v_c = -D_{ss}^T v_t \tag{3.18}$$

Now, the tree and co-tree, each divided as specified above, are illustrated in Figure 3.9. The tree, $\mathcal{T}$, consists entirely of an independent set of capacitors. These are divided into two sets $\mathcal{T}_1, \mathcal{T}_2$. The co-tree, $\mathcal{L}$, consists entirely of an independent set of inductors, also divided into two sets $\mathcal{L}_1, \mathcal{L}_2$. These divisions are arbitrary, except that the elements in $\mathcal{L}_2$ do not make fundamental loops with elements in $\mathcal{T}_1$.

The capacitor charges in $\mathcal{T}_1$, denoted q_{C1}, and the inductor fluxes in $\mathcal{L}_2$, denoted λ_{L2}, are the generalized coordinates. The capacitor voltages in $\mathcal{T}_2$, v_{C2}, and the inductor currents in $\mathcal{L}_1$, i_{L1}, are the generalized velocities. Equation (3.18) then can be written as

$$\begin{bmatrix} i_{C1} \\ i_{C2} \end{bmatrix} = \begin{bmatrix} D_{ss11} & 0 \\ D_{ss21} & D_{ss22} \end{bmatrix} \begin{bmatrix} i_{L1} \\ i_{L2} \end{bmatrix} \tag{3.19}$$

$$\begin{bmatrix} v_{L1} \\ v_{L2} \end{bmatrix} = -\begin{bmatrix} D_{ss11}^T & D_{ss21}^T \\ 0 & D_{ss22}^T \end{bmatrix} \begin{bmatrix} v_{C1} \\ v_{C2} \end{bmatrix} \tag{3.20}$$

Now, take the upper part of (3.19) and lower part of (3.20) to provide a relationship between the coordinate derivatives and the generalized velocities

$$\frac{d}{dt}\begin{bmatrix} q_{C1} \\ \lambda_{L2} \end{bmatrix} = V \begin{bmatrix} i_{L1} \\ v_{C2} \end{bmatrix} \tag{3.21}$$

where

$$V = \begin{bmatrix} D_{ss11} & 0 \\ 0 & -D_{ss22}^T \end{bmatrix} \tag{3.22}$$

The remaining equations are

$$i_{C2} = D_{ss21} i_{L1} + D_{ss22} i_{L2} \tag{3.23}$$

$$v_{L1} = -D_{ss11}^T v_{C1} - D_{ss21}^T v_{C2} \tag{3.24}$$

Equation (3.21) relates the coordinate velocities and the generalized velocities. We shall refer to (3.21) as the *velocity transformation relation* and to the matrix V as the *velocity transformation matrix*. Moreover, (3.21) is to be used to uniquely establish the generalized velocities as a function of the coordinate velocities. This is trivially accomplished when V has an inverse, but when that is not the case, we led to several important results which are discussed in the following paragraphs. Note that Equations (3.23) and (3.24) are the network loop and node equations, respectively.

3.5.3 *Other Forms of Lagrange Equations*

3.5.3.1 Generalized Lagrange Equations

Consider, as in Remark 3.20, the case without excess or nonconservative elements. The Lagrangian is assumed to take the form:

$$L(p,q) = W^*(p) - Z(q) + q^T G\, p \tag{3.25}$$

where

$$p = \begin{bmatrix} i_{L1} \\ v_{C2} \end{bmatrix}, \quad q = \begin{bmatrix} q_{C1} \\ \lambda_{L2} \end{bmatrix} \tag{3.26}$$

and

$$W^*(p) = U^*_{\mathcal{L}_1}(i_{L1}) + T^*_{\mathcal{T}_2}(v_{C2}) \tag{3.27}$$

$$Z(q) = U_{\mathcal{L}_2}(\lambda_{L2}) + T_{\mathcal{T}_1}(q_{C1}) \tag{3.28}$$

Note the cross product term appearing in (3.25). It is unusual, but necessary to address general circuits. The special case considered now lends insight into its meaning. Early treatments of circuit models via Lagrange methods such as [138, 104] do not contain a cross product term. However, [54, 117, 150] do.

Consider Poincarè equations (3.7), (3.8). In the present case, (3.7) is given by (3.21) and (3.8) can be divided into two sets of equations

$$\frac{d}{dt}\left(\frac{\partial L}{\partial i_{L1}}\right) - \frac{\partial L}{\partial q_{C1}} D_{ss11} = 0 \tag{3.29}$$

$$\frac{d}{dt}\left(\frac{\partial L}{\partial v_{C2}}\right) - \frac{\partial L}{\partial \lambda_{L2}} D_{ss22}^T = 0 \tag{3.30}$$

A few simple examples verify that the first of these equations is a set of voltage loop equations, while the second is a set of current nodal equations.

The key to deriving the generalized Lagrange equations is understanding the relationship between the coordinate time derivatives $\dot{q}$ and the quasi-velocities p as given by (3.21) and (3.22), or simply

$$\dot{q} = Vp \tag{3.31}$$

Suppose rank $V = k \leq \min(m, n)$ where m is the dimension of q and n is the dimension of p. Then, V can be factored to the form $V = LR, L \in R^{m \times k}, R \in R^{k \times n}$. L and R are both of full rank k, and consequently, L possesses and left inverse L^l, and R possesses a right inverse R^r. Consequently, V has a pseudo-inverse $V^+ = R^r L^l$, such that $VV^+V = V$.

Now, the matrices $(I - R^r R)$ and $\left(I - LL^l\right)$ can be similarly rank-factored

$$\left(I - R^r R\right) = \Delta\Phi, \quad \Delta \in R^{n \times (n-k)}, \Phi \in R^{(n-k) \times n}$$

$$\left(I - LL^l\right) = \Gamma\Sigma, \quad \Gamma \in R^{m \times (m-k)}, \Sigma \in R^{(m-k) \times m}$$

The solution properties of (3.31) for p in terms of $\dot{q}$ can be summarized in terms of these matrices. A solution of (3.31) exists if and only if the *compatibility constraints* hold

$$\Sigma\dot{q} = 0 \tag{3.32}$$

If (3.32) holds, then all solutions of (3.31) are of the form:

$$p = V^+\dot{q} + \Delta\dot{w} \tag{3.33}$$

where $w(t)$ is an arbitrary differentiable $(n-k)$ vector of *quasi-coordinates*. Notice that (3.32) and (3.33) can be combined to yield

$$\begin{bmatrix} p \\ 0 \end{bmatrix} = \begin{bmatrix} V^+ & \Delta \\ \Sigma & 0 \end{bmatrix} \begin{bmatrix} \dot{q} \\ \dot{w} \end{bmatrix} \tag{3.34}$$

Its inverse is

$$\begin{bmatrix} \dot{q} \\ \dot{w} \end{bmatrix} = \begin{bmatrix} V & \Gamma \\ \Phi & 0 \end{bmatrix} \begin{bmatrix} p \\ 0 \end{bmatrix} \tag{3.35}$$

As noted in [117], if the velocity transformation matrix V has a right inverse, then there are no compatibility constraints, and if V has a left inverse, then no additional coordinates are required.

For convenience in the following discussion, the following matrices are defined:

$$\mathcal{A} = \left[\begin{array}{c|c} V & \Gamma \\ \Phi & 0 \end{array} \right] = [\tilde{V} \mid \Lambda] \tag{3.36}$$

$$\mathcal{A}^{-1} = \begin{bmatrix} V^+ & \Delta \\ \Sigma & 0 \end{bmatrix} = \left[\begin{array}{c} \tilde{V}^l \\ \hline \Psi \end{array} \right] \tag{3.37}$$

Proposition 3.21 (Generalized Lagrange Equations) *Consider a network without excess elements and having generalized coordinates q_{C1}, λ_{L2}, generalized velocities i_{L1}, v_{C2}, and velocity transformation matrix as given in (3.22) and Lagrangian (3.25), then the* generalized Lagrange equations *are*

$$\dot{\tilde{q}} = \tilde{V} p, \quad \tilde{V} = \begin{bmatrix} D_{ss11} & 0 \\ 0 & -D_{ss22}^T \\ \Phi_1 & \Phi_2 \end{bmatrix} \tag{3.38}$$

$$\frac{d}{dt}\frac{\partial L}{\partial p} - \frac{\partial L}{\partial \tilde{q}} \tilde{V} = Q \tag{3.39}$$

where $\tilde{q}^T = \begin{bmatrix} q^T & w^T \end{bmatrix}$ *and*

$$L(p, \tilde{q}) = W^*(p) - Z(q) + \tilde{q}^T G p \tag{3.40}$$

with

$$G = \begin{bmatrix} 0 & 0 \\ -D_{ss22}^{+} D_{ss21} & 0 \\ \Delta_2^T D_{ss21} & 0 \end{bmatrix} \tag{3.41}$$

Proof Equation (3.39) is readily reduced by direct computation of the two relations

$$\frac{d}{dt}\left(\frac{\partial L}{\partial i_{L1}}\right) - \frac{\partial L}{\partial q_{C1}} D_{ss11} - \frac{\partial L}{\partial w}\Phi_1 = Q_1 \tag{3.42}$$

$$\frac{d}{dt}\left(\frac{\partial L}{\partial v_{C2}}\right) + \frac{\partial L}{\partial \lambda_{L2}} D_{ss22}^T - \frac{\partial L}{\partial w}\Phi_2 = Q_2 \tag{3.43}$$

In the absence of the cross product term, the first is a set of voltage loop equations and the second is a set of current nodal equations. The cross product term is necessary, however, to accommodate modeling of complex network topologies. Consequently, the first step is to determine choices for G such that the loop nodal distinctions are preserved. First, expand the cross product term so that the Lagrangian is in the form:

$$L(p,q) = W^*(p) - Z(q) + \begin{bmatrix} q_{C1}^T & \lambda_{L2}^T & w^T \end{bmatrix} \begin{bmatrix} G_{11} & G_{12} \\ G_{21} & G_{22} \\ G_{31} & G_{32} \end{bmatrix} \begin{bmatrix} i_{L1} \\ v_{C2} \end{bmatrix}$$

Notice that the coordinates w appear only in the cross product term. The first two terms represent "kinetic" energy minus "potential" energy and depend only on the original generalized coordinates and velocities. A somewhat tedious calculation shows that in order to insure that the loop nodal properties are retained with the addition of the cross product term in the Lagrangian, G must satisfy the following condition [117]:

$$\tilde{V}G = \frac{1}{2}(M + N)$$

where M is the skew symmetric matrix

$$M = \begin{bmatrix} 0 & -D_{ss21}^T \\ D_{ss21} & 0 \end{bmatrix}$$

and N is an arbitrary symmetric matrix. If N is chosen to be

$$N = \begin{bmatrix} 0 & D_{ss21}^T \\ D_{ss21} & 0 \end{bmatrix}$$

Then, using a left inverse of $\tilde{V}$, G is obtained as (3.41).

Now, as above, the generalized velocities, i_{L1}, v_{C2}, comprise a set of n variables, the generalized coordinates, q_{C1}, λ_{L2}, comprise a set of m variables, and w consists of $n + m - k$ quasi-coordinates (recall k is the rank of V). Thus, there are a total of $2n + 2m - k$ variables. Equation (3.38), along with Equations (3.42), (3.43) (or, (3.39)), comprise a complete set of $2n + 2m - k$ first-order differential equations that define the evolution of the network. □

Remark 3.22 (Generalized Force Expressions) The generalized Lagrange equations (3.39) reproduce the loop equations that specify the generalized velocity v_{L1} and the node equations that specify the generalized velocity i_{C2}, i.e., from (3.16) and (3.17)

$$i_{C2} = D_{ss21} i_{L1} + D_{ss22} i_{L2} + D_{sn21} i_{J2} + D_{sn22} i_{R2}$$

$$v_{L1} = -D_{ss11}^T v_{C1} - D_{ss21}^T v_{C2} - D_{ns11}^T v_{E1} - D_{ns21}^T v_{R1}$$

The last two terms of each of these relations come from nonconservative source and resistance elements and must be produced by the generalized force Q. Thus, define

$$Q(p,t) = \begin{bmatrix} Q_1 \\ Q_2 \end{bmatrix} = \begin{bmatrix} D_{sn21} i_{J2} + D_{sn22} i_{R2} \\ -D_{ns11}^T v_{E1} - D_{ns21}^T v_{R1} \end{bmatrix}$$

A more convenient form of the generalized forces can be obtained by defining potential functions for the nonconservative elements. Following Cherry [44] define

1. *resistor content*: $G_{R_{\mathcal{T}_1}}\left(i_{R_1}\right) = \int v_{R_1}^T\left(i_{R_1}\right) di_{R_1}$
2. *voltage source content*: $G_{E_{\mathcal{T}_1}}\left(i_{E_1}, t\right) = \int v_{E_1}^T(t)\, di_{E_1}$
3. *resistor co-content*: $G_{R_{\mathcal{L}_2}}^*\left(v_{R_2}\right) = \int i_{R_2}^T\left(v_{R_2}\right) dv_{R_2}$
4. *current source co-content*: $G_{J_{\mathcal{L}_2}}^*\left(v_{J_2}, t\right) = \int i_{J_2}^T(t)\, dv_{J_2}$

In terms of these functions, the generalized forces can be expressed in the very convenient form:

$$Q_1\left(i_{L_1}, t\right) = \frac{\partial}{\partial i_{L_1}}\left[G_{E_{\mathcal{T}_1}}\left(D_{ns1} i_{L_1}, t\right) + G_{R_{\mathcal{T}_1}}\left(D_{ns2} i_{L_1}\right)\right] \tag{3.44}$$

$$Q_2\left(v_{C_2}, t\right) = \frac{\partial}{\partial v_{C_2}}\left[G_{J_{\mathcal{L}_2}}^*\left(-D_{sn1}^T v_{C_2}, t\right) + G_{R_{\mathcal{L}_2}}^*\left(-D_{sn1}^T v_{C_2}\right)\right] \tag{3.45}$$

Remark 3.23 (Reduced Equations) Equations (3.42) and (3.43) can be further evaluated to reveal interesting structure in the system equations. Consider the Lagrangian (3.41). Notice that the cross product term reduces to

$$\tilde{q}^T G p = -\lambda_{L2}^T D_{ss22}^+ D_{ss21} i_{L1} + w^T \Delta_2^T D_{ss21} i_{L1}$$

Thus, (3.42) and (3.43) become

$$\frac{d}{dt}\left(\frac{\partial W^*}{\partial i_{L1}}\right) + \left(v_{C2}^T D_{ss22} D_{ss22}^+ D_{ss21} + \left(i_{L1}^T \Phi_1^T + v_{C2}^T \Phi_2^T\right) \Delta_2^T D_{ss21}\right) \\ - \frac{\partial Z(q)}{\partial q_{C1}} D_{ss11} - i_{L1}^T D_{ss21}^T \Delta_2 \Phi_1 = Q_1$$

$$\frac{d}{dt}\left(\frac{\partial W^*}{\partial v_{C2}}\right) - \frac{\partial Z(q)}{\partial \lambda_{L2}} D_{ss22}^T - \left(D_{ss22}^+ D_{ss21} i_{L1}\right)^T D_{ss22}^T - \left(\Delta_2^T D_{ss21} i_{L1}\right)^T \Phi_2 = Q_2$$

Now, Equation (3.38) provides

$$\dot{\lambda}_{L2} = -D_{ss22}^T v_{C2} \quad \dot{w} = \Phi_1 i_{L1} + \Phi_2 v_{C2}$$

so that after substitution and some rearrangement, we obtain

$$\frac{d}{dt}\left(\frac{\partial W^*}{\partial i_{L1}}\right) + v_{C2}^T \left[D_{ss22} D_{ss22}^+ D_{ss21} + \Phi_2^T \Delta_2^T D_{ss21}\right] \\ + i_{L1}^T \left[\Phi_1^T \Delta_2^T D_{ss21} - D_{ss21}^T \Delta_2 \Phi_1\right] - \frac{\partial Z(q)}{\partial q_{C1}} D_{ss11} = Q_1$$

$$\frac{d}{dt}\left(\frac{\partial W^*}{\partial v_{C2}}\right) + i_{L1}^T \left[-D_{ss21}^T \left(D_{ss22} D_{ss22}^+\right)^T - D_{ss21}^T \Delta_2 \Phi_2\right] - \frac{\partial Z(q)}{\partial \lambda_{L2}} D_{ss22}^T = Q_2$$

Recall Equations (3.36) and (3.37) and note that $\mathcal{A}^{-1}\mathcal{A} = I$. Thus,

$$\begin{bmatrix} V^+ & \Delta \\ \Sigma & 0 \end{bmatrix} \begin{bmatrix} V & \Gamma \\ \Phi & 0 \end{bmatrix} = \begin{bmatrix} V^+ V + \Delta \Phi & V^+ \Gamma \\ \Sigma V & \Sigma \Gamma \end{bmatrix} = I_{2n+2m-k}$$

Take the upper left $(n+m) \times (n+m)$ block and write it as

$$V^+ V + \Delta \Phi = \begin{bmatrix} D_{ss11}^+ D_{ss11} & 0 \\ 0 & \left(D_{ss22}^T\right)^+ D_{ss22}^T \end{bmatrix} + \begin{bmatrix} \Delta_1 \Phi_1 & \Delta_1 \Phi_2 \\ \Delta_2 \Phi_1 & \Delta_2 \Phi_2 \end{bmatrix} = I_{n+m}$$

From which it can be seen that

$$\begin{aligned} &D_{ss11}^+ D_{ss11} + \Delta_1 \Phi_1 = I_n \\ &\Delta_1 \Phi_2 = 0_{n \times m} \\ &\Delta_2 \Phi_1 = 0_{n \times n} \\ &\left(D_{ss22}^T\right)^+ D_{ss22}^T + \Delta_2 \Phi_2 = I_m \end{aligned}$$

These relations lead to the simplified equations

$$\frac{d}{dt}\left(\frac{\partial W^*}{\partial i_{L1}}\right) + v_{C2}^T D_{ss21} - \frac{\partial Z(q)}{\partial q_{C1}} D_{ss11} = Q_1 \tag{3.46}$$

$$\frac{d}{dt}\left(\frac{\partial W^*}{\partial v_{C2}}\right) - i_{L1}^T D_{ss21}^T - \frac{\partial Z(q)}{\partial \lambda_{L2}} D_{ss22}^T = Q_2 \tag{3.47}$$

It is interesting to observe that the i_{L1} and v_{C2} equations, (3.46) and (3.47), respectively, are connected via an antisymmetric matrix. This interaction is conservative and similar to the gyroscopic force interaction in mechanical systems.

Example 3.24 Linear RLC circuit. Consider the circuit in Figure 3.10. All capacitors, inductors, and resistors are assumed linear. The selected tree and co-tree are shown in Figure 3.11.

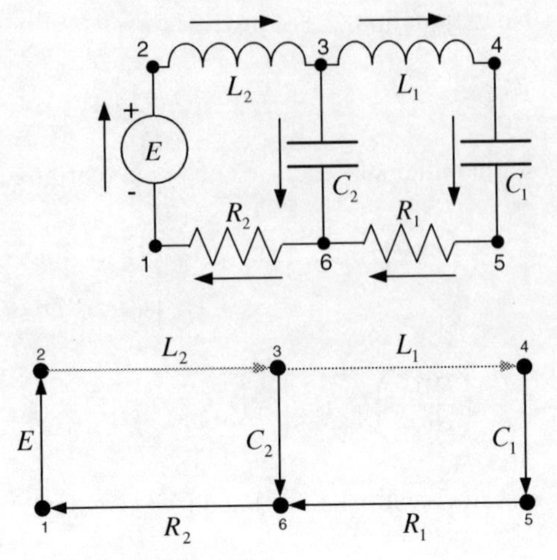

Fig. 3.10 The linear RLC network includes a voltage source and resistors

Fig. 3.11 The graph shows the selected tree (black) and the chords (gray)

The tree and co-tree divisions are chosen to be

$$\mathcal{T}_1 = \{C_1, C_2, E, R_1, R_2\}, \mathcal{T}_2 = \{\}$$

$$\mathcal{L}_1 = \{L_1, L_2\}, \mathcal{L}_2 = \{\}$$

Thus, the network generalized coordinates are the capacitor charges, q_{C_1}, q_{C_2}, and the generalized velocities are the inductor currents, i_{L_1}, i_{L_2}. The dynamic transformation relations are

$$\begin{bmatrix} i_{C_1} \\ i_{C_2} \\ i_E \\ i_{R_1} \\ i_{R_2} \end{bmatrix} = \begin{bmatrix} 1 & 0 \\ -1 & 1 \\ 0 & 1 \\ 1 & 0 \\ 0 & 1 \end{bmatrix} \begin{bmatrix} i_{L_1} \\ i_{L_2} \end{bmatrix}, \quad \begin{bmatrix} v_{L_1} \\ v_{L_2} \end{bmatrix} = -\begin{bmatrix} 1 & -1 & 0 & 1 & 0 \\ 0 & 1 & 1 & 0 & 1 \end{bmatrix} \begin{bmatrix} v_{C_1} \\ v_{C_2} \\ v_E \\ v_{R_1} \\ v_{R_2} \end{bmatrix}$$

From the first two loop equations,

$$\dot{q} = Vp \Rightarrow \frac{d}{dt} \begin{bmatrix} q_{C_1} \\ q_{C_2} \end{bmatrix} = \begin{bmatrix} 1 & 0 \\ -1 & 1 \end{bmatrix} \begin{bmatrix} i_{L_1} \\ i_{L_2} \end{bmatrix}$$

For this setup, the Lagrangian is

$$L\left(i_{L_1}, i_{L_2}, q_{C_1}, q_{C_2}\right) = U_L^*\left(i_{L_1}, i_{L_2}\right) - T_C\left(q_{C_1}, q_{C_2}\right)$$

$$U_L^*\left(i_{L_1}, i_{L_2}\right) = \frac{1}{2}\left(L_1 i_{L_1}^2 + L_2 i_{L_2}^2\right), \quad T_C\left(q_{C_1}, q_{C_2}\right) = \frac{1}{2}\left(\frac{q_{C_1}^2}{C_1} + \frac{q_{C_2}^2}{C_2}\right)$$

From which generalized Lagrange equations are derived

$$L_1 \frac{d}{dt} i_{L_1} - \frac{q_{C_1}}{C_1} + \frac{q_{C_2}}{C_2} = v_E - \frac{i_{L_1}}{R_1} + \frac{i_{L_2}}{R_2}$$

$$L_2 \frac{d}{dt} i_{L_2} - \frac{q_{C_2}}{C_2} = -v_E - \frac{i_{L_2}}{R_2}$$

Example 3.25 Example 3.15 Revisited. Consider once again the circuit in Figure 3.7. Select a tree and co-tree

$$\mathcal{T} = \{E, R, C_1, C_2\}, \quad \mathcal{L} = \{L_1, L_2\}$$

There are several ways to partition $\mathcal{T}$ and $\mathcal{L}$. Consider the following four partitions:

1. $\mathcal{T}_1 = \{E, R, C_1, C_2\}, \mathcal{T}_2 = \{\} \quad \mathcal{L}_1 = \{L_1, L_2\}, \mathcal{L}_2 = \{\}$
2. $\mathcal{T}_1 = \{E, R\}, \mathcal{T}_2 = \{C_1, C_2\} \quad \mathcal{L}_1 = \{\}, \mathcal{L}_2 = \{L_1, L_2\}$
3. $\mathcal{T}_1 = \{E, R, C_1\}, \mathcal{T}_2 = \{C_2\} \quad \mathcal{L}_1 = \{L_1\}, \mathcal{L}_2 = \{L_2\}$
4. $\mathcal{T}_1 = \{E, R, C_2\}, \mathcal{T}_2 = \{C_1\} \quad \mathcal{L}_1 = \{L_2\}, \mathcal{L}_1 = \{L_2\}$

Partitions 1. and 2. are classical. Partition 1. corresponds to the choice of capacitor charges q_{C1}, q_{C2} as generalized coordinates and inductor currents i_{L1}, i_{L2} as generalized velocities. Alternatively, Partition 2. corresponds to choosing inductor flux linkages $\lambda_{L1}, \lambda_{L2}$ as coordinates and capacitor voltages v_{C1}, v_{C2} as velocities.

Partitions 3. and 4. are unusual. They are different because the coordinates are mixed capacitor charge and inductor flux linkage. With partition 3., the coordinates are q_{C1}, λ_{L2} and the corresponding velocities are i_{L1}, v_{C2}. Similarly, Partition 4. produces coordinates q_{C2}, λ_{L1} and velocities i_{L2}, v_{C1}.

Consider Partition 3. The dynamic transformation relations are

$$\begin{bmatrix} i_E \\ i_R \\ i_{C1} \\ i_{C2} \end{bmatrix} = \begin{bmatrix} 1 & 0 \\ 1 & -1 \\ 1 & 0 \\ 0 & 1 \end{bmatrix} \begin{bmatrix} i_{L1} \\ i_{L2} \end{bmatrix}, \quad \begin{bmatrix} v_{L1} \\ v_{L2} \end{bmatrix} = \begin{bmatrix} -1 & -1 & -1 & 0 \\ 0 & 1 & 0 & -1 \end{bmatrix} \begin{bmatrix} v_E \\ v_R \\ v_{C1} \\ v_{C2} \end{bmatrix}$$

The Lagrangian is

$$L\left(i_{L1}, v_{C2}, q_{C1}, \lambda_{L2}\right) = U_{L_1}^*\left(i_{L1}\right) + T_{C_2}^*\left(v_{C2}\right) - U_{L_2}\left(\lambda_{L2}\right) - T_{C_1}\left(q_{c1}\right)$$

where

$$U_L^*(i) = \frac{1}{2} L i^2, \, U_L(\lambda) = \frac{1}{2} \frac{\lambda^2}{L}, \, T_C^*(v) = \frac{1}{2} C v^2, \, T_C(q) = \frac{1}{2} \frac{q^2}{C}$$

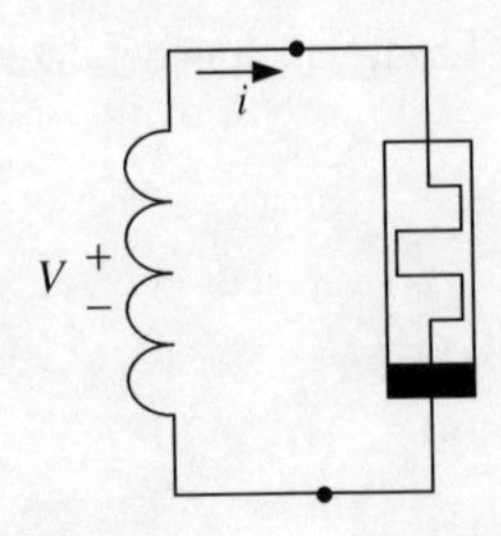

Fig. 3.12 A simple memristor circuit

Example 3.26 Memristor Examples

Recall that the full model for the charge-controlled memristor is

$$v = M(q)\,i, \quad \frac{dq}{dt} = i \tag{3.48}$$

and for the flux-controlled memristor,

$$i = W(\lambda)\,v, \quad \frac{d\lambda}{dt} = v \tag{3.49}$$

Note that in the linear case, i.e., M, W are constant, the differential equations are unnecessary and the memristor is essentially a resistor. The interesting case is the nonlinear memristor. Consider the circuit in Figure 3.12.

Place the current-controlled inductor in $\mathcal{L}_1$ and the charge-controlled memristor in $\mathcal{T}_1$. Thus, there are no generalized coordinates and the only generalized velocity is the inductor current, i_L. Define a single quasi-coordinate, q_L,

$$\frac{dq_L}{dt} = i_L \tag{3.50}$$

The Lagrangian is

$$L = U^*(i_L) = \frac{1}{2}L_1 i_L^2$$

Note that the incremental work done by the memristor is $\delta\mathcal{W} = v_L\,di_L$ so that

$$L_1\frac{di_L}{dt} = -M(q_L) \tag{3.51}$$

Equations (3.2) and (3.3) form a complete description of the circuit.

As a slightly more complex example, consider the circuit in Figure 3.13. In this case, the inductor and capacitor are considered to be linear, the voltage source is constant, and the memristor is flux-controlled. Thus, the tree and co-tree can be defined and partitioned in the following way:

$$\mathcal{T}_1 = \{E\}, \mathcal{T}_2 = \{C\}$$

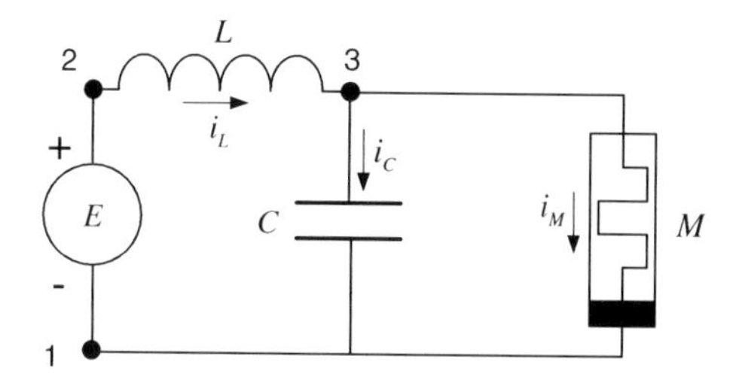

Fig. 3.13 A four element memristor circuit

$$\mathcal{L}_1 = \{L\}, \mathcal{L}_2 = \{M\}$$

The state variables are the two generalized velocities v_C and i_L, so the Lagrangian is

$$L = U_L^*(i_L) + T_C^*(v_C) = \tfrac{1}{2}L_1 i_L^2 + \tfrac{1}{2}C_1 v_C^2$$

Now, the flux-controlled memristor is characterized by the work done by the voltage source $\delta \mathcal{W}_E = V_E \, di_L$ and by the memristor $\delta \mathcal{W}_M = i_M \, dv_C = W(\lambda_M) \, v_C \, dv_C$. Thus, the Lagrange equations are

$$L\frac{di_L}{dt} = V_E, \quad C\frac{dv_C}{dt} = W(\lambda_M) \, v_C \tag{3.52}$$

Again, notice that Equations (3.49) and (3.52) provide a complete description of the network.

3.5.3.2 Brayton–Moser Equations

If all capacitors are voltage-controlled and all inductors are current-controlled, then it is possible to place all capacitors in $\mathcal{T}_2$ and all inductors in $\mathcal{L}_1$. In this case, the capacitor voltages and inductor currents are viewed as generalized velocities and no generalized coordinates are explicitly identified. Thus, q is vacuous and all of the generalized coordinates are in fact quasi-coordinates, w, defined via

$$\dot{w} = p \tag{3.53}$$

so that $\Phi = I$ and $\Delta = I$. In this case, the Lagrangian assumes the form:

$$L = W^*(p) + w^T G p$$

and Lagrange equations are

$$\frac{d}{dt}\frac{\partial L}{\partial p} - \frac{\partial L}{\partial w} = Q \tag{3.54}$$

Upon further evaluation, these become

$$\left[\frac{\partial^2 W^*(p)}{\partial p^2}\right]\frac{dp}{dt} + p^T\left(G - G^T\right) = \left[\,Q_1\left(i_{L_1}, t\right)\; Q_2\left(v_{C_2}, t\right)\right] \tag{3.55}$$

with

$$G = \left[\,\Delta_2^T D_{ss21}\; 0\,\right] = \begin{bmatrix} 0 & 0 \\ D_{ss21} & 0 \end{bmatrix} \tag{3.56}$$

In partitioned form, the Lagrange equations are reduced to

$$\left[\frac{\partial^2 U^*_{\mathcal{L}_1}\left(i_{L_1}\right)}{\partial {i_{L_1}}^2}\right]\frac{di_{L_1}}{dt} + v_{C_2}^T D_{ss21} = Q_1\left(i_{L_1}, t\right) \tag{3.57}$$

$$\left[\frac{\partial^2 T^*_{\mathcal{T}_2}\left(v_{C_2}\right)}{\partial {i_{L_1}}^2}\right]\frac{dv_{C_2}}{dt} - i_{L_1}^T D_{ss21}^T = Q_2\left(v_{C_2}, t\right) \tag{3.58}$$

Now, define the *mixed potential function*

$$\begin{aligned} P\left(i_{L_1}, v_{C_2}, t\right) &= v_{C_2}^T D_{ss21} i_{L_1} - G^*_{J_{\mathcal{L}_2}}\left(-D_{sn1}^T v_{C_2}, t\right) \\ &\quad - G^*_{R_{\mathcal{L}_2}}\left(-D_{sn1}^T v_{C_2}\right) + G_{E_{\mathcal{T}_1}}\left(D_{ns1} i_{L_1}, t\right) + G_{R_{\mathcal{T}_1}}\left(D_{ns2} i_{L_1}\right) \end{aligned} \tag{3.59}$$

Then, Equation (3.55) can be written as

$$\left[\frac{\partial^2 W^*(p)}{\partial p^2}\right]\frac{dp}{dt} = J\frac{\partial P}{\partial p}, \quad J = \operatorname{diag}\left(I_{n_{L_1}}, I_{n_{C_2}}\right) \tag{3.60}$$

These are the Brayton–Moser equations [31]. They can also be written in partitioned form:

$$\left[\frac{\partial^2 U^*_{\mathcal{L}_1}\left(i_{L_1}\right)}{\partial {i_{L_1}}^2}\right]\frac{di_{L_1}}{dt} = \frac{\partial P}{\partial i_{L_1}} \tag{3.61}$$

$$\left[\frac{\partial^2 T^*_{\mathcal{T}_2}\left(v_{C_2}\right)}{\partial {i_{L_1}}^2}\right]\frac{dv_{C_2}}{dt} = -\frac{\partial P}{\partial v_{C_2}} \tag{3.62}$$

Example 3.27 Degenerate RLC circuit. As an example of the special case described above, consider the three-node, four-branch circuit in Figure 3.14. All capacitors, inductors, and resistors are assumed linear. The selected tree and co-tree are shown in Figure 3.15, i.e.,

$$\begin{aligned} \mathcal{T} &= \{E, C\} \\ \mathcal{L} &= \{R, L\} \end{aligned}$$

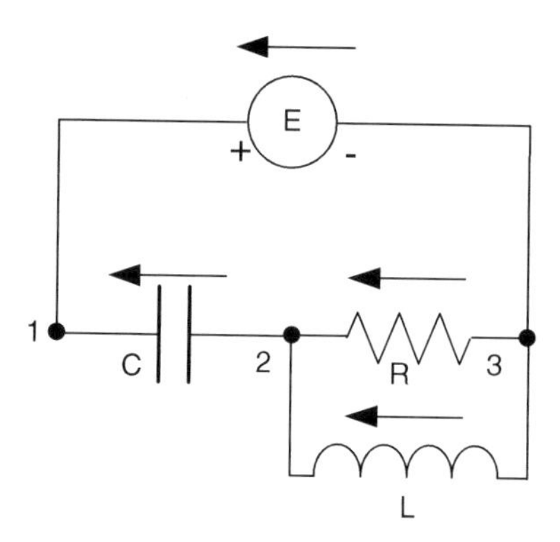

Fig. 3.14 The 4-element RLC network includes a voltage source

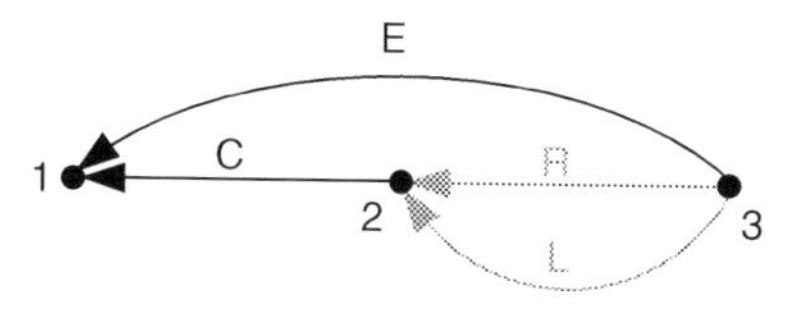

Fig. 3.15 The graph shows the selected tree (black) and the chords (gray)

Tree and co-tree partitions that satisfy the criteria are

$$\mathcal{T}_1 = \{E\}, \mathcal{T}_2 = \{C\}$$

$$\mathcal{L}_1 = \{L\}, \mathcal{L}_2 = \{R\}$$

The dynamic transformation relations are

$$\begin{aligned} \begin{bmatrix} i_C \\ i_E \end{bmatrix} &= \begin{bmatrix} D_{ss21} & D_{sn21} \\ D_{ns11} & D_{nn12} \end{bmatrix} \begin{bmatrix} i_L \\ i_R \end{bmatrix} \\ &= \begin{bmatrix} 1 & 1 \\ -1 & -1 \end{bmatrix} \begin{bmatrix} i_L \\ i_R \end{bmatrix} \end{aligned}$$

$$\begin{bmatrix} v_L \\ v_R \end{bmatrix} = \begin{bmatrix} -1 & 1 \\ -1 & 1 \end{bmatrix} \begin{bmatrix} v_C \\ v_E \end{bmatrix}$$

Note that in this arrangement, the state variables are the quasi-velocities v_C and i_L. A complete set of equations can be obtained in the form of Equations (3.57) and (3.58).

3.5.3.3 Generalized Hamilton Equations

In the usual way, the Hamiltonian, $H(\pi, \tilde{q})$, can be obtained by Legendre transformation of the Lagrangian, $L(p, \tilde{q})$. The variable π is the *generalized momentum* defined by

$$\pi \triangleq \frac{\partial L(p, \tilde{q})}{\partial p} \tag{3.63}$$

It is assumed that (3.63) admits a solution $p = \varphi(\pi, \tilde{q})$ for all admissible $\tilde{q}$. Then,

$$H(\pi, \tilde{q}) \triangleq \left[\pi p - L(p, \tilde{q})\right]_{p \to \varphi(\pi, \tilde{q})} \tag{3.64}$$

The transformation defined by Equations (3.63) and (3.64) possesses the properties

$$p = \frac{\partial H}{\partial \pi} \tag{3.65}$$

$$\frac{\partial H}{\partial \tilde{q}} = -\frac{\partial L}{\partial \tilde{q}} \tag{3.66}$$

In view of (3.63) and (3.64), the generalized Lagrange equations, Equations (3.38), (3.39), and (3.65), can be written

$$\frac{d\pi}{dt} = -\frac{\partial H}{\partial \tilde{q}}\tilde{V} + Q \tag{3.67}$$

$$\frac{d\tilde{q}}{dt} = \tilde{V}\frac{\partial H}{\partial \pi} \tag{3.68}$$

$$\lambda = \frac{\partial H}{\partial \tilde{q}}\Lambda \tag{3.69}$$

These constitute *Hamilton equations* for a general nonlinear circuit.

The Hamiltonian can be explicitly obtained from its definition (3.64) using the Lagrangian (3.40). The momentum is

$$\pi = \frac{\partial L(p, \tilde{q})}{\partial p} = \frac{\partial W^*(p)}{\partial p} + \tilde{q}^T G \tag{3.70}$$

and the Hamiltonian is

$$H(\pi, \tilde{q}) = \frac{\partial W^*(p)}{\partial p}p - W^*(p) + Z(q) \tag{3.71}$$

Note that

$$W\left(\frac{\partial W^*}{\partial p}\right) = \frac{\partial W^*}{\partial p}p - W^*(p) \tag{3.72}$$

so that using (3.70), the Hamiltonian is

$$H(\pi,\tilde{q}) = W(\pi - \tilde{q}G) + Z(q) \tag{3.73}$$

Note that

$$\dot{H}(\pi,\tilde{q}) = \dot{\pi}\frac{\partial H}{\partial \pi} + \frac{\partial H}{\partial \tilde{q}}\dot{\tilde{q}} = Qp \tag{3.74}$$

There is an interesting alternative Hamiltonian approach. Define the momentum

$$\mu = \frac{\partial W^*(p)}{\partial p} = \pi - \tilde{q}^T G \tag{3.75}$$

In terms of μ, the Hamiltonian is

$$\tilde{H}(\mu,\tilde{q}) = H(\pi,\tilde{q})|_{\pi\to\mu+\tilde{q}^T G} = W(\mu) + Z(q) \tag{3.76}$$

Now, Hamilton equations, (3.67) and (3.68), can be rewritten in the form:

$$\frac{d\mu}{dt} = \left(\frac{\partial W(\mu)}{\partial \mu}\right)^T \left(G^T\tilde{V} - \tilde{V}^T G\right) - \frac{\partial Z(q)}{\partial \tilde{q}}\tilde{V} + Q \tag{3.77}$$

$$\frac{d\tilde{q}}{dt} = \tilde{V}\frac{\partial W(\mu)}{\partial \mu} \tag{3.78}$$

Recall

$$\left(G^T\tilde{V} - \tilde{V}^T G\right) = M^T$$

and note that since $Z(q)$ does not depend on the quasi-coordinates w, Equation (3.77) is independent of w and there is no need to compute them using (3.78). That is, the quasi-coordinates are ignorable and (3.77) and (3.78) can be reduced to

$$\frac{d\mu}{dt} = \left(\frac{\partial W(\mu)}{\partial \mu}\right)^T M^T - \frac{\partial Z(q)}{\partial q}V + Q \tag{3.79}$$

$$\frac{dq}{dt} = V\frac{\partial W(\mu)}{\partial \mu} \tag{3.80}$$

These equations provide a complete and minimal state description of the system. Note also that $\tilde{H} = W(\mu) + Z(q)$ is the total energy of the system.

3.5.4 Excess Elements

Example 3.28 RLC System with Excess Elements. Consider the four-node, six-branch circuit in Figure 3.16. All capacitors, inductors, and resistors are assumed linear. Notice that the network has one excess capacitor and one excess inductor.

The selected tree and co-tree are shown in Figure 3.17, i.e.,

$$\mathcal{T} = \{a, c, e\}$$
$$\mathcal{L} = \{b, d, f\}$$

Now, voltage source c and inductor e must be assigned to $\mathcal{T}_1$ and capacitor b, and resistor d must be assigned to $\mathcal{L}_2$. Consequently, capacitor a must be assigned to $\mathcal{T}_2$ to avoid fundamental loops formed by chords of $\mathcal{L}_2$. Inductor f can be placed either in $\mathcal{L}_1$ or in $\mathcal{L}_2$. For now, it is placed in $\mathcal{L}_1$. Thus,

$$\mathcal{T}_1 = \{c, e\}, \quad \mathcal{T}_2 = \{a\}$$
$$\mathcal{L}_1 = \{f\}, \quad \mathcal{L}_2 = \{b, d\}$$

The chord–branch current relationship is

$$\begin{bmatrix} i_a \\ i_c \\ i_e \end{bmatrix} = \begin{bmatrix} 1 & 1 & -1 \\ -1 & -1 & 0 \\ -1 & 0 & 0 \end{bmatrix} \begin{bmatrix} i_f \\ i_d \\ i_b \end{bmatrix}$$

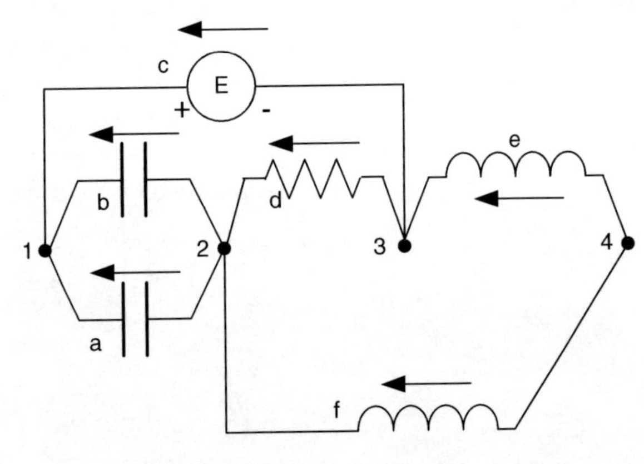

Fig. 3.16 The 6-element network includes 2 excess elements

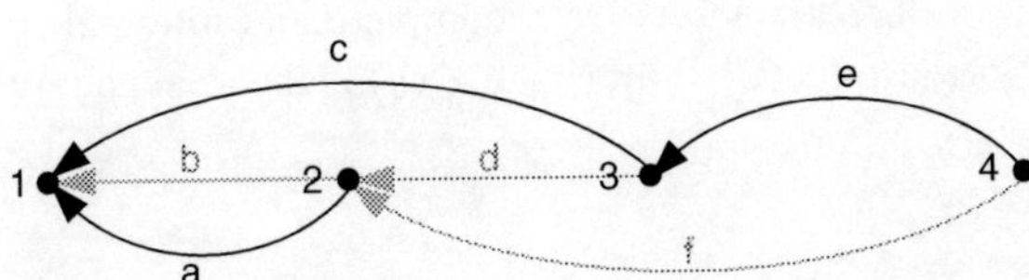

Fig. 3.17 The graph shows the selected tree (black) and the chords (gray)

and the branch–chord voltage relationship is

$$\begin{bmatrix} v_f \\ v_d \\ v_b \end{bmatrix} = \begin{bmatrix} -1 & 1 & 1 \\ -1 & 1 & 0 \\ 1 & 0 & 0 \end{bmatrix} \begin{bmatrix} v_a \\ v_c \\ v_e \end{bmatrix}$$

With this partition structure, the single generalized coordinate is the flux linkage of inductor f, λ_f. The corresponding generalized velocity is the capacitor a voltage v_a.

3.6 Coupled Circuits and Electromechanical Devices

Systems in which electrical and mechanical components interact are considered in this section. Electromechanical transducers are central to electric power systems—the most obviously important being generators and motors. A direct approach to developing the equations describing the behavior of such devices is to apply Lagrange equations in some form. The generalized Lagrange equations are particularly convenient for the systems of interest herein where the mechanical dynamics are rather simple and the complexity arises only in their integration with the electrical dynamics. The key ideas for application of the generalized equations to such systems will be illustrated with a series of examples.

Each example involves a coupled system that may include only electrical or both electrical and mechanical components. The system interacts with the external world by energy exchange through multiple ports. The first issue is to identify the idealized system energy storage potential function and the constitutive relations that can be derived from it. Then, the generalized Lagrange equations are derived for the nonideal system.

Example 3.29 Solenoid. As a simple example, first consider the ideal solenoid shown in Figure 3.18. The kinematic relationships are

$$\frac{d\lambda(t)}{dt} = e(t), \quad \frac{dx(t)}{dt} = v(t) \tag{3.81}$$

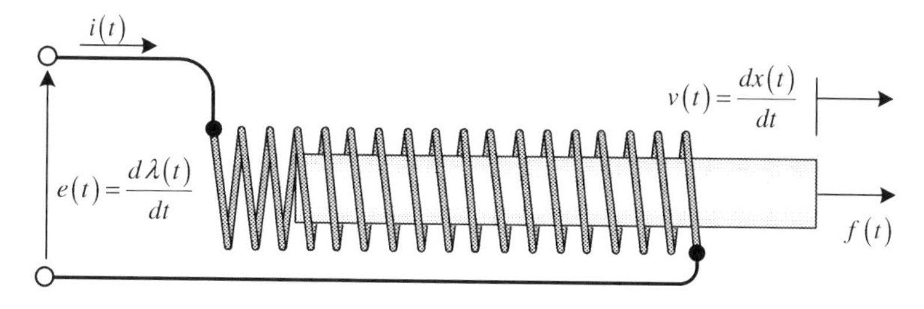

Fig. 3.18 An ideal solenoid with movable core

In addition, there are two constitutive relations

$$i = \phi_i(\lambda, x), \quad f = \phi_f(\lambda, x)$$

The work done by the solenoid in an infinitesimal time dt is

$$\delta W = ei\,dt + fv\,dt = i\,d\lambda + f\,dx$$

Consider the displacement of the solenoid from one static position to another. Assuming a lossless system, the work delivered is stored as magnetic energy, $U_m(\lambda, x)$. Thus,

$$dU_m(\lambda, x) = i\,d\lambda + f\,dx \tag{3.82}$$

or

$$U_m(\lambda, x) = \int_{\mathcal{P}} (i\,d\lambda + f\,dx) \tag{3.83}$$

where $\mathcal{P}$ is any path from $(0, 0)$ to (λ, x). Choose $\mathcal{P}$ to be composed of two segments: first, a line along the x-axis from $(0, 0)$ to $(0, x)$ and second, a line parallel to the λ-axis from $(0, x)$ to (λ, x). Then, the integral is reduced to

$$U_m(\lambda, x) = \int_0^{\lambda} \phi_i(\lambda, x)\,d\lambda$$

Assuming the solenoid constitutive relation is linear in current, $i = \lambda / L(x)$, U_m is reduced to

$$U_m(\lambda, x) = \frac{\lambda^2}{2L(x)}$$

Note that from (3.82),

$$i = \frac{\partial U_m}{\partial \lambda}, \quad f = \frac{\partial U_m}{\partial x}$$

Thus, if the magnetic energy function $U_m(\lambda, x)$ is known, the constitutive equations can be obtained from it. Consequently, in this example, it is

$$f = -\frac{\lambda^2 L'(x)}{2L^2(x)}$$

Now consider the nonideal solenoid in Figure 3.19 for which Lagrange equations will be derived.

The electrical part of the system consists of the simple circuit shown in Figure 3.20. The voltage source and the resistance form the tree, and as required, these elements are placed in $\mathcal{T}_1$. The only chord is the inductor, and it must be placed in

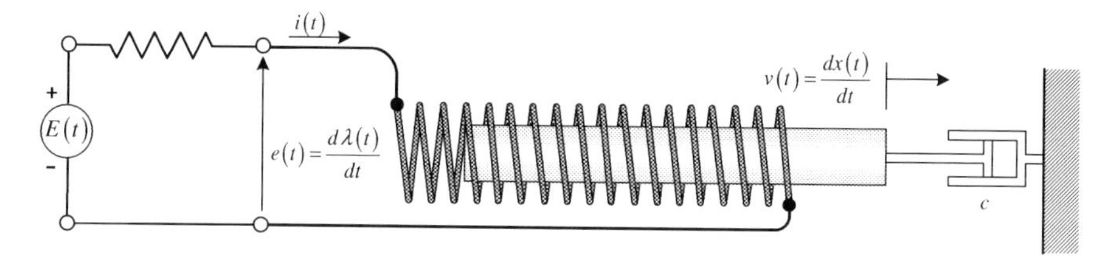

Fig. 3.19 A nonideal solenoid that includes electrical resistance and a mechanical damper

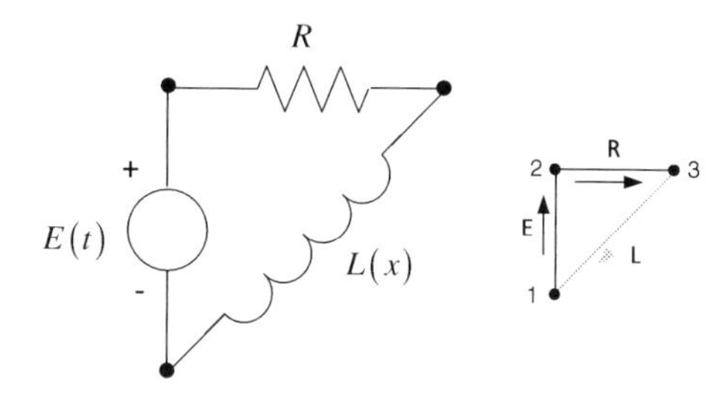

Fig. 3.20 The electrical part of the system consists of a three-element loop

$\mathcal{L}_1$ so as not to form loops with elements in $\mathcal{T}_1$. The only state variable in this setup is the inductor current, i, which is a generalized velocity. The only coordinate is the corresponding quasi-coordinate w, which is ignorable.

The mechanical subsystem contributes the coordinates x and its corresponding velocity, v

$$\frac{dx}{dt} = v \tag{3.84}$$

The Lagrangian is

$$L(x, i, v) = U_m^*(x, i) + T^*(v) - V(x)$$

with

$$U_m^*(i) = \tfrac{1}{2}L(x)\,i^2,\; T^*(v) = \tfrac{1}{2}mv^2,\;\; V(x) = \tfrac{1}{2}kx^2$$

The total infinitesimal work done by the external forces is

$$\delta W = (E(t) - Ri)\,\delta q - c\,v\,\delta x$$

Thus, the generalized Lagrange equations in the matrix form are

$$\begin{bmatrix} L(x) & 0 \\ 0 & m \end{bmatrix} \frac{d}{dt}\begin{bmatrix} i \\ v \end{bmatrix} + \begin{bmatrix} R & L_x(x)\,i \\ -\frac{1}{2}L_x(x)\,i & c \end{bmatrix}\begin{bmatrix} i \\ v \end{bmatrix} + \begin{bmatrix} 0 \\ k \end{bmatrix} x = \begin{bmatrix} E(t) \\ 0 \end{bmatrix}$$

Note that

$$\begin{bmatrix} 0 & L_x(x)\,i \\ -\frac{1}{2}L_x(x)\,i & 0 \end{bmatrix}\begin{bmatrix} i \\ v \end{bmatrix} = \begin{bmatrix} \frac{1}{2}L_x(x)\,v & \frac{1}{2}L_x(x)\,i \\ -\frac{1}{2}L_x(x)\,i & 0 \end{bmatrix}\begin{bmatrix} i \\ v \end{bmatrix}$$

So that the Lagrange equations can be written

$$\begin{bmatrix} L(x) & 0 \\ 0 & m \end{bmatrix}\frac{d}{dt}\begin{bmatrix} i \\ v \end{bmatrix} + \begin{bmatrix} \frac{1}{2}L_x(x)\,v + R & \frac{1}{2}L_x(x)\,i \\ -\frac{1}{2}L_x(x)\,i & c \end{bmatrix}\begin{bmatrix} i \\ v \end{bmatrix} + \begin{bmatrix} 0 \\ k \end{bmatrix}x = \begin{bmatrix} E(t) \\ 0 \end{bmatrix} \tag{3.85}$$

Equations (3.81) and (3.85) provide a complete characterization of the nonideal solenoid. The form of the equations (3.85) is very useful as the inertia, stiffness, gyroscopic, and dissipative terms are readily identifiable.

Example 3.30 Mutually Coupled Inductors

Mutually coupled inductors play an important role in motors and other devices. Consider the mutually coupled ideal inductors shown in Figure 3.21. The flux linkages of each coil depend on the currents in both coils

$$\lambda_1 = \phi_1(i_1, i_2)$$

$$\lambda_2 = \phi_2(i_1, i_2)$$

Assume, for now, that these relationships are invertible so that

$$i_1 = \varphi_1(\lambda_1, \lambda_2)$$

$$i_2 = \varphi_2(\lambda_1, \lambda_2)$$

The instantaneous power flowing into the pair of coils is

$$P = e_1 i_1 + e_2 i_2$$

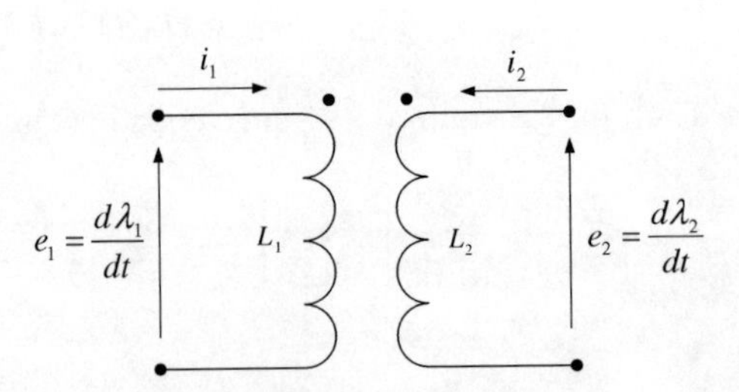

Fig. 3.21 Two mutually coupled inductors

Thus, the work done by the system in an infinitesimal time interval dt is

$$\delta U = e_1 i_1 \, dt + e_2 i_2 \, dt = i_1 \, d\lambda_1 + i_2 \, d\lambda_2 \tag{3.86}$$

Since the ideal inductors are conservative, the infinitesimal work equals the increase in magnetic stored energy in the pair of coils. Now, Equation (3.86) is an exact differential, $dU(\lambda_1, \lambda_2)$, only if the constitutive relations satisfy the integrability conditions

$$\frac{\partial \varphi_1(\lambda_1, \lambda_2)}{\partial \lambda_2} = \frac{\partial \varphi_2(\lambda_1, \lambda_2)}{\partial \lambda_1}$$

If this is the case, the magnetic energy function $U(\lambda_1, \lambda_2)$ exists for coupled pair.

If $U(\lambda_1, \lambda_2)$ is known, then the constitutive equations can be obtained by differentiation,

$$i_1 = \frac{\partial U(\lambda_1, \lambda_2)}{\partial \lambda_1}, \quad i_2 = \frac{\partial U(\lambda_1, \lambda_2)}{\partial \lambda_2}$$

Similarly, the co-energy $U^*(i_1, i_2)$ exists provided the integrability conditions hold

$$\frac{\partial \phi_1(i_1, i_2)}{\partial i_2} = \frac{\partial \phi_2(i_1, i_2)}{\partial i_1}$$

In this case, and if U^* is known, the constitutive relations can be obtained in the form:

$$\lambda_1 = \frac{\partial U^*(i_1, i_2)}{\partial i_1}, \quad \lambda_2 = \frac{\partial U^*(i_1, i_2)}{\partial i_2}$$

Furthermore, both U and U^* exist and they are related by Legendre transformation, example

$$U^*(i_1, i_2) = \lambda_1 i_1 + \lambda_2 i_2 - U(\lambda_1, \lambda_2)$$

Consider the case where the constitutive relations are linear, i.e.,

$$\begin{aligned} \lambda_1 &= L_{11} i_1 + L_{12} i_2 \\ \lambda_2 &= L_{21} i_1 + L_{22} i_2 \end{aligned}$$

Integrability requires $L_{12} = L_{21}$, that is, the inductance matrix is symmetric. Consequently,

$$\begin{bmatrix} \lambda_1 \\ \lambda_2 \end{bmatrix} = \begin{bmatrix} L_{11} & L_{12} \\ L_{12} & L_{22} \end{bmatrix} \begin{bmatrix} i_1 \\ i_2 \end{bmatrix}$$

and

$$\begin{bmatrix} i_1 \\ i_2 \end{bmatrix} = \begin{bmatrix} \Gamma_{11} & \Gamma_{12} \\ \Gamma_{12} & \Gamma_{22} \end{bmatrix} \begin{bmatrix} \lambda_1 \\ \lambda_2 \end{bmatrix}, \quad \begin{bmatrix} \Gamma_{11} & \Gamma_{12} \\ \Gamma_{12} & \Gamma_{22} \end{bmatrix} = \begin{bmatrix} L_{11} & L_{12} \\ L_{12} & L_{22} \end{bmatrix}^{-1}$$

With the constitutive relations known, it is a simple matter to integrate along any path to find the energy and co-energy functions. For example, to find the energy function, take any path C from $(0, 0)$ to (λ_1, λ_2)

$$U(\lambda_1, \lambda_2) = \frac{1}{2}\begin{bmatrix}\lambda_1 & \lambda_2\end{bmatrix}\begin{bmatrix}\Gamma_{11} & \Gamma_{12} \\ \Gamma_{12} & \Gamma_{22}\end{bmatrix}\begin{bmatrix}\lambda_1 \\ \lambda_2\end{bmatrix} \tag{3.87}$$

Similarly, integrate or use Legendre transformation to obtain

$$U^*(i_1, i_2) = \frac{1}{2}\begin{bmatrix}i_1 & i_2\end{bmatrix}\begin{bmatrix}L_{11} & L_{12} \\ L_{12} & L_{22}\end{bmatrix}\begin{bmatrix}i_1 \\ i_2\end{bmatrix} \tag{3.88}$$

Now consider the system shown in Figure 3.22. The network is composed of two circuits coupled by the mutual inductance. All elements are assumed to have linear constitutive relations. Each circuit has its own graph as illustrated in Figure 3.23. The left circuit tree and co-tree are subdivided in accordance with the criteria in Section 3.5.2 as follows:

$$\mathcal{T}_1 = \{E, R_1, L_3\}, \mathcal{T}_2 = \{\}, \mathcal{L}_1 = \{L_1\}, \mathcal{L}_2 = \{\}$$

Similarly, the right circuit tree and co-tree are divided as follows:

$$\mathcal{T}_1 = \{\}, \mathcal{T}_2 = \{C_2\}, \mathcal{L}_1 = \{L_2\}, \mathcal{L}_2 = \{R_2\}$$

The significance of these tree and co-tree divisions is that they organize the choice of state variables. Note that L_3 is an excess element in the left circuit. From the left circuit, inductor L_1 contributes a velocity i_{L_1}. From the right circuit, L_2 contributes

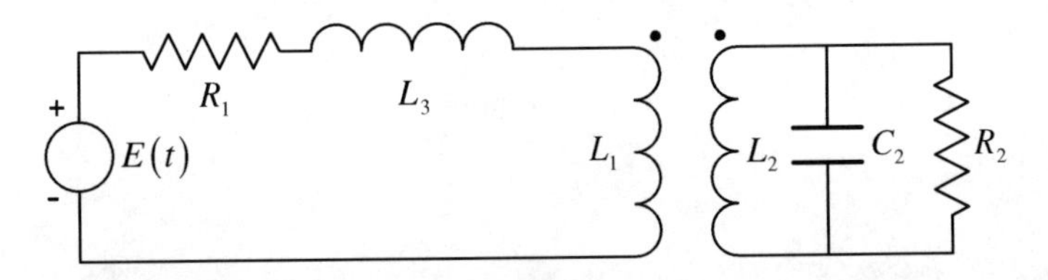

Fig. 3.22 Two circuits coupled through a mutual inductance

Fig. 3.23 Directed graphs for the two distinct circuits in Figure 3.22

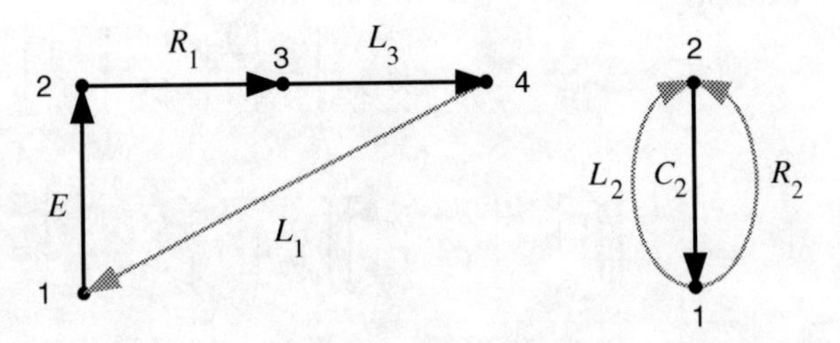

a velocity i_{L_2} and C_2 contributes a velocity v_{C_2}. As in the case of Section 3.5.3.2, all generalized coordinates are quasi-coordinates, w, with $\dot{w} = p$, i.e.,

$$\frac{d}{dt}\begin{bmatrix} w_1 \\ w_2 \\ w_3 \end{bmatrix} = \begin{bmatrix} i_{L_1} \\ i_{L_2} \\ v_{C_2} \end{bmatrix} \tag{3.89}$$

Consider the assembly of the Lagrangian. In this case, since all storage elements contribute velocity states, the Lagrangian involves only co-energy functions. The circuit cross-coupling terms, $w^T Gp$, can be assembled independently for the left and right circuits and then combined. First, the dynamic transformation relations for the two circuits, respectively, are

$$\begin{bmatrix} i_E \\ i_{R_1} \\ i_{L_3} \end{bmatrix} = \begin{bmatrix} 1 \\ 1 \\ 1 \end{bmatrix} i_{L_1}, \quad i_{C_2} = \begin{bmatrix} 1 & 1 \end{bmatrix} \begin{bmatrix} i_{L_2} \\ i_{R_2} \end{bmatrix}$$

Notice that for the left circuit, D_{ss21} is null, so the cross product term is absent. On the other hand, for the right circuit, the cross product term is $w_3 i_{L_2}$. Consequently, the Lagrangian is

$$L\left(i_{L_1}, i_{L_2}, i_{C_2}, w_3\right) = U^*_{L_1L_2}\left(i_{L_1}, i_{L_2}\right) + U^*_{L_3}\left(i_{L_1}\right) + T^*_{C_2}\left(v_{C_2}\right) + w_3 i_{L_2} \tag{3.90}$$

where

$$U^*_{L_1L_2}\left(i_{L_1}, i_{L_2}\right) = \frac{1}{2}\begin{bmatrix} i_{L_1} & i_{L_2} \end{bmatrix}\begin{bmatrix} L_{11} & L_{12} \\ L_{12} & L_{22} \end{bmatrix}\begin{bmatrix} i_{L_1} \\ i_{L_2} \end{bmatrix}$$

$$U^*_{L_3}\left(i_{L_1}\right) = \frac{1}{2}L_3 i^2_{L_1}, \quad T^*_{C_2}\left(v_{C_2}\right) = \frac{1}{2}C_2 v^2_{C_2}$$

The generalized forces are easily determined

$$Q = \begin{bmatrix} E(t) - R_1 i_{L_1} & 0 & -v_{C_2}/R_2 \end{bmatrix} \tag{3.91}$$

Thus, Lagrange equations are obtained in the form of Equation (3.55)

$$\begin{bmatrix} (L_{11} + L_3) & L_{12} & 0 \\ L_{12} & L_{22} & 0 \\ 0 & 0 & C_2 \end{bmatrix} \frac{d}{dt}\begin{bmatrix} i_{L_1} \\ i_{L_2} \\ v_{C_2} \end{bmatrix} + \begin{bmatrix} R_1 & 0 & 0 \\ 0 & 0 & 1 \\ 0 & -1 & R_2^{-1} \end{bmatrix}\begin{bmatrix} i_{L_1} \\ i_{L_2} \\ v_{C_2} \end{bmatrix} = \begin{bmatrix} E(t) \\ 0 \\ 0 \end{bmatrix} \tag{3.92}$$

Equation (3.92) is a complete representation of the inductor-coupled network. All three of the quasi-coordinates can be considered ignorable.

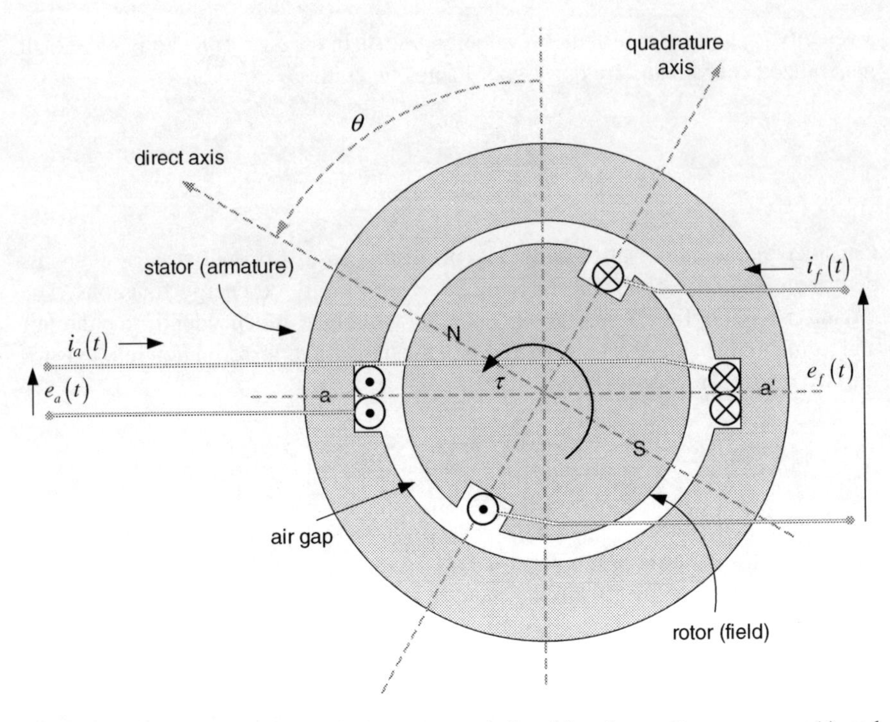

Fig. 3.24 A motor consisting of a single armature winding driven by a voltage source, $e_a(t)$, and a single field winding with a voltage source, $e_f(t)$

Example 3.31 Single-Phase Motor. Consider the ideal electric motor shown in Figure 3.24. Because the rotor is free to rotate, the mutual inductance will vary as a periodic function of rotor angle θ. The system can be viewed as having three ports through which energy can be provided to or withdrawn from the system. They include the electrical inputs to the rotor and stator and the mechanical input to the rotor shaft.

The kinematic relations are

$$\frac{d\lambda_a(t)}{dt} = e_a(t), \quad \frac{d\lambda_f(t)}{dt} = e_f(t), \quad \frac{d\theta(t)}{dt} = \omega(t) \tag{3.93}$$

The constitutive relations for the coupled inductors now take the form:

$$\begin{aligned} \lambda_a &= \phi_a(i_a, i_f, \theta) \\ \lambda_f &= \phi_f(i_a, i_f, \theta) \end{aligned} \tag{3.94}$$

or

$$\begin{aligned} i_a &= \varphi_a(\lambda_a, \lambda_f, \theta) \\ i_f &= \varphi_f(\lambda_a, \lambda_f, \theta) \end{aligned} \tag{3.95}$$

Now, the instantaneous power delivered to the system is

$$P(t) = e_a(t)\, i_a(t) + e_f(t)\, i_f(t) + \tau(t)\, \omega(t)$$

Thus, the work done in an infinitesimal time, dt, is

$$\delta W = e_a i_a dt + e_f i_f dt + \tau \omega dt = i_a d\lambda_a + i_f d\lambda_f + \tau d\theta \tag{3.96}$$

Consider the work done in displacing the rotor from one static position to another. Since the system is conservative, the resultant energy input is stored as magnetic energy. Thus, δU is an exact differential, and there exists a potential function $U\left(\lambda_a, \lambda_f, \theta\right)$ that can be obtained by integrating (3.96) over an arbitrary path $\mathcal{P}$ beginning at $(0, 0, 0)$ and terminating at $\left(\lambda_a, \lambda_f, \theta\right)$

$$U\left(\lambda_a, \lambda_f, \theta\right) = \int_{\mathcal{P}} \left(i_a d\lambda_a + i_f d\lambda_f + \tau d\theta\right) \tag{3.97}$$

Formally, the differential (3.96) is exact only if the integrability condition is satisfied, $J = J^T$ where

$$J = \begin{bmatrix} \frac{\partial \varphi_a}{\partial \lambda_a} & \frac{\partial \varphi_a}{\partial \lambda_f} & \frac{\partial \varphi_a}{\partial \theta} \\ \frac{\partial \varphi_f}{\partial \lambda_a} & \frac{\partial \varphi_f}{\partial \lambda_f} & \frac{\partial \varphi_f}{\partial \theta} \\ \frac{\partial \tau}{\partial \lambda_a} & \frac{\partial \tau}{\partial \lambda_f} & \frac{\partial \tau}{\partial \theta} \end{bmatrix} \tag{3.98}$$

Suppose the constitutive relations are linear in the currents and flux linkages so that (3.95) takes the form:

$$\begin{bmatrix} i_a \\ i_f \end{bmatrix} = \begin{bmatrix} \Gamma_{11}(\theta) & \Gamma_{12}(\theta) \\ \Gamma_{12}(\theta) & \Gamma_{22}(\theta) \end{bmatrix} \begin{bmatrix} \lambda_a \\ \lambda_f \end{bmatrix} \tag{3.99}$$

Notice that in the absence of a magnetic field, the rotor can be displaced to the desired position, θ, without any effort, i.e., $\tau = 0$. Thus, integrate first along the θ-axis, so that $\lambda_a = 0$, $\lambda_f = 0$, from the origin to the desired θ. Then, with θ constant, integrate in the λ_a direction with $\lambda_f = 0$ to the desired λ_a. Finally, with θ and λ_a constant, integrate in the λ_f direction to the desired λ_f. The result is

$$U\left(\lambda_a, \lambda_f, \theta\right) = \frac{1}{2} \begin{bmatrix} \lambda_a & \lambda_f \end{bmatrix} \begin{bmatrix} \Gamma_{11}(\theta) & \Gamma_{12}(\theta) \\ \Gamma_{12}(\theta) & \Gamma_{22}(\theta) \end{bmatrix} \begin{bmatrix} \lambda_a \\ \lambda_f \end{bmatrix} \tag{3.100}$$

As usual, Legendre transformation leads to the co-energy

$$U^*\left(i_a, i_f, \theta\right) = \frac{1}{2} \begin{bmatrix} i_a & i_f \end{bmatrix} \begin{bmatrix} L_{11}(\theta) & L_{12}(\theta) \\ L_{12}(\theta) & L_{22}(\theta) \end{bmatrix} \begin{bmatrix} i_a \\ i_f \end{bmatrix} \tag{3.101}$$

The constitutive relations can be recovered from the energy functions once they are known. In this case, the mechanical torque can be obtained

$$\tau\left(\lambda_a,\lambda_f,\theta\right)=\frac{\partial U\left(\lambda_a,\lambda_f,\theta\right)}{\partial\theta}=\frac{1}{2}\left[\lambda_a\;\lambda_f\right]\begin{bmatrix}\partial\Gamma_{11}(\theta)/\partial\theta & \partial\Gamma_{12}(\theta)/\partial\theta\\ \partial\Gamma_{12}(\theta)/\partial\theta & \partial\Gamma_{22}(\theta)/\partial\theta\end{bmatrix}\begin{bmatrix}\lambda_a\\ \lambda_f\end{bmatrix} \tag{3.102}$$

$$\tau\left(i_a,i_f,\theta\right)=-\frac{U^*\left(i_a,i_f,\theta\right)}{\partial\theta}=-\frac{1}{2}\left[i_a\;i_f\right]\begin{bmatrix}\partial L_{11}(\theta)/\partial\theta & \partial L_{12}(\theta)/\partial\theta\\ \partial L_{12}(\theta)/\partial\theta & \partial L_{22}(\theta)/\partial\theta\end{bmatrix}\begin{bmatrix}i_a\\ i_f\end{bmatrix} \tag{3.103}$$

Consider the nonideal motor shown in Figure 3.25 with electrical resistance and load torque, T_{load}. The electric network consists of two coupled circuits. The armature circuit consists of a single loop made up of the voltage source, $E_a(t)$, the resistor, R_a, and the coupled inductor, L_a. The field is similar, composed of the voltage source, $E_f(t)$. the resistor, R_f, and the coupled inductor, L_f. The armature circuit is divided

$$\mathcal{T}_1=\{E_a,R_a\},\mathcal{T}_2=\{\},\mathcal{L}_1=\{L_a\},\mathcal{L}_2=\{\}$$

and the field circuit divided similarly

$$\mathcal{T}_1=\left\{E_f,R_f\right\},\mathcal{T}_2=\{\},\mathcal{L}_1=\left\{L_f\right\},\mathcal{L}_2=\{\}$$

Thus, the armature circuit contributes a velocity, i_{L_a}, and the field circuit contributes a velocity, i_{L_f}. The coordinates are quasi-coordinates characterized by

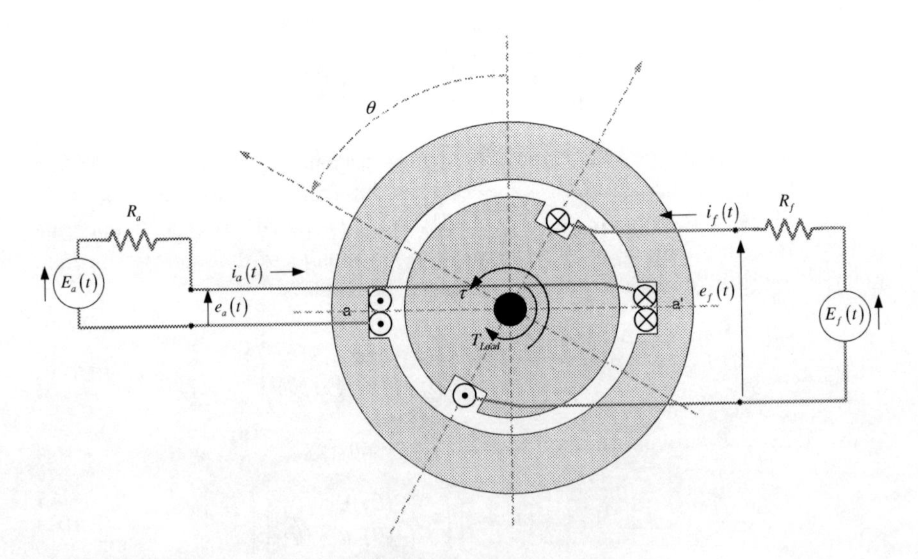

Fig. 3.25 A single-phase motor with resistance and load

$$\frac{d}{dt}\begin{bmatrix} w_1 \\ w_2 \end{bmatrix} = \begin{bmatrix} i_{L_a} \\ i_{L_f} \end{bmatrix} \tag{3.104}$$

The mechanical subsystem contributes the generalized velocity, the angular velocity of the rotor and shaft, ω, and the corresponding coordinate, θ. Since there are no capacitors in either the armature or field circuits, a cross product will not appear in the Lagrangian which takes the form:

$$L\left(i_a, i_f, \omega, \theta\right) = U^*\left(i_a, i_f, \theta\right) + T^*(\omega) \tag{3.105}$$

$$U^*\left(i_a, i_f, \theta\right) = \frac{1}{2}\begin{bmatrix} i_a & i_f \end{bmatrix}\begin{bmatrix} L_{11}(\theta) & L_{12}(\theta) \\ L_{12}(\theta) & L_{22}(\theta) \end{bmatrix}\begin{bmatrix} i_a \\ i_f \end{bmatrix}, \quad T^*(\omega) = \frac{1}{2}J\omega^2 \tag{3.106}$$

The Lagrange equations are

$$M(\theta)\frac{d}{dt}\begin{bmatrix} i_a \\ i_f \\ \omega \end{bmatrix} + C\left(i_a, i_f, \omega, \theta\right)\begin{bmatrix} i_a \\ i_f \\ \omega \end{bmatrix} = \begin{bmatrix} E_a(t) \\ E_f(t) \\ -T_{Load}(t) \end{bmatrix} \tag{3.107}$$

with

$$M(\theta) = \begin{bmatrix} L_{11}(\theta) & L_{12}(\theta) & 0 \\ L_{12}(\theta) & L_{22}(\theta) & 0 \\ 0 & 0 & J \end{bmatrix} \tag{3.108}$$

$$C\left(i_a, i_f, \omega, \theta\right) = \begin{bmatrix} R_a + \frac{1}{2}\omega L_{\theta,11} & \frac{1}{2}\omega L_{\theta,12} & \frac{1}{2}\left(i_a L_{\theta,11} + i_f L_{\theta,12}\right) \\ \frac{1}{2}\omega L_{\theta,12} & R_f + \frac{1}{2}\omega L_{\theta,22} & \frac{1}{2}\left(i_a L_{\theta,12} + i_f L_{\theta,22}\right) \\ -\frac{1}{2}\left(i_a L_{\theta,11} + i_f L_{\theta,12}\right) & -\frac{1}{2}\left(i_a L_{\theta,12} + i_f L_{\theta,22}\right) & 0 \end{bmatrix} \tag{3.109}$$

Equation (3.107) along with the kinematic relation

$$\frac{d\theta}{dt} = \omega \tag{3.110}$$

forms a complete system of equations. Notice that the matrix $C\left(i_a, i_f, \omega, \theta\right)$ is written in a form such that the symmetric (dissipative) and antisymmetric (conservative) elements are easily identified.

The concepts developed in the above example will be expanded in Chapter 4 to obtain models for multi-phase AC motors and generators.

Chapter 4
AC Power Systems

> *"The enchanting charms of this sublime science reveal only to those who have the courage to go deeply into it."*
>
> —Carl Friedrich Gauss

4.1 Introduction

For our purposes, an AC network is simply a network, in which all voltage and current sources are sinusoidal with a common frequency. An example is shown in Figure 4.1. Most power systems involve AC circuits because AC systems provide considerable benefits over DC in terms of cost to build, efficiency, and safety.

The chapter begins with a discussion of networks that are simple combinations of linear, one-port elements, and then introduces the concepts of real and reactive power. Multi-port models are introduced along with the reciprocity theorem. Models for single-phase machines, transformers, and transmission lines are discussed.

A discussion of three-phase AC networks follows with a detailed examination of models for synchronous machines, induction machines, and permanent magnet synchronous machines. Again, the modeling framework is Lagrange equations. Balanced operation of AC networks is discussed and, in this context, simplified machine models are derived. Attention then turns to the network power flow equations after which the classical simplified model is derived for balanced AC power systems.

H.G. Kwatny and K. Miu-Miller, *Power System Dynamics and Control*,
Control Engineering, DOI 10.1007/978-0-8176-4674-5_4

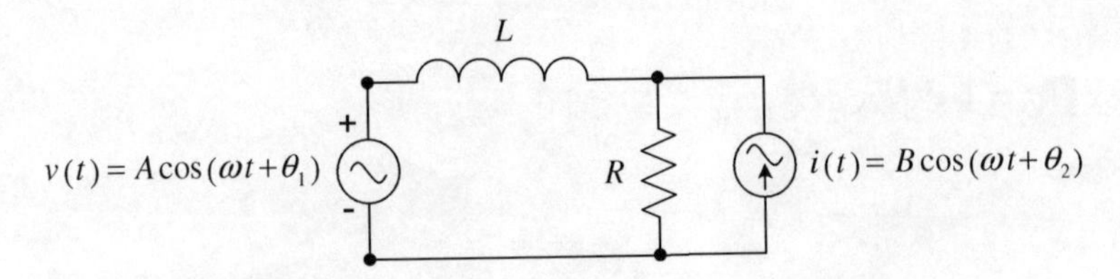

Fig. 4.1 This simple network illustrates AC voltage and current sources

4.2 Basics Concepts of AC Networks

4.2.1 Impedance Models of Linear Networks

We will examine some basic characteristics of networks composed of linear time-invariant components – resistor, capacitors, and inductors – in addition to controlled voltage and current sources. Consider the source supplied voltages or currents as inputs and all dependent voltages and currents as outputs. Then, as with any linear, time-invariant system, one can define a transfer matrix relating inputs to outputs [7, 41].

Specifically, consider the linear two-terminal or one-port network shown in Figure 4.2. Assume the network has no internal sources. Two terminals constitute a port if the current entering one-terminal equals the current leaving the other. The port current into the network responds to the applied voltage. Accordingly, we can express the input–output relationship in terms of a transfer function $Y(s)$

$$I(s) = Y(s)\,V(s)$$

where s is the Laplace variable.

Now, suppose

$$v(t) = \sqrt{2}\,\mathbf{V}e^{j\omega t} \Rightarrow V(s) = \frac{\sqrt{2}\,\mathbf{V}}{s - j\omega},$$

where $\mathbf{V}$ is a complex number and ω is real. The scale factor $\sqrt{2}$ is introduced for reasons to be discussed below. It follows that we can compute the current

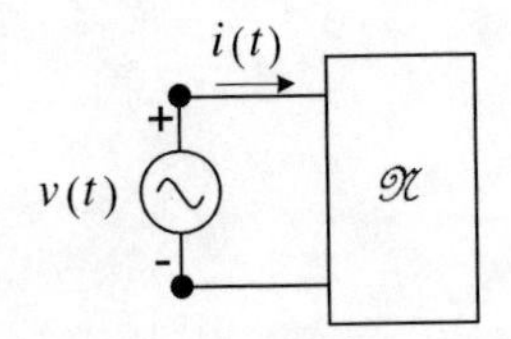

Fig. 4.2 A linear one-port network with voltage source

$$I(s) = Y(s)\frac{\sqrt{2}\,\mathbf{V}}{s - j\omega} = \frac{\sqrt{2}\,\mathbf{V}Y(j\omega)}{s - j\omega} + \textit{transient terms}$$

where the last expression is obtained from partial fraction expansion. If we assume the network is stable, then as $t \to \infty$ the transient vanishes and we obtain the *steady-state* forced, periodic solution

$$\lim_{t\to\infty} i(t) = \sqrt{2}\,\mathbf{I}e^{j\omega t} = \sqrt{2}\,\mathbf{V}Y(j\omega)\,e^{j\omega t}$$

Consequently, we have

$$\mathbf{I} = Y(j\omega)\mathbf{V},\ \ \mathbf{V} = Z(j\omega)\mathbf{I},$$

where $Z(j\omega) = Y^{-1}(j\omega)$ is called the *impedance* of the one-port network and $Y(j\omega)$ is the *admittance*.

Example 4.1 Simple calculations lead to the following impedances for the elementary linear R, L, C components:

- Resistor

$$v(t) = Ri(t) \Rightarrow Z_R = R$$

- Inductor

$$L\frac{di(t)}{dt} = v(t) \Rightarrow Z_L = j\omega L$$

- Capacitor

$$C\frac{dv(t)}{dt} = i(t) \Rightarrow Z_C = \frac{1}{j\omega C}$$

It is often necessary to compute the aggregate impedance of networks of elements with known impedances. We will derive simple formulas for series and parallel combinations elements. These can be applied to resolve arbitrary networks.

From KCL, each of the three elements shown in the series combination of Figure 4.3 has the same current. Using this fact and applying KVL,

$$\mathbf{V} = (Z_1 + Z_2 + Z_3)\,\mathbf{I}$$

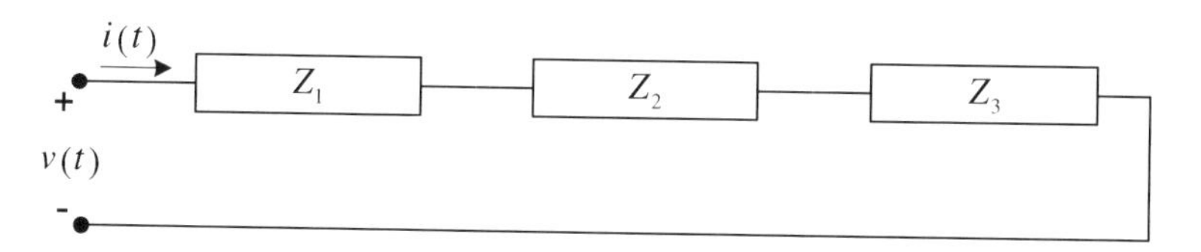

Fig. 4.3 A series combination is easily addressed in terms of impedances

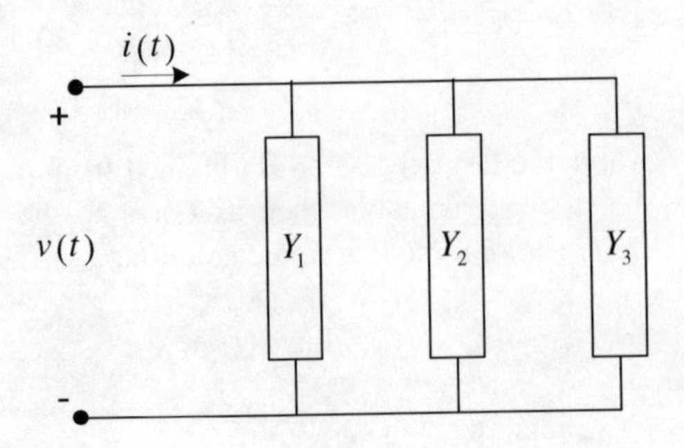

Fig. 4.4 A parallel combination is easily addressed in terms of admittances

Consequently, the equivalent impedance for the series connection is

$$Z = Z_1 + Z_2 + Z_3$$

By KVL, each element in the parallel combination in Figure 4.4 has the same voltage. KCL implies

$$\mathbf{I} = (Y_1 + Y_2 + Y_3)\,\mathbf{V}$$

The equivalent impedance for the parallel connection is

$$Z = \frac{1}{\frac{1}{Z_1} + \frac{1}{Z_2} + \frac{1}{Z_3}}$$

4.2.2 *Active and Reactive Power*

In this section, we focus on *steady-state* analysis of AC circuits as described above. By this, we mean that all currents and voltages have achieved a periodic behavior, so that each voltage or current has the following form

$$x(t) = X_{\max} \cos(\omega t + \theta) \tag{4.1}$$

where the *amplitude* X_{max}, *frequency* ω, and *phase* θ are all real numbers. The steady-state approximation will be relaxed later to admit slowly varying amplitude, frequency, and phase, where 'slowly varying' will be understood relative to the period $2\pi/\omega$.

Using Euler's formula, we can also write

$$x(t) = \mathrm{Re} X_{\max} e^{j(\omega t + \theta)} = \mathrm{Re}\sqrt{2}\mathbf{X} e^{j\omega t} \tag{4.2}$$

where

$$\mathbf{X} = \frac{X_{\max}}{\sqrt{2}} e^{j\theta} \tag{4.3}$$

The quantity $\mathbf{X}$ is the *phasor* representation for the sinusoid $x(t)$. Now, let us compute the root mean square value of $x(t)$:

$$X_{rms} = \left[\frac{\omega}{2\pi} \int_0^{2\pi/\omega} \left(X_{\max} \cos(\omega t + \theta)\right)^2 dt \right]^{\frac{1}{2}} = \frac{X_{\max}}{\sqrt{2}} = |\mathbf{X}| \tag{4.4}$$

Thus, we see one benefit of the scale factor $\sqrt{2}$.

Consider, once again, the network of Figure 4.2. Suppose the applied voltage is

$$v(t) = V_{\max} \cos(\omega t + \theta_v)$$

and the resulting current is

$$i(t) = I_{\max} \cos(\omega t + \theta_i)$$

The instantaneous power is

$$p(t) = v(t)\, i(t) = V_{\max} I_{\max} \cos(\omega t + \theta_v) \cos(\omega t + \theta_i)$$

Now, the average power injection over one period is

$$P = \frac{\omega}{2\pi} \int_0^{2\pi/\omega} V_{\max} I_{\max} \cos(\omega t + \theta_v) \cos(\omega t + \theta_i)\, dt \tag{4.5}$$

which evaluates to

$$P = \frac{V_{\max} I_{\max}}{2} \cos\phi = |\mathbf{V}|\, |\mathbf{I}| \cos\phi \tag{4.6}$$

where $\phi = \theta_v - \theta_i$. P is called the *real power* or *active power*. The factor $\cos\phi$ in (4.6) is called the *power factor*. For more details and examples, see [27].

Notice that we could write

$$P = \mathrm{Re}\, |\mathbf{V}|\, e^{j\theta_v}\, |\mathbf{I}|\, e^{-j\theta_i} = \mathrm{Re}\mathbf{V}\mathbf{I}^*$$

The quantity $\mathbf{V}\mathbf{I}^*$ is defined as the *complex power*, S

$$S = \mathbf{V}\mathbf{I}^* = |\mathbf{V}|\, |\mathbf{I}|\, e^{j\phi} = P + jQ \tag{4.7}$$

where $Q = \mathrm{Im}\mathbf{V}\mathbf{I}^*$ is called the *reactive power*.

Example 4.2 Consider a load with impedance Z, then

$$S = \mathbf{VI}^* = Z\mathbf{II}^* = Z|\mathbf{I}|^2$$

from which we can obtain

$$P = |\mathbf{I}|^2\,|Z|\cos\angle Z,\quad Q = |\mathbf{I}|^2\,|Z|\sin\angle Z$$

Now, suppose $i(t) = \sqrt{2}\,|\mathbf{I}|\cos(\omega t)$, in which case $v(t) = \sqrt{2}\,|Z|\,|\mathbf{I}|\cos(\omega t + \angle Z)$. Then, the instantaneous power is

$$\begin{aligned} p(t) &= v(t)\,i(t) = 2\,|Z|\,|\mathbf{I}|^2\cos(\omega t + \angle Z)\cos(\omega t) \\ &= |Z|\,|\mathbf{I}|^2\,(\cos(2\omega t + \angle Z) + \cos(\angle Z)) \\ &= |Z|\,|\mathbf{I}|^2\,(\cos(\angle Z) + \cos(2\omega t)\cos(\angle Z) - \sin(2\omega t)\sin(\angle Z)) \\ &= P\,(1 + \cos(2\omega t)) - Q\sin(2\omega t) \end{aligned}$$

Example 4.3 Consider the previous Example 4.2 in which the load corresponds to linear resistance, inductance, or capacitance, with corresponding impedances

$$Z_R = R, Z_L = \omega L e^{j\pi/2}, Z_C = \frac{e^{-j\pi/2}}{\omega C}$$

The absorbed real and reactive powers for each case are:

1. resistor $P = R|\mathbf{I}|^2, \quad Q = 0$
2. inductor $P = 0, \quad Q = \omega L|\mathbf{I}|^2$
3. capacitor $P = 0, \quad Q = -|\mathbf{I}|^2/\omega C.$

From this, we see that a resistor absorbs real power, whereas the inductor absorbs reactive power. On the other hand, the capacitor actually exports reactive power.

Remark 4.4 (*Implication of Reactive Power*) An essential aspect of the complex power S is shown in Figure 4.5 which summarizes the relationships in (4.7). Clearly, if the voltage magnitude $|V|$ and active power P are fixed, then increased reactive power Q results in increased current. This, in turn, means that equipment must be employed to accommodate the higher current requirements. Higher current also implies increased I^2R losses in generation and transmission equipment. As a result, reactive power is a critically important consideration in the design and operation of power systems.

An important property of AC circuits is *conservation of instantaneous power* which can be derived from KVL and KCL (see Theorem 3.13). Similarly, it can be shown that conservation of active power and conservation of reactive power hold true.

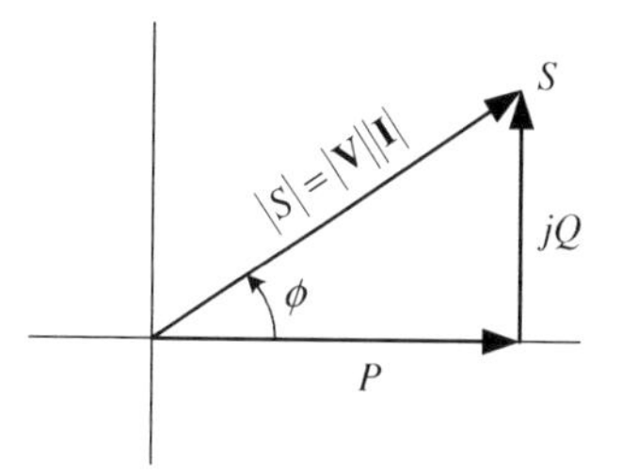

Fig. 4.5 The reactive power phasor shown as the sum of its parts

Proposition 4.5 (Conservation of Real and Reactive Power) *Consider a network with n branches. Let* $\mathbf{V}_i$, I_i, $i = 1, \ldots, n$ *denote the branch voltage phasors and branch current phasors, respectively. Then,*

$$\sum_{i=1}^{n} P_i = 0$$

$$\sum_{i=1}^{n} Q_i = 0$$

Proof The voltage phasors satisfy KVL and the current phasors satisfy KCL. If A is the network incident matrix, KCL implies

$$A\mathbf{I} = 0 \Rightarrow A\mathbf{I}^* = 0,$$

so that Tellegen's theorem implies

$$\sum_{i=1}^{n} \mathbf{V}_i \mathbf{I}_i^* = 0$$

so that

$$\sum_{i=1}^{n} S_i = 0 \Rightarrow \sum_{i=1}^{n} P_i = 0 \wedge \sum_{i=1}^{n} Q_i = 0$$

■

4.2.3 *Multi-port Networks*

In this section, we will be particularly concerned with models of *four-terminal* or *two-port* networks. However, we first discuss some properties of general multi-port networks. As in any multi-port device [165], we associate two variables with each port, in our case voltage and current. One of the variables at each port is treated as an input and the other as an output so the two-port device is a two input – two output dynamical system. If the device is linear, we can model it with a 2×2 transfer matrix. If, in addition, the device is stable and the inputs are sinusoids with a common frequency, then we can focus on its steady state, periodic response.

Example 4.6 Consider the network in Figure 4.6. The relationship between the port voltages and the currents can be written as

$$\begin{bmatrix} \mathbf{I}_1 \\ \mathbf{I}_2 \end{bmatrix} = \begin{bmatrix} \frac{1}{j\omega L} & -\frac{1}{j\omega L} \\ \frac{1}{j\omega L} & \frac{1}{R} - \frac{1}{j\omega L} \end{bmatrix} \begin{bmatrix} \mathbf{V}_1 \\ \mathbf{V}_2 \end{bmatrix}$$

or in reverse

$$\begin{bmatrix} \mathbf{V}_1 \\ \mathbf{V}_2 \end{bmatrix} = \begin{bmatrix} j\omega L - R & R \\ -R & R \end{bmatrix} \begin{bmatrix} \mathbf{I}_1 \\ \mathbf{I}_2 \end{bmatrix}$$

The first equation relates the voltages and currents with an impedance matrix, whereas the second uses its inverse, an admittance matrix. These relations can be generalized to arbitrary multi-ports composed of passive elements.

Example 4.7 A similar example that represents the simplest model of an AC transmission line is the network in Figure 4.7. In this case, the relationship between the port voltages and currents is

$$\begin{bmatrix} \mathbf{I}_1 \\ \mathbf{I}_2 \end{bmatrix} = \begin{bmatrix} \frac{1}{R+j\omega L} & -\frac{1}{R+j\omega L} \\ -\frac{1}{R+j\omega L} & \frac{1}{R+j\omega L} \end{bmatrix} \begin{bmatrix} \mathbf{V}_1 \\ \mathbf{V}_2 \end{bmatrix}$$

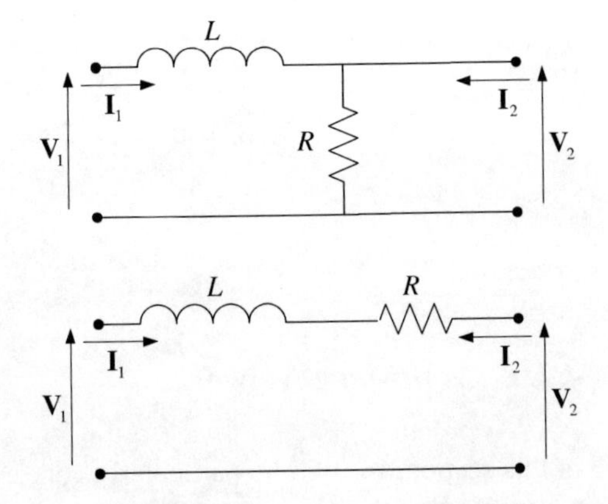

Fig. 4.6 A simple two-port network composed of an inductor and resistance

Fig. 4.7 A simple two-port network composed of an inductor and resistance

The power injected into port 1 is

$$\begin{aligned} P_1 = \mathrm{Re}\mathbf{V}_1\mathbf{I}_1^* &= \tfrac{R}{R^2+L^2\omega^2}\left(|\mathbf{V}_1|^2 - |\mathbf{V}_1|\,|\mathbf{V}_2|\cos(\theta_1-\theta_2)\right) \\ &+ \tfrac{L\omega}{R^2+L^2\omega^2}|\mathbf{V}_1|\,|\mathbf{V}_2|\sin(\theta_1-\theta_2) \\ Q_1 = \mathrm{Im}\mathbf{V}_1\mathbf{I}_1^* &= \tfrac{L\omega}{R^2+L^2\omega^2}\left(|\mathbf{V}_1|^2 - |\mathbf{V}_1|\,|\mathbf{V}_2|\cos(\theta_1-\theta_2)\right) \\ &- \tfrac{R}{R^2+L^2\omega^2}|\mathbf{V}_1|\,|\mathbf{V}_2|\sin(\theta_1-\theta_2) \end{aligned}$$

And for port 2 the injected power is

$$\begin{aligned} P_2 &= \tfrac{R}{R^2+L^2\omega^2}\left(|\mathbf{V}_2|^2 - |\mathbf{V}_1|\,|\mathbf{V}_2|\cos(\theta_1-\theta_2)\right) \\ &- \tfrac{L\omega}{R^2+L^2\omega^2}|\mathbf{V}_1|\,|\mathbf{V}_2|\sin(\theta_1-\theta_2) \\ Q_2 &= \tfrac{L\omega}{R^2+L^2\omega^2}\left(|\mathbf{V}_2|^2 - |\mathbf{V}_1|\,|\mathbf{V}_2|\cos(\theta_1-\theta_2)\right) \\ &+ \tfrac{R}{R^2+L^2\omega^2}|\mathbf{V}_1|\,|\mathbf{V}_2|\sin(\theta_1-\theta_2) \end{aligned}$$

For a general n-port, linear, passive circuit, we define the vector $\mathbf{V}$ of port voltages and vector $\mathbf{I}$ of port currents. Typically, we can write

$$\mathbf{I} = \mathbf{Y}\mathbf{V}, \quad \mathbf{V} = \mathbf{Z}\mathbf{I}, \quad \mathbf{Y} = \mathbf{Z}^{-1} \tag{4.8}$$

where $\mathbf{Z}$ and $\mathbf{Y}$ are, respectively, the network impedance and admittance matrices.

Remark 4.8 There are situations in which only $\mathbf{Z}$ exists and not $\mathbf{Y}$ or vice versa. For example, if the inductor is absent in the two-port of Example 4.6, then the admittance model is

$$\begin{bmatrix} \mathbf{V}_1 \\ \mathbf{V}_2 \end{bmatrix} = \begin{bmatrix} R & R \\ R & R \end{bmatrix} \begin{bmatrix} \mathbf{I}_1 \\ \mathbf{I}_2 \end{bmatrix}$$

Thus, we see that the impedance matrix $\mathbf{Z}$ exists but it is singular so the admittance matrix does not exist. Physically, this makes perfect sense. We could place a current source on both ports, but certainly not voltage sources.

On the other hand, it is possible to place a current source on one port and a voltage source on the other. For example, suppose a voltage source is applied to port 1 and current source to port 2. Then

$$\begin{bmatrix} \mathbf{I}_1 \\ \mathbf{V}_2 \end{bmatrix} = \begin{bmatrix} 1 & -1/R \\ 0 & 1 \end{bmatrix} \begin{bmatrix} \mathbf{I}_2 \\ \mathbf{V}_1 \end{bmatrix}$$

Thus, we have a ‘mixed’ representation.

Reciprocity is an important notion in circuit theory. It is discussed in the literature under several variations. We adopt the formulation in [43]. Consider an m-port network with two different excitation signals as shown in Figure 4.8. At each port, either variable, voltage or current may be the excitation and the other the response.

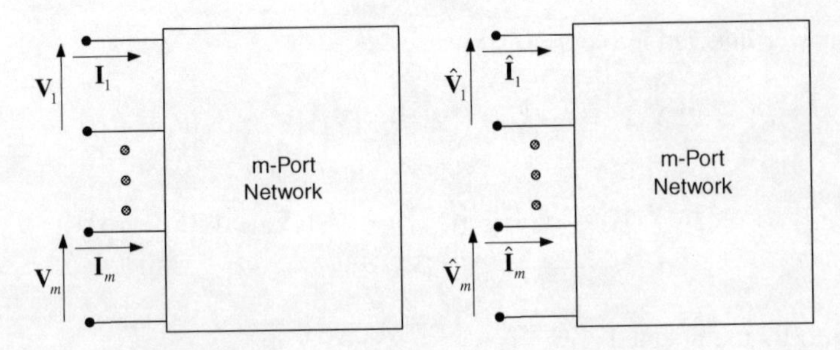

Fig. 4.8 An m-port network shown with two different excitations

Definition 4.9 (*Reciprocal Network*) An m-port network is *reciprocal* if

$$\sum_{k=1}^{m} \mathbf{I}_k \hat{\mathbf{V}}_k = \sum_{k=1}^{m} \hat{\mathbf{I}}_k \mathbf{V}_k \tag{4.9}$$

The meaning of the reciprocal property is most easily understood in terms of a 2-port network in which case (4.9) reduces to

$$\mathbf{I}_1 \hat{\mathbf{V}}_1 + \mathbf{I}_2 \hat{\mathbf{V}}_2 = \hat{\mathbf{I}}_1 \mathbf{V}_1 + \hat{\mathbf{I}}_2 \mathbf{V}_2 \tag{4.10}$$

Now, set

$$\mathbf{V}_1 = \hat{\mathbf{V}}_2 = \mathbf{E}, \quad \mathbf{V}_2 = \hat{\mathbf{V}}_1 = 0$$

So that (4.10) yields

$$\hat{\mathbf{I}}_1 = \mathbf{I}_2$$

In words, this means the following: Place a voltage source $\mathbf{E}$ at port 1 and short circuit port 2. This will produce a current $\mathbf{I}_2 = \mathbf{I}$ at port 2. Now interchange the procedure and place the source $\mathbf{E}$ at port 2 while short circuiting port 1. This will produce a current at port 1, $\hat{\mathbf{I}}_1 = \mathbf{I}$.

Alternatively, place a current source at port 1 and open circuit port 2. Then, reverse the procedure and place the current source at port 2 while open circuiting port 1:

$$\mathbf{I}_1 = \hat{\mathbf{I}}_2 = \mathbf{I}, \quad \mathbf{I}_2 = \hat{\mathbf{I}}_1 = 0$$

Then (4.10) yields

$$\mathbf{V}_2 = \hat{\mathbf{V}}_1$$

Theorem 4.10 (Reciprocity Theorem) *An m-port network with well-defined admittance matrix (impedance matrix) is reciprocal if and only if its admittance matrix (impedance matrix) is symmetric.*

Proof Sufficiency is proved in [43]. To establish necessity note that reciprocity requires

$$\mathbf{I}^T\hat{\mathbf{V}} = \hat{\mathbf{I}}^T\mathbf{V} \Rightarrow \mathbf{V}^T Y^T \hat{\mathbf{V}} = \hat{\mathbf{V}}^T Y^T \mathbf{V}$$

This must be true for all $\mathbf{V}$ and $\hat{\mathbf{V}}$. Thus, we conclude that $Y = Y^T$. ■

Remark 4.11 (*Two-Port Reciprocity*) The classical reciprocity result for two-port networks is often derived using Tellegen's theorem. Consider a two-port network with m internal branches as shown in Figure 4.9. Suppose that the m internal branches are characterized by the voltage and current variables $\mathbf{V}_1, \ldots, \mathbf{V}_m$ and $\mathbf{I}_1, \ldots, \mathbf{I}_m$ in the left excitation case and $\hat{\mathbf{V}}_1, \ldots, \hat{\mathbf{V}}_m, \hat{\mathbf{I}}_1, \ldots, \hat{\mathbf{I}}_m$ in the right. Assume that the two-port has linear, passive branch components, each characterized by an admittance model, of the form $\mathbf{I}_j = Y_j\mathbf{V}_j$. The port variables are designated $\mathbf{V}_{m+1}, \mathbf{I}_{m+1}, \mathbf{V}_{m+2}, \mathbf{I}_{m+2}$ and $\hat{\mathbf{V}}_{m+1}, \hat{\mathbf{I}}_{m+1}, \hat{\mathbf{V}}_{m+2}, \hat{\mathbf{I}}_{m+2}$, respectively.

Now, since $\mathbf{I}$, $\hat{\mathbf{I}}$, and $\mathbf{V}$, $\hat{\mathbf{V}}$ satisfy the KCL and KVL requirements of Tellegen's theorem, we can pair $\mathbf{V}$ and $\hat{\mathbf{I}}$ to obtain

$$\sum_{j=1}^{m} \mathbf{V}_j\hat{\mathbf{I}}_j + \mathbf{V}_{m+1}\hat{\mathbf{I}}_{m+1} + \mathbf{V}_{m+2}\hat{\mathbf{I}}_{m+2} = 0$$

and pair $\hat{\mathbf{V}}$ and $\mathbf{I}$ to obtain

$$\sum_{j=1}^{m} \hat{\mathbf{V}}_j\mathbf{I}_j + \hat{\mathbf{V}}_{m+1}\mathbf{I}_{m+1} + \hat{\mathbf{V}}_{m+2}\mathbf{I}_{m+2} = 0$$

Now, $\mathbf{V}_{m+2} = 0$, $\hat{\mathbf{V}}_{m+1} = 0$, so we have

$$\sum_{j=1}^{m} \mathbf{V}_j\hat{\mathbf{I}}_j + \mathbf{V}_{m+1}\hat{\mathbf{I}}_{m+1} = 0, \quad \sum_{j=1}^{m} \hat{\mathbf{V}}_j\mathbf{I}_j + \hat{\mathbf{V}}_{m+2}\mathbf{I}_{m+2} = 0$$

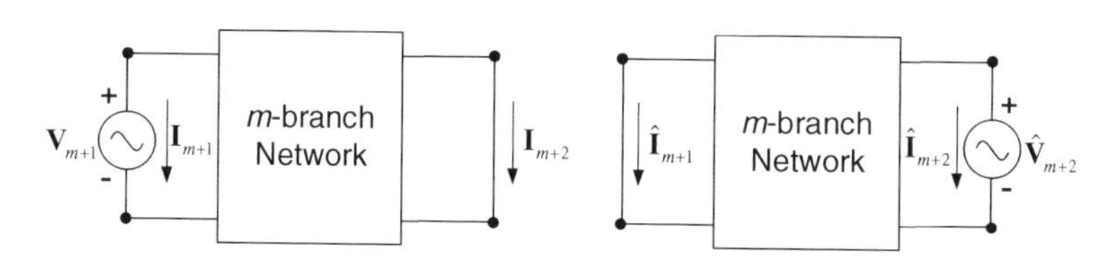

Fig. 4.9 A 2-port network with m internal branches is shown with two different excitation configurations

and using the constitutive relations for the internal branches, $\hat{\mathbf{I}}_j = Y_j \hat{\mathbf{V}}_j$, and $\mathbf{I}_j = Y_j \mathbf{V}_j$

$$\sum_{j=1}^{m} Y_j \mathbf{V}_j \hat{\mathbf{V}}_j + \mathbf{V}_{m+1} \hat{\mathbf{I}}_{m+1} = 0, \quad \sum_{j=1}^{m} Y_j \mathbf{V}_j \hat{\mathbf{V}}_j + \hat{\mathbf{V}}_{m+2} \mathbf{I}_{m+2}$$

Consequently,

$$\mathbf{V}_{m+1} \hat{\mathbf{I}}_{m+1} = \hat{\mathbf{V}}_{m+2} \mathbf{I}_{m+2}$$

Suppose $\mathbf{V}_{m+1} = \hat{\mathbf{V}}_{m+2}$, then

$$\hat{\mathbf{I}}_{m+1} = \mathbf{I}_{m+2}$$

Thus, we have reciprocity of the port short circuit currents.

Remark 4.12 Consider a network with n nodes and m branches. Assume all branches have a well-defined admittance. Choose two of the n nodes in order to create a port. For convenience, suppose the two nodes are $n-1$ and n. Denote the port current by $\mathbf{I}_{m+1}$ and the voltage across it (from node n to node $n-1$) by $\mathbf{V}_{m+1}$. KCL is now modified to

$$A^T \mathbf{I} = \begin{bmatrix} 0_{(n-2)\times 1} \\ \mathbf{I}_{m+1} \\ -\mathbf{I}_{m+1} \end{bmatrix}$$

and KVL to

$$\mathbf{V} = A\mathbf{v} = A \begin{bmatrix} v_1 \\ \vdots \\ v_{n-1} \\ \mathbf{V}_{m+1} + v_{n-1} \end{bmatrix}$$

Now,

$$\mathbf{I} = \operatorname{diag}(Y_1, \ldots, Y_m)\, \mathbf{V}$$

So, premultiplying by A^T and using KCL, KVL

$$\begin{bmatrix} 0_{(n-2)\times 1} \\ \mathbf{I}_{m+1} \\ -\mathbf{I}_{m+1} \end{bmatrix} = \begin{bmatrix} B_{(n-2)\times(n-2)} & C_{(n-2)\times 2} \\ C^T_{(n-2)\times 2} & D_{2\times 2} \end{bmatrix} \begin{bmatrix} v_1 \\ \vdots \\ \mathbf{V}_{m+1} + v_n \\ v_n \end{bmatrix}$$

where the matrix on the right is a partitioning of the symmetric matrix $A^T \operatorname{diag}(Y_1, \ldots, Y_m) A$; which implies that B and D are symmetric matrices.

4.2.4 Single-Phase Machines

Consider the motor of Example 3.31. Equation (3.107) can be written with mechanical and electrical dynamics separated

$$L(\theta)\frac{d\mathbf{i}}{dt} + \frac{\partial L(\theta)}{\partial\theta}\mathbf{i}\omega + \mathbf{R}\mathbf{i} = \mathbf{v} \tag{4.11}$$

$$J\frac{d\omega}{dt} + \frac{1}{2}\mathbf{i}^T\frac{\partial L(\theta)}{\partial\theta}\mathbf{i} = -T_L \tag{4.12}$$

with

$$\mathbf{i} = \begin{bmatrix} i_a \\ i_f \end{bmatrix}, \quad \mathbf{v} = \begin{bmatrix} v_a \\ v_f \end{bmatrix}$$

and

$$L(\theta) = \begin{bmatrix} L_s & M_f\cos(\theta) \\ M_f\cos(\theta) & L_f \end{bmatrix}, \quad \mathbf{R} = \begin{bmatrix} R & 0 \\ 0 & R_f \end{bmatrix}$$

Expanding (4.11) and (4.12) yields

$$\begin{bmatrix} L_s & M_f\cos(\theta) \\ M_f\cos(\theta) & L_f \end{bmatrix}\frac{d}{dt}\begin{bmatrix} i_a \\ i_f \end{bmatrix} + \begin{bmatrix} R & -\omega M_f\sin(\theta) \\ -\omega M_f\sin(\theta) & R_f \end{bmatrix}\begin{bmatrix} i_a \\ i_f \end{bmatrix} = \begin{bmatrix} v_a \\ v_f \end{bmatrix}$$

$$J\frac{d\omega}{dt} - M_f\sin(\theta)\, i_a i_f = -T_L$$

Now, consider the steady operation of the motor under constant load torque, T_L, and with AC stator voltage,

$$v_a(t) = V_a\cos(\omega_s t)$$

where V_a is the supply voltage magnitude and ω_s is the network synchronous frequency. In steady operation, it is assumed that the field current, i_f, is constant. In general, the synchronous machine field voltage, v_f, is used to regulate some variable, such as power factor in the motoring case or terminal voltage in the generating case, that results in a constant field current. These *excitation* control systems will be addressed below. It is also assumed that the mechanical dynamics evolve on a timescale much larger than the synchronous period $2\pi/\omega_s$.

In order to efficiently study the steady behavior, assume that the motor angular velocity, ω, is close to the synchronous frequency, ω_s, so that

$$\omega = \omega_s + v, \quad |v| \ll \omega_s$$

Under these conditions, the equations of motion can be rewritten

$$\frac{di_a}{dt} = -\left(\frac{R}{L_s}\right) i_a + (\omega_s + \upsilon)\, i_f \frac{M_f}{L_s} \sin\left(\omega_s t - \frac{\pi}{2} + \delta\right) + \frac{1}{L_s} V_a \cos(\omega_s t) \quad (4.13)$$

$$\frac{d\upsilon}{dt} = \left(i_f \frac{M_f}{J}\right) \sin\left(\omega_s t - \frac{\pi}{2} + \delta\right) i_a - \frac{1}{J} T_L \quad (4.14)$$

$$\frac{d\delta}{dt} = \upsilon \quad (4.15)$$

Note that the mechanical variable θ and its velocity ω have now been replaced by δ and υ, respectively. Recall that θ represents the counterclockwise rotation of the rotor d-axis relative to the stator a-axis. The $\pi/2$ terms shift the rotation with reference to the rotor q-axis. The variable δ should be viewed as the rotor angle relative to a frame rotating with synchronous velocity ω_s. The strategy to solve these equations is to use the *method of averaging* as described in [84]. The process is as follows:

1. Solve the short timescale equation, (4.13), for $i_a(t)$. This is accomplished by assuming the field current, i_f, is constant, treating the slowly varying υ as a small parameter and generating the solution as a function of υ. The result in the limit $\upsilon \to 0$ is a periodic, steady solution with frequency ω_s.
2. Substitute the steady solution for i_a into the long timescale equation (4.14) and average the right-hand side over time. Solve the averaged equation for the speed deviation, $\upsilon(t)$.

Remark 4.13 Averaging. The basis for the application of averaging is that the synchronous rotor speed ω_s is large when compared to the mechanical time constants. Thus, once the short time periodic behavior of the phase current is obtained and substituted into (4.14), the mechanical equations, (4.14) and (4.15), can be rewritten by changing the timescale according to the relation $t = \varepsilon\tau$, where $\varepsilon = \omega_s^{-1}$ is treated as a small parameter. In this way, (4.14) and (4.15) take the form

$$\frac{dx}{d\tau} = \varepsilon f(x, \tau, \varepsilon) \quad (4.16)$$

with $f(x, \tau, \varepsilon)$ almost periodic in τ. Define the *averaged* function

$$f_0(x) = \lim_{T \to \infty} \frac{1}{T} \int_0^T f(x, \tau, 0)\, d\tau \quad (4.17)$$

and the *averaged* system

$$\frac{dx}{d\tau} = \varepsilon f_0(x) \quad (4.18)$$

As shown in [84], there exists an almost periodic transformation $x \to y$ that is near identity for small ϵ, such that (4.16) transforms to

$$\frac{dy}{d\tau} = \varepsilon \left[f_0 (y) + f_1 (y, \tau, \varepsilon) \right]$$

with $f_1 (y, \tau, 0) \equiv 0$. The significance of this is that the averaged system (4.18) is the same as the original system (4.16) up through terms of order ϵ.

The steady solution for the stator current, Equation (4.13), is

$$\bar{i}_a = \frac{-i_f M_f \omega_s \left(R \cos(\omega_s t + \delta) + \omega_s L_s sin(\omega_s t + \delta)\right) + V_a \left(R \cos\left(\omega_s t\right) + \omega_s L_s \sin\left(\omega_s t\right)\right)}{\left(R^2 + L_s^2 \omega_s^2\right)} \tag{4.19}$$

Substituting this result in (4.14) and averaging yields

$$\frac{dv}{dt} = \frac{i_f M_f}{2J} \frac{i_f M_f R \omega_s - R V_a \cos(\delta) + \omega_s L_s V_a \sin(\delta)}{\left(R^2 + \omega_s^2 L_s^2\right)} - \frac{1}{J} T_L \tag{4.20}$$

With some tedious algebra, Equation (4.18) can be written

$$\begin{aligned} \bar{i}_a &= \frac{\left(-i_f M_f \omega_s e^{-j\delta} + V_a\right) e^{-j\omega_s t}}{2(R - j\omega_s L_s)} + \frac{\left(-i_f M_f \omega_s e^{j\delta} + V_a\right) e^{j\omega_s t}}{2(R + j\omega_s L_s)} \\ &= \mathrm{Re} \frac{\left(-i_f M_f \omega_s e^{j\delta} + V_a\right)}{(R + j\omega_s L_s)} e^{j\omega_s t} \end{aligned}$$

Thus, a phasor representation of (4.18) is

$$\mathbf{I}_a = \frac{\mathbf{V}_a - \mathbf{E}_a}{(R + j\omega_s L_s)}, \quad \mathbf{E}_a = \frac{i_f M_f \omega_s}{\sqrt{2}} e^{j\delta} \tag{4.21}$$

Equation (4.21) leads to the equivalent circuit shown in Figure 4.10.

Remark 4.14 Motor Operation. Consider the case where the resistance R is negligible. Then Equations (4.20) and (4.21) reduce to

$$\frac{dv}{dt} = \frac{i_f M_f}{2J} \frac{V_a \sin(\delta)}{\omega_s L_s} - \frac{1}{J} T_L \tag{4.22}$$

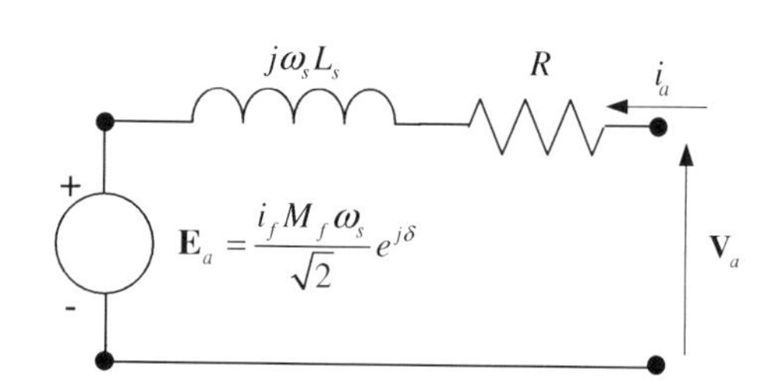

Fig. 4.10 Equivalent circuit for the single-phase synchronous motor

$$\mathbf{I}_a = \frac{\mathbf{V}_a - \mathbf{E}_a}{j\omega_s L_s} \tag{4.23}$$

From (4.22), it is clear that the electrical torque is

$$\tau = \frac{i_f M_f V_a \sin(\delta)}{2\omega_s L_s} \tag{4.24}$$

Furthermore, in steady operation, $\dot{\upsilon} = 0$ so

$$\frac{i_f M_f V_a \sin(\delta)}{2\omega_s L_s} = T_L \tag{4.25}$$

Thus, for given constant load torque, the rotor angle δ can be obtained from (4.25). Note also that the mechanical power delivered to the rotor shaft is

$$P = \omega_s \tau = \frac{i_f M_f V_a \sin(\delta)}{2L_s} \tag{4.26}$$

The electrical power delivered to the rotor can be obtained using (4.23)

$$\mathbf{S}_r = -\mathbf{E}_a \mathbf{I}_a^* = \frac{i_f M_f \omega_s}{2} e^{j\delta} \frac{V_a - i_f M_f \omega_s e^{-j\delta}}{-j\omega_s L_s}$$

which can be reduced to

$$\mathbf{S}_r = P + jQ = \frac{i_f M_f V_a \sin\delta}{2L_s} + j\left(\frac{(i_f M_f \omega_s)^2 - V_a i_f M_f \omega_s \cos\delta}{2\omega_s L_s}\right) \tag{4.27}$$

Note that Equation (4.27) is consistent with Equation (4.26).

Remark 4.15 The Single-phase Generator. The single-phase synchronous motor model given above could serve equally well as a model for the single-phase generator. However, it is convention to reverse the positive direction of the stator current, i_a as shown in Figure 4.11. Thus, the appropriate sign changes are required in the motor equations.

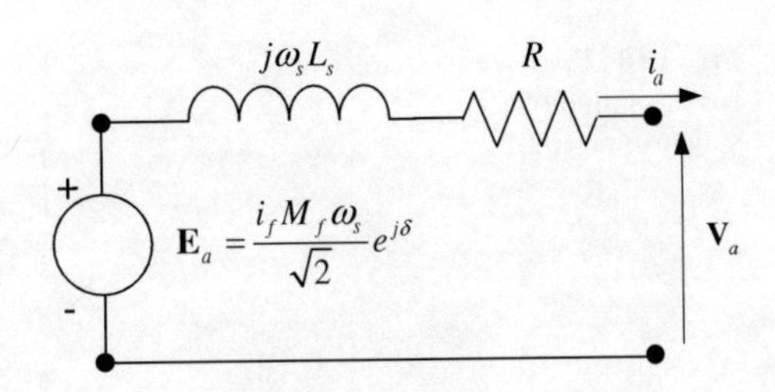

Fig. 4.11 Equivalent circuit for the single-phase synchronous generator

The equivalent circuit phasor relation is

$$\mathbf{I}_a = \frac{\mathbf{E}_a - \mathbf{V}_a}{(R + j\omega_s L_s)} \tag{4.28}$$

and the instantaneous power supplied to the generator terminals is

$$\mathbf{S}_a = P_a + jQ_a$$

with

$$P_a = \frac{R\left(|\mathbf{V}_a|\,|\mathbf{E}_a|\cos\delta - |\mathbf{V}_a|^2\right) + \omega_s L_s\,|\mathbf{V}_a|\,|\mathbf{E}_a|\sin\delta}{\left(R^2 + (\omega_s L_s)^2\right)} \tag{4.29}$$

$$Q_a = \frac{\omega_s L_s\,|\mathbf{V}_a|\,|\mathbf{E}_a|\cos\delta - R\,|\mathbf{V}_a|\,|\mathbf{E}_a|\sin\delta - \omega_s L_s |\mathbf{V}_a|^2}{\left(R^2 + (\omega_s L_s)^2\right)} \tag{4.30}$$

In many large generators the resistance is negligible, in which case (4.29) and (4.30) reduce to

$$P_a = \frac{|\mathbf{V}_a|\,|\mathbf{E}_a|\sin\delta}{\omega_s L_s}, \quad Q_a = \frac{|\mathbf{V}_a|\,|\mathbf{E}_a|\cos\delta - |\mathbf{V}_a|^2}{\omega_s L_s} \tag{4.31}$$

The situation examined above for both the motor and the generator is valid for slow variations of the terminal voltage (relative to synchronous frequency, ω_s).

4.2.5 *Transmission Lines and Transformers*

Here, we will discuss two essential two-port devices – transmission lines and transformers. A transmission line can be viewed as a two-port as shown in Figure 4.12. The transmission line is a continuous insulated conductor with relatively small diameter and long length. It can be modeled by a partial differential equation with time and one space dimension as independent variables—denoted t and x, respectively. An infinitesimal element is shown in Figure 4.12. The element includes a series resistance c, a series inductance l, a shunt resistance modeled by its susceptance (admittance) g, and a shunt capacitance c.

It follows that the voltage and current along the line satisfy the partial differential equations

$$\frac{\partial \mathbf{V}(x)}{\partial x} = -(r + j\omega l)\,\mathbf{I}(x), \quad \frac{\partial \mathbf{I}(x)}{\partial x} = -(g + j\omega c)\,\mathbf{V}(x) \tag{4.32}$$

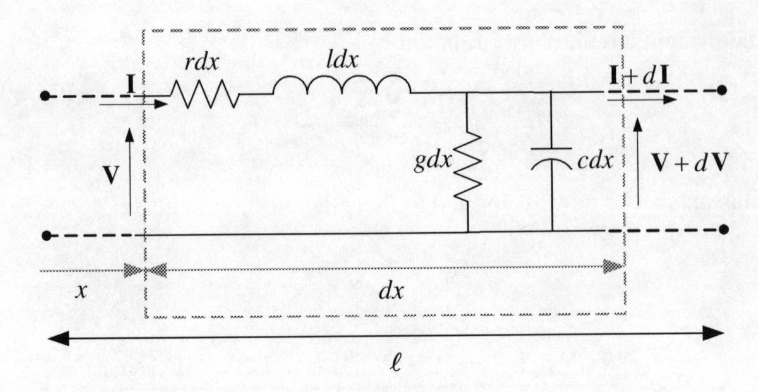

Fig. 4.12 An infinitesimal segment of a transmission line two-port of length ℓ

It is easy to show that Equations (4.32) can be combined to yield the *Telegraph* equations

$$\frac{\partial^2 \mathbf{V}(x)}{\partial x^2} = \gamma^2 \mathbf{V}(x), \quad \frac{\partial^2 \mathbf{I}(x)}{\partial x^2} = \gamma^2 \mathbf{I}(x) \tag{4.33}$$

with

$$\gamma = \sqrt{(r + j\omega l)(g + j\omega c)}$$

The solution of the first of Equations (4.33) is

$$\mathbf{V}(x) = ae^{-\gamma x} + be^{\gamma x} \tag{4.34}$$

where the constants a and b need to be determined from boundary conditions. Now compute

$$\mathbf{I}(x) = \frac{-1}{(r + j\omega l)} \frac{\partial \mathbf{V}(x)}{\partial x} = \frac{1}{Z_0}\left(ae^{-\gamma x} - be^{\gamma x}\right) \tag{4.35}$$

where

$$Z_0 = \sqrt{\frac{r + j\omega l}{g + j\omega c}}$$

Z_0 is called the *characteristic impedance* of the line. Now, suppose the voltage and current at the boundary $x = 0$ are specified as $\mathbf{V}_1, \mathbf{I}_1$, respectively. Then

$$\mathbf{V}_1 = a + b, \quad \mathbf{I}_1 = (a - b)/Z_0$$

from which a, b are obtained. Substitute these results into Equations (4.34) and (4.35) with $x = l$ and simplify to obtain

$$\mathbf{V}_2 = \cosh(\gamma\ell)\,\mathbf{V}_1 - Z_0 \sinh(\gamma\ell)\,\mathbf{I}_1 \tag{4.36}$$

$$\mathbf{I}_2 = \frac{1}{Z_0}\sinh(\gamma\ell)\,\mathbf{V}_1 - \cosh(\gamma\ell)\,\mathbf{I}_1 \tag{4.37}$$

where $\mathbf{V}_2, \mathbf{I}_2$ are the terminal voltage and current with sign convention as shown in Figure 4.13.

The two-port model can be interpreted in several ways. One commonly used model is the **Π**–equivalent circuit shown in Figure 4.13. By computing the matrix parameters for the **Π**–circuit and comparing them with the coefficients in Equations (4.36) and (4.37), the following parameters are identified:

$$Z = Z_0 \sinh(\gamma\ell), \quad Y = \frac{1}{Z_0}\tanh\left(\frac{\gamma\ell}{2}\right)$$

A little algebra shows that the **Π**–equivalent circuit for the transmission line can be modeled by the relation

$$\begin{bmatrix}\mathbf{I}_1\\ \mathbf{I}_2\end{bmatrix} = Y_\Pi \begin{bmatrix}\mathbf{V}_1\\ \mathbf{V}_2\end{bmatrix}, \quad Y_\Pi = \begin{bmatrix} Y + Z^{-1} & -Z^{-1}\\ -Z^{-1} & Y + Z^{-1}\end{bmatrix} \tag{4.38}$$

An ideal transformer was illustrated in Example 2.11. A two-port network schematic of the ideal transformer is shown in Figure 4.14. The two windings are shown as coils. Since the actual coils could be wound clockwise or counterclockwise, dots are used to identify the positive voltage side. The ideal transformer has a simple functional model $V_2 = V_1/n$, where $n = N_1/N - 2$, the *turns ratio*. The ideal transformer neglects several important elements that are relevant to varying degrees in real transformers. The equivalent circuit for a nonideal transformer shown in Figure 4.15 includes the leakage reactance, jX_1, jX_2, and winding losses R_1, R_2. Also, included are core losses, R_c, and magnetizing reactance jX_m.

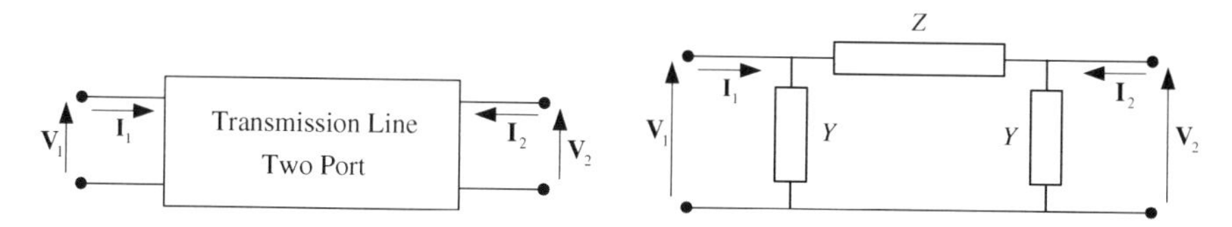

Fig. 4.13 Transmission line as a two-port. General form on the left and **Π**–equivalent circuit on the right

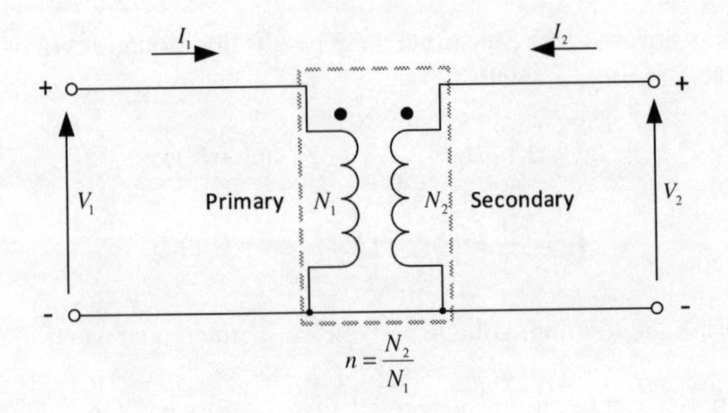

Fig. 4.14 The circuit diagram for an ideal transformer

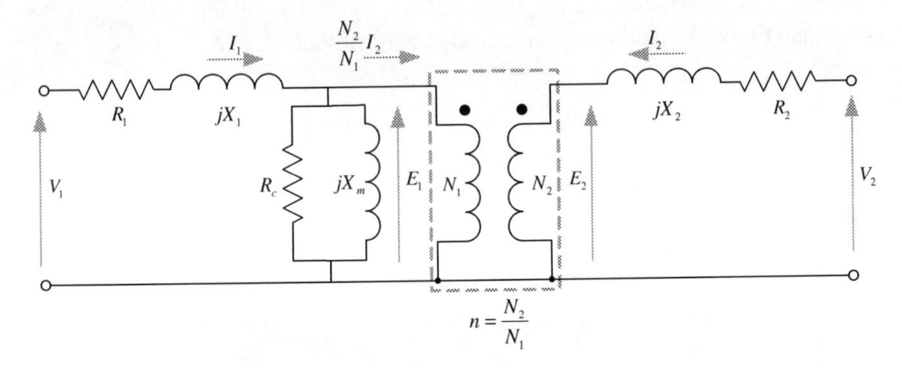

Fig. 4.15 The circuit diagram for a nonideal ideal transformer

4.2.5.1 Per-Unit Normalization

Single-phase and balanced three-phase[1] power networks are conveniently analyzed using a single-phase equivalent network defined using normalized variables with *per-unit* dimensions. Using per-unit dimensions has three chief advantages: 1) Systems containing transformers and operating with multiple voltage levels are significantly simplified, 2) many network components have physical parameters that vary within a relatively narrow range when expressed in per-unit dimension with respect to their rating, and 3) numerical computations are more reliable. Thus, it is common practice using a *per-phase, per-unit* equivalent circuit for power system analysis. The system is composed of one- and multi-port elements.

Normalization means that the variables are scaled in accordance with specified base values that are appropriate for the system. Then, for any specific quantity

$$\text{quantity in per unit} = \frac{\text{quantity in SI units}}{\text{base value of quantity}}$$

[1] A definition and discussion of balanced three-phase power network can be found in Section 4.4.

Base units need to be defined in a consistent fashion. In particular, we want to insure that the complex relation, Ohms law,

$$V = ZI$$

remains true after the scaling. Consequently, it is necessary to choose real base values V_b, Z_b, I_b that satisfy

$$V_b = Z_b I_b$$

so that

$$\frac{V}{V_b} = \frac{Z}{Z_b}\frac{I}{I_b}$$

Thus, the scaled quantities satisfy

$$V_{p.u.} = Z_{p.u.} I_{p.u.}$$

Similarly, choose a power base S_b to preserve the relation $S = VI^*$

$$S_b = V_b I_b$$

yielding

$$S_{p.u.} = V_{p.u.} I^*_{p.u.}$$

Note that from this relation, it follows that

$$P_{p.u.} = \frac{P}{S_b}, \ Q_{p.u.} = \frac{Q}{S_b}$$

Similarly,

$$Z_{p.u.} = R_{p.u.} + jX_{p.u.}, \ R_{p.u.} = \frac{R}{Z_b}, \ X_{p.u.} = \frac{X}{Z_b}$$

Consider the conservative transformer shown in Figure 4.16. It represents a simple circuit that operates with two voltage levels. The transformer model is shown including the magnetizing inductance X_m, the primary side inductance X_1, and the secondary side inductance X_2. A set of base quantities are chosen for the left side of the network: $V_{1b}, I_{1b}, Z_{1b}, S_{1b}$. Now choose the secondary side base quantities as follows:

$$V_{2b} = nV_{1b}, \ I_{2b} = \frac{1}{n}I_{1b}, \ Z_{2b} = n^2 Z_{1b}, \ S_{2b} = S_{1b}$$

The equivalent per-unit model is shown in Figure 4.17.

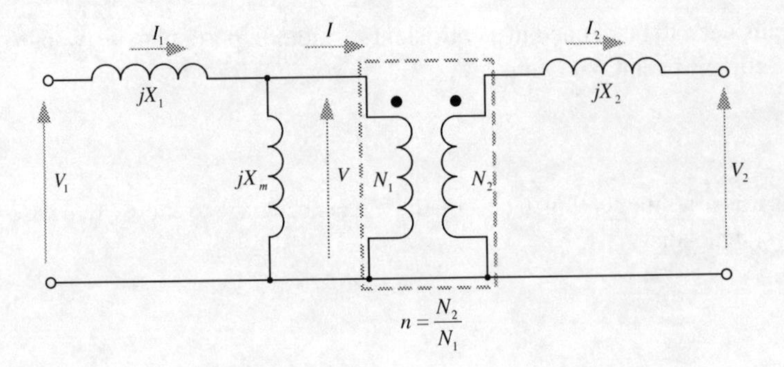

Fig. 4.16 A conservative transformer is a simple circuit with two voltage levels

Fig. 4.17 Per-unit representation of the above transformer

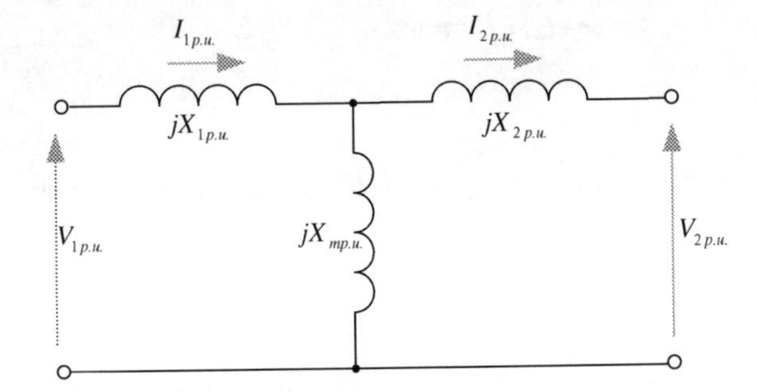

4.3 Three-Phase AC Systems

In most modern systems, electricity is generated and transported as three-phase alternating current. Ideally, each phase carries identical, sinusoidal current and voltage waveforms with each phase separated by precisely 120 degrees. Such systems have advantages over alternatives such as DC or single-phase AC systems.

4.3.1 Principles of Three-Phase Transmission

4.3.1.1 The Single-Line Diagram

Complex power systems are frequently portrayed graphically as a *single-line (one-line) diagram.* Multi-port elements are connected via single lines that join a port of one element to a port of a second element. A typical single-line diagram is illustrated in Figure 4.18.

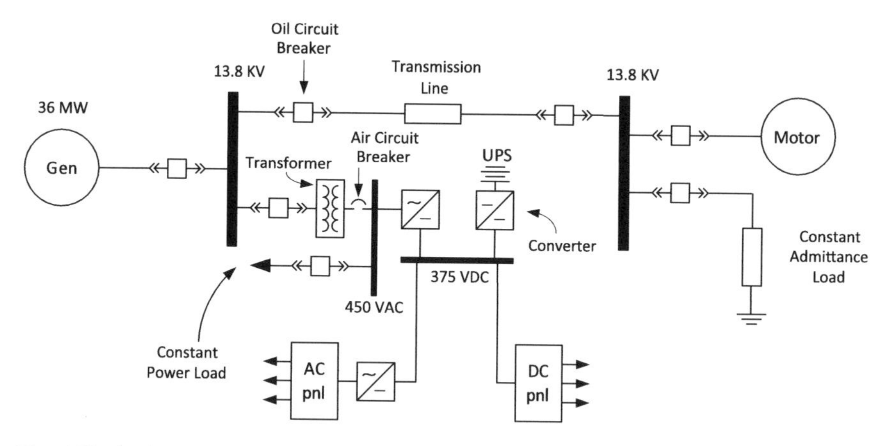

Fig. 4.18 A single-line diagram for a simple power network shows the graphical symbol for frequently used components

4.3.2 Three-Phase Synchronous Machines

Synchronous motors and generators have similar principles of operation and construction. In a generator, mechanical power is supplied to the rotor shaft and electrical power is discharged from the electrical terminals. Motors operate in reverse with electrical power supplied to the device through the electrical terminals and mechanical power delivered by the rotor shaft. The machine consists of a stationary body, the *stator*, and a rotating body, the *rotor*, as shown in Figure 4.19. In comparison with Figure 3.24, note the addition of the b and c phase coils. The three coils are spaced 120 degrees apart.

The analysis proceeds along the lines of Example 3.31 except now there are four coupled circuits instead of two. The following assumptions will be made:

1. Each stator circuit a, b, c has a terminal voltage, $v_a(t)$, $v_b(t)$, $v_c(t)$ and each circuit has the same resistance, R. The corresponding phase currents are $i_a(t)$, $i_b(t)$, $i_c(t)$. For the motor, positive current is into the machine positive voltage terminal and for the generator, positive current is out.
2. The rotor circuit is driven by a voltage source, $v_f(t)$, and has resistance, R_f. Again, for the generator case, the positive current direction is out of the positive terminal.
3. Each of the stator coil windings is distributed around the inner circumference of the stator, so that the induced flux density is sinusoidally distributed around the *air gap* between the rotor and stator (for more details see [27, 82]).
4. The rotor has an external mechanical torque, T, acting positively in the θ direction. Thus, T is positive when the machine acts as a generator and negative when the machine acts as a motor.
5. The rotor is round. Consequently, the inductances take the form:

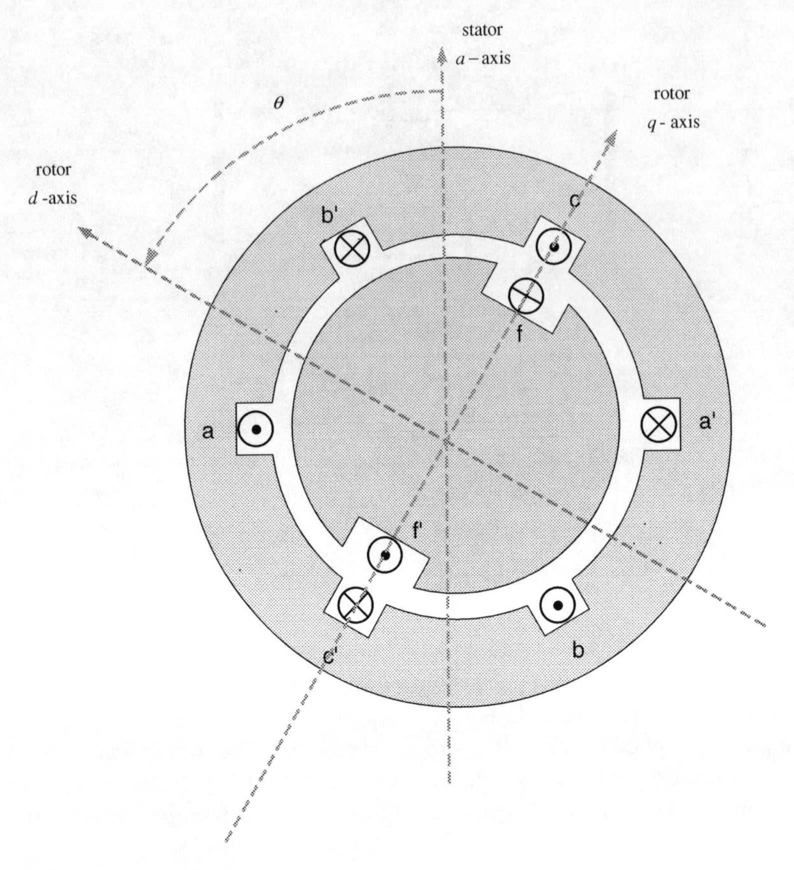

Fig. 4.19 A three-phase machine

(a) *stator coil self-inductances* are independent of rotor position. They are all the same and designated $L_s > 0$,

$$L_s = L_{aa} = L_{bb} = L_{cc}$$

(b) *stator coil mutual inductances* are also independent of θ and equal to each other with designation $M_s > 0$

$$M_s = -L_{ab} = -L_{bc} = -L_{ca}$$

(c) *field inductances* The field self-inductance is a constant denoted $L_f > 0$

$$L_{ff} = L_f$$

The field mutual inductances with each of the stator coils are periodically θ-dependent, with $M_f > 0$, they can be expressed

$$L_{af} = M_f \cos(\theta), L_{bf} = M_f \cos\left(\theta - \frac{2\pi}{3}\right), L_{cf} = M_f \cos\left(\theta + \frac{2\pi}{3}\right)$$

4.3.2.1 The Lagrange Equations

In the case of three phases, the generalized velocities expand to $p = \left(i_a, i_b, i_c, i_f, \omega\right)$. As before, all coordinates are ignorable except the rotor angle, θ. The magnetic co-energy of Equation (3.106) generalizes to

$$U^*\left(i_a, i_b, i_c, i_f, \theta\right) = \tfrac{1}{2}\left[\, i_a \; i_b \; i_c \; i_f \,\right] L(\theta) \begin{bmatrix} i_a \\ i_b \\ i_c \\ i_f \end{bmatrix} = \tfrac{1}{2}\mathbf{i}^T L(\theta)\,\mathbf{i} \tag{4.39}$$

where

$$L(\theta) = \begin{bmatrix} L_s & -M_s & -M_s & M_f\cos(\theta) \\ -M_s & L_s & -M_s & M_f\cos\left(\theta - \frac{2\pi}{3}\right) \\ -M_s & -M_s & L_s & M_f\cos\left(\theta + \frac{2\pi}{3}\right) \\ M_f\cos(\theta) & M_f\cos\left(\theta - \frac{2\pi}{3}\right) & M_f\cos\left(\theta + \frac{2\pi}{3}\right) & L_f \end{bmatrix} \tag{4.40}$$

The Lagrangian is

$$L\left(i_a, i_b, i_c, i_f, \omega, \theta\right) = U^*\left(i_a, i_b, i_c, i_f, \theta\right) + T^*(\omega) \tag{4.41}$$

where $T^*(\omega)$ is the rotor kinetic co-energy as defined in (3.106). The generalized force vector, Q, can be obtained from the dissipation function

$$\mathcal{D}\left(i_a, i_b, i_c, i_f, \omega\right) = \tfrac{1}{2}\mathbf{i}^T\mathbf{R}\,\mathbf{i} + \mathbf{i}^T\mathbf{v} + T\omega, \quad \mathbf{v} = \left[\, v_a \; v_b \; v_c \; v_f \,\right] \tag{4.42}$$

$$\mathbf{R} = \operatorname{diag}\left(R, R, R, R_f\right) \tag{4.43}$$

$$Q = \frac{\partial \mathcal{D}}{\partial p} \tag{4.44}$$

Lagrange's equations (3.39) are obtained in the form

$$\frac{d\left(L(\theta)\,\mathbf{i}\right)}{dt} + \mathbf{R}\mathbf{i} = \mathbf{v} \tag{4.45}$$

$$J\frac{d\omega}{dt} - \frac{1}{2}\mathbf{i}^T\frac{dL(\theta)}{d\theta}\mathbf{i} = T \tag{4.46}$$

The instantaneous electromagnetic torque is

$$\tau = -\frac{1}{2}\mathbf{i}^T\frac{dL(\theta)}{d\theta}\mathbf{i} = i_f i_a M_f \sin(\theta) + i_f i_b M_f \sin\left(\theta - \frac{2\pi}{3}\right) + i_f i_c M_f \sin\left(\theta + \frac{2\pi}{3}\right) \tag{4.47}$$

In steady state, constant speed operation, of course, $T = \tau$.

These equations can be greatly simplified by transforming the stator electrical variables to the rotor coordinates. That will be done in the next section.

4.3.2.2 The Blondel-Park Transformation

The *Park transformation*, or *Blondel transformation*, is used to transforms the stator *abc* voltages, currents, and flux linkages to rotor *dqo* coordinates. Thus,

$$\begin{bmatrix} i_d \\ i_q \\ i_o \end{bmatrix} = \sqrt{\frac{2}{3}} \begin{bmatrix} \cos\theta & \cos\left(\theta - \frac{2\pi}{3}\right) & \cos\left(\theta + \frac{2\pi}{3}\right) \\ \sin\theta & \sin\left(\theta - \frac{2\pi}{3}\right) & \sin\left(\theta + \frac{2\pi}{3}\right) \\ \frac{1}{\sqrt{2}} & \frac{1}{\sqrt{2}} & \frac{1}{\sqrt{2}} \end{bmatrix} \begin{bmatrix} i_a \\ i_b \\ i_c \end{bmatrix} \tag{4.48}$$

In general

$$\mathbf{i}_{dqo} = B\mathbf{i}_{abc}, \mathbf{v}_{dqo} = B\mathbf{v}_{abc}, \lambda_{dqo} = B\lambda_{abc}$$

The inverse transformation is

$$B^{-1} = B^T = \sqrt{\frac{2}{3}} \begin{bmatrix} \cos\theta & \sin\theta & \frac{1}{\sqrt{2}} \\ \cos\left(\theta - \frac{2\pi}{3}\right) & \sin\left(\theta - \frac{2\pi}{3}\right) & \frac{1}{\sqrt{2}} \\ \cos\left(\theta + \frac{2\pi}{3}\right) & \sin\left(\theta + \frac{2\pi}{3}\right) & \frac{1}{\sqrt{2}} \end{bmatrix} \tag{4.49}$$

Now, after a somewhat tedious computation the Lagrange equations can be obtained in the form

$$\begin{bmatrix} L_d & 0 & 0 & L_{fd} \\ 0 & L_q & 0 & 0 \\ 0 & 0 & L_o & 0 \\ L_{fd} & 0 & 0 & L_f \end{bmatrix} \frac{d}{dt} \begin{bmatrix} i_d \\ i_q \\ i_o \\ i_f \end{bmatrix} = - \begin{bmatrix} R & \omega L_q & 0 & 0 \\ -\omega L_d & R & 0 & -\omega L_{fd} \\ 0 & 0 & R & 0 \\ 0 & 0 & 0 & R_f \end{bmatrix} \begin{bmatrix} i_d \\ i_q \\ i_o \\ i_f \end{bmatrix} + \begin{bmatrix} v_d \\ v_q \\ v_o \\ v_f \end{bmatrix} \tag{4.50}$$

$$J\frac{d\omega}{dt} = L_{fd} i_f i_q + T \tag{4.51}$$

Where

$$L_d = L_q \equiv L_s + M_s, L_o \equiv L_s - 2M_s, L_{fd} \equiv \sqrt{\tfrac{3}{2}} M_f$$

The electrical and mechanical equations can be combined to yield

$$\begin{bmatrix} L_d & 0 & 0 & L_{fd} & 0 \\ 0 & L_q & 0 & 0 & 0 \\ 0 & 0 & L_o & 0 & 0 \\ L_{fd} & 0 & 0 & L_f & 0 \\ 0 & 0 & 0 & 0 & J \end{bmatrix} \frac{d}{dt} \begin{bmatrix} i_d \\ i_q \\ i_o \\ i_f \\ \omega \end{bmatrix} = - \begin{bmatrix} R & \omega L_q & 0 & 0 & 0 \\ -\omega L_d & R & 0 & 0 & -L_{fd} i_f \\ 0 & 0 & R & 0 & 0 \\ 0 & 0 & 0 & R_f & 0 \\ 0 & L_{fd} i_f & 0 & 0 & 0 \end{bmatrix} \begin{bmatrix} i_d \\ i_q \\ i_o \\ i_f \\ \omega \end{bmatrix} + \begin{bmatrix} v_d \\ v_q \\ v_o \\ v_f \\ T \end{bmatrix} \tag{4.52}$$

The result of this transformation is the removal of the θ-dependency. Equation (4.52) does not contain θ at all. It is also of note that the electromagnetic torque imposed on the rotor is

$$\tau(t) = L_{fd}\, i_f(t)\, i_q(t) \tag{4.53}$$

Remark 4.16 (*Synchronous operation*) Consider steady, synchronous operation which means constant rotor speed at angular velocity, ω_s, constant δ with $\theta = \omega_s t + \delta$, the currents i_d and i_q are constant, and balanced phase voltage and currents so that $i_o = 0$. The inverse Park transformation yields

$$i_a = \sqrt{\frac{2}{3}}\, i_d \cos(\omega_s t + \delta) + \sqrt{\frac{2}{3}}\, i_q \sin(\omega_s t + \delta)$$

In phasor form, this equation may be written

$$\mathbf{I}_a = \left(I_q + jI_d\right) e^{j\delta} \tag{4.54}$$

where I_d, I_q are the constants $I_q = i_q / \sqrt{3},\ I_d = i_d / \sqrt{3}$. The geometry of this relationship is exhibited in Figure 4.20.

Similarly, the voltage relationships can be derived

$$v_a = \sqrt{\tfrac{2}{3}}\, v_d \cos(\omega_s t + \delta) + \sqrt{\tfrac{2}{3}}\, v_q \sin(\omega_s t + \delta)$$

and

$$\mathbf{V}_a = \left(V_q + jV_d\right) e^{j\delta} \tag{4.55}$$

with $V_q = v_q / \sqrt{3},\ V_d = v_d / \sqrt{3}$.

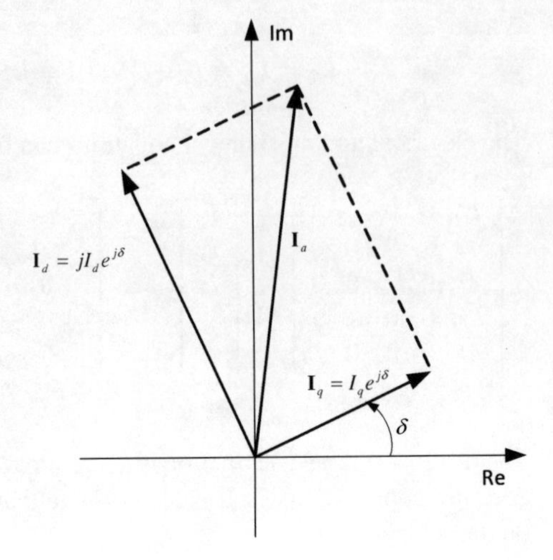

Fig. 4.20 The geometry of the phasor relationship

Remark 4.17 (*Salient rotor*) The above analysis is specifically for a round rotor synchronous motor. In the event of a salient pole machine, the upper left 3×3 block of the matrix $L(\theta)$ in Equation (4.40) is

$$L_{11}(\theta) = \begin{bmatrix} L_s + L_m \cos 2\theta & -M_s - L_m \cos 2\left(\theta + \frac{\pi}{6}\right) & -M_s - L_m \cos 2\left(\theta + \frac{5\pi}{6}\right) \\ -M_s - L_m \cos 2\left(\theta + \frac{\pi}{6}\right) & L_s + L_m \cos 2\left(\theta - \frac{2\pi}{3}\right) & -M_s - L_m \cos 2\left(\theta - \frac{\pi}{2}\right) \\ -M_s - L_m \cos 2\left(\theta + \frac{5\pi}{6}\right) & -M_s - L_m \cos 2\left(\theta + \frac{\pi}{2}\right) & L_s + L_m \cos 2\left(\theta + \frac{2\pi}{3}\right) \end{bmatrix}$$

with $L_m > 0$. Applying the Park transformation to L_{11} reduces it to

$$BL_{11}B^{-1} = \begin{bmatrix} L_d & 0 & 0 \\ 0 & L_q & 0 \\ 0 & 0 & L_o \end{bmatrix}$$

However, in this case,

$$L_d = L_s + M_s + 3/2 L_m, \quad L_q = L_s + M_s - 3/2 L_m, \quad L_o = L_s + 2M_s$$

The important observation is that whereas in the round rotor case, $L_d = L_q$, that is not the case with a salient rotor for which $L_d > L_q$.

4.3.2.3 Induction Machines

Induction machines are either of the wound rotor or squirrel cage type. In either case, the rotor typically has multiple windings which are short-circuited so that the excitation results from rotor winding currents induced by the rotating stator field. In the wound rotor type, the rotor windings are closed through resistors via slip rings. Squirrel cage devices have the rotor terminals directly connected. Thus, a model for a squirrel cage motor with three stator phases and two rotor windings is obtained by repeating the analysis above after adding a second rotor winding, identical and orthogonally placed to the first, and with both field voltages set equal to zero. As a result, the dynamic equations are

$$\begin{bmatrix} L_d & 0 & 0 & L_{fd} & 0 \\ 0 & L_q & 0 & 0 & L_{fd} \\ 0 & 0 & L_o & 0 & 0 \\ L_{fd} & 0 & 0 & L_f & 0 \\ 0 & L_{fd} & 0 & 0 & L_f \end{bmatrix} \frac{d}{dt} \begin{bmatrix} i_d \\ i_q \\ i_o \\ i_{f_1} \\ i_{f_2} \end{bmatrix} = - \begin{bmatrix} R & \omega L_d & 0 & 0 & -\omega L_{fd} \\ -\omega L_q & R & 0 & \omega L_{fd} & 0 \\ 0 & 0 & R & 0 & 0 \\ 0 & 0 & 0 & R_f & 0 \\ 0 & 0 & 0 & 0 & R_f \end{bmatrix} \begin{bmatrix} i_d \\ i_q \\ i_o \\ i_{f_1} \\ i_{f_2} \end{bmatrix} + \begin{bmatrix} v_d \\ v_q \\ v_o \\ 0 \\ 0 \end{bmatrix} \tag{4.56}$$

$$J \frac{d\omega}{dt} = L_{fd} \left(i_{f_1} i_q - i_{f_2} i_d \right) + T \tag{4.57}$$

As before, combining the electrical and mechanical equations can lead to insights into the conservative and nonconservative internal forces

$$\begin{bmatrix} L_d & 0 & 0 & L_{fd} & 0 & 0 \\ 0 & L_q & 0 & 0 & L_{fd} & 0 \\ 0 & 0 & L_o & 0 & 0 & 0 \\ L_{fd} & 0 & 0 & L_f & 0 & 0 \\ 0 & L_{fd} & 0 & 0 & L_f & 0 \\ 0 & 0 & 0 & 0 & 0 & J \end{bmatrix} \frac{d}{dt} \begin{bmatrix} i_d \\ i_q \\ i_o \\ i_{f_1} \\ i_{f_2} \\ \omega \end{bmatrix} = - \begin{bmatrix} R & \omega L_d & 0 & 0 & 0 & -L_{fd} i_{f_2} \\ -\omega L_q & R & 0 & 0 & 0 & L_{fd} i_{f_1} \\ 0 & 0 & R & 0 & 0 & 0 \\ 0 & 0 & 0 & R_f & 0 & 0 \\ 0 & 0 & 0 & 0 & R_f & 0 \\ L_{fd} i_{f_2} & -L_{fd} i_{f_1} & 0 & 0 & 0 & 0 \end{bmatrix} \begin{bmatrix} i_d \\ i_q \\ i_o \\ i_{f_1} \\ i_{f_2} \\ \omega \end{bmatrix} + \begin{bmatrix} v_d \\ v_q \\ v_o \\ 0 \\ 0 \\ T \end{bmatrix} \tag{4.58}$$

4.3.2.4 Permanent Magnet Synchronous Machines

In permanent magnet synchronous machines, the field coil of the synchronous device is replaced by permanent magnets. Thus, there is no field circuit and one degree of freedom is eliminated. The generalized velocity vector is now $p = [i_a, i_b, i_c, \omega]^T$. If the constant permanent magnet field intensity in flux linkages is λ_f, then the relevant magnetic co-energy reduces to

$$U^* (i_a, i_b, i_c, \theta) = \tfrac{1}{2} \mathbf{i}^T L_S \mathbf{i} + \lambda_f \left[\cos(\theta) \ \cos\left(\theta - \tfrac{2\pi}{3}\right) \ \cos\left(\theta + \tfrac{2\pi}{3}\right) \right] \mathbf{i} \tag{4.59}$$

$$\mathbf{i} = \begin{bmatrix} i_a \\ i_b \\ i_c \end{bmatrix}, \quad L_{\mathbf{S}} = \begin{bmatrix} L_s & -M_s & -M_s \\ -M_s & L_s & -M_s \\ -M_s & -M_s & L_s \end{bmatrix} \tag{4.60}$$

The dissipation function is

$$\mathcal{D}(i_a, i_b, i_c, \omega, t) = \tfrac{1}{2}\mathbf{i}^T R_{\mathbf{S}}\mathbf{i} + \mathbf{v}^T\mathbf{i} + T\omega, \quad R_{\mathbf{S}} = \mathrm{diag}(R, R, R) \tag{4.61}$$

The Lagrangian is

$$L(i_a, i_b, i_c, \omega, \theta) = U^*(i_a, i_b, i_c, \theta) + T^*(\omega) \tag{4.62}$$

from which the Lagrange equations are obtained

$$L_{\mathbf{S}}\frac{d\mathbf{i}}{dt} - \lambda_f \begin{bmatrix} \sin(\theta) \\ \sin\left(\theta - \frac{2\pi}{3}\right) \\ \sin\left(\theta + \frac{2\pi}{3}\right) \end{bmatrix} \omega = \mathbf{v} \tag{4.63}$$

$$J\frac{d\omega}{dt} + \lambda_f \left[\sin(\theta) \ \sin\left(\theta - \tfrac{2\pi}{3}\right) \ \sin\left(\theta + \tfrac{2\pi}{3}\right) \right] \mathbf{i} = T \tag{4.64}$$

Using Park's transformation, the current coordinates can be converted from i_a, i_b, i_c to i_d, i_q, i_o and combining the electrical and mechanical equations to obtain

$$\begin{bmatrix} L_d & 0 & 0 & 0 \\ 0 & L_q & 0 & 0 \\ 0 & 0 & L_0 & 0 \\ 0 & 0 & 0 & J \end{bmatrix} \frac{d}{dt} \begin{bmatrix} i_d \\ i_q \\ i_o \\ \omega \end{bmatrix} = \begin{bmatrix} R & -\omega L_q & 0 & 0 \\ \omega L_d & R & 0 & -\sqrt{3/2}\,\lambda_f \\ 0 & 0 & R & 0 \\ 0 & \sqrt{3/2}\,\lambda_f & 0 & 0 \end{bmatrix} \begin{bmatrix} i_d \\ i_q \\ i_o \\ \omega \end{bmatrix} + \begin{bmatrix} v_d \\ v_q \\ v_o \\ T \end{bmatrix} \tag{4.65}$$

The electromagnetic torque induced on the rotor is

$$\tau = \sqrt{3/2}\,\lambda_f i_q \tag{4.66}$$

4.4 Balanced Three-Phase AC Power Networks

A balanced three-phase AC circuit is essentially three separate, but equivalent, circuits as illustrated in Figure 4.21. In the circuit shown, the three sinusoidal voltages have the same magnitude are separated in phase by $2\pi/3$ rad, and the line and load admittances are identical.

Because they are balanced, they can be depicted with a single-line diagram.

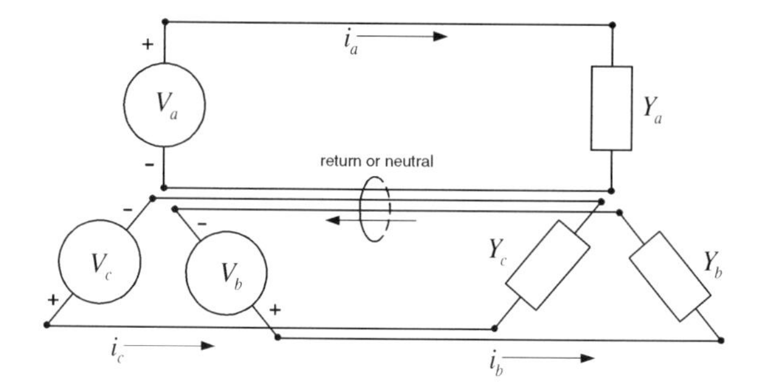

Fig. 4.21 A three-phase circuit with generator and loads

4.4.1 Synchronous Generator in Steady State

Consider the three-phase synchronous machine of Equations. (4.45) and (4.46), or (4.50) and (4.51), operating as a generator. Suppose, for now, that the system operates in the *steady-state* condition where the speed $\omega = \omega_s$ and the field current i_f are both constant. Also, the system is *balanced* which means the three-phase voltages and currents are identical sinusoids except they are separated in phase by $2\pi/3$ rad. Thus,

$$i_a + i_b + i_c = 0, \quad v_a + v_b + v_c = 0 \tag{4.67}$$

It is also true that the total instantaneous electric power output is constant

$$P = i_d v_d + i_q v_q + i_o v_o = i_a v_a + i_b v_b + i_c v_c = \text{constant} \tag{4.68}$$

Equation (4.45) can be written

$$L(\theta)\frac{d\mathbf{i}}{dt} + \frac{\partial L(\theta)}{\partial \theta}\mathbf{i}\omega + \mathbf{R}\mathbf{i} = \mathbf{v} \tag{4.69}$$

with

$$\frac{\partial L(\theta)}{\partial \theta} = \begin{bmatrix} 0 & 0 & 0 & -M_f \sin(\theta) \\ 0 & 0 & 0 & -M_f \sin\left(\theta - \frac{2\pi}{3}\right) \\ 0 & 0 & 0 & -M_f \sin\left(\theta + \frac{2\pi}{3}\right) \\ -M_f \sin(\theta) & -M_f \sin\left(\theta - \frac{2\pi}{3}\right) & -M_f \sin\left(\theta + \frac{2\pi}{3}\right) & 0 \end{bmatrix} \tag{4.70}$$

Notice that the first (a-phase) of these equations can be written

$$L_s\frac{di_a}{dt} + M_s\left(\frac{di_b}{dt} + \frac{di_c}{dt}\right) + M_f\frac{di_f}{dt} - M_f \sin(\theta)\,\omega_s i_f + Ri_a = v_a$$

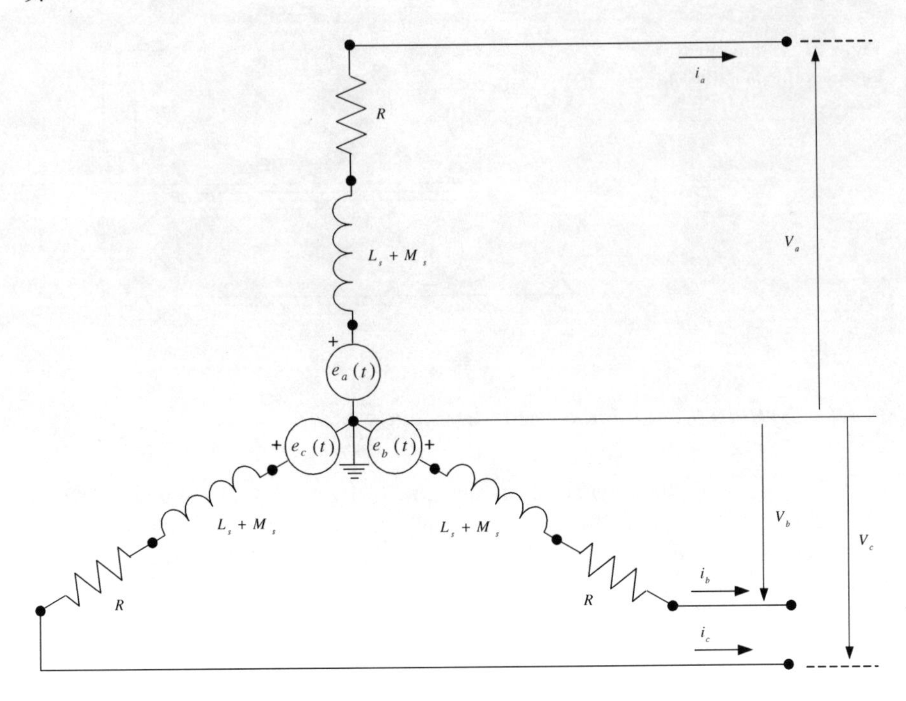

Fig. 4.22 Circuit diagram for a three-phase generator

Since i_f is constant and $i_a = -(i_b + i_c)$ this simplifies to

$$(L_s + M_s)\frac{di_a}{dt} - M_f \sin(\theta)\,\omega_s i_f + Ri_a = v_a \tag{4.71}$$

Reversing the current sign convention, the generator relation is

$$-(L_s + M_s)\frac{di_a}{dt} + M_f \sin(\theta)\,\omega_s i_f - Ri_a = v_a \tag{4.72}$$

A circuit equivalent to Equation (4.72) for all three phases is shown in Figure 4.22. The voltage, $e_a(t)$, is defined by

$$e_a(t) = \omega_s M_f i_f \cos(\omega_s t + \delta)\,,\ \mathrm{E}_a = \frac{\omega_s M_f i_f}{\sqrt{2}} e^{j\delta} = E_a e^{j\delta} \tag{4.73}$$

4.4.2 *Synchronous Machine Simplified Dynamic Model*

Once again, consider a synchronous machine in balanced operation but not in steady-state condition. In this case, the speed can vary with time but it will assume that ω

remains close to ω_s. Once again suppose $\omega = \omega_s + \upsilon$ with $|\upsilon| \ll \omega_s$. Recall that balanced operation implies that $i_o = 0$, $v_o = 0$. Equation (4.52) can be reduced to the three electrical equations,

$$L_d \frac{di_d}{dt} + L_{fd} \frac{di_f}{dt} = -Ri_d - \omega L_q i_q + v_d \tag{4.74}$$

$$L_q \frac{di_q}{dt} = \omega L_d i_d - Ri_q + \omega L_{fd} i_f + v_q \tag{4.75}$$

$$L_{fd} \frac{di_d}{dt} + L_f \frac{di_f}{dt} = -R_f i_f + v_f \tag{4.76}$$

and one mechanical equation

$$J \frac{d\upsilon}{dt} = L_{fd} i_f i_q + T \tag{4.77}$$

Now, suppose $di_d/dt, di_q/dt, di_f/dt$ are small compared to $\omega_s i_d, \omega_s i_q, \omega_s i_f$. Then Equations (4.74) and (4.75) reduce to

$$0 = -Ri_d - \omega_s L_q i_q + v_d \tag{4.78}$$

$$0 = \omega_s L_d i_d - Ri_q - \omega_s L_{fd} i_f + v_q \tag{4.79}$$

These are the steady, synchronous motion equations, so the results and definitions of Remark 4.16 apply. Thus, Equations (4.78) and (4.79) can be rewritten as the equivalent complex relation

$$\left(V_q + jV_d\right) e^{j\delta} = R\left(I_q + jI_d\right) e^{j\delta} - \omega_s L_d I_d e^{j\delta} + j\omega_s L_q I_q e^{j\delta} + \omega_s L_{fd} I_f e^{j\delta}$$

or

$$\mathbf{V}_a = R\mathbf{I}_a - \omega_s \left(L_d I_d - jL_q I_q\right) e^{j\delta} + \mathbf{E}_a \tag{4.80}$$

Equation (4.80) can be used to solve for the real variables I_d, I_q if the voltages E_a, V_a are known. This then allows determination of the real and reactive power flows into the machine. To do this, first choose the rotor angle as a reference and set it arbitrarily to zero, $\delta = 0$. Relative to this reference, suppose the angle of the terminal voltage $\mathbf{V}_a$ is θ. Then

$$\mathbf{E}_a = E_a, \; \mathbf{V}_a = V_a e^{j\theta} = V_a \cos\theta + jV_a \sin\theta$$

and (4.80) can be written

$$E_a = V_a \cos\theta + jV_a \sin\theta - R\left(I_q - jI_d\right) - j\omega_s \left(L_q I_q + jL_d I_d\right) \tag{4.81}$$

in which all variables are real. Equating real and imaginary parts of (4.81) leads to two real equations

$$E_a = V_a \cos\theta - RI_q + \omega_s L_d I_d$$

$$0 = V_a \sin\theta + RI_d - \omega_s L_q I_q$$

These equations, linear in I_d, I_q, are easily solved to yield

$$I_d = \frac{L_q \omega_s (E_a - V_a \cos\theta) + RV_a \sin\theta}{L_d L_q \omega_s^2 - R^2} \tag{4.82}$$

$$I_q = \frac{-L_d \omega_s V_a \sin\theta + R(V_a \cos\theta - E_a)}{L_d L_q \omega_s^2 - R^2} \tag{4.83}$$

The electric power delivered to the motor terminals, $\mathbf{S}_a = \mathbf{V}_a \mathbf{I}_a^*$, and the power delivered to the rotor, $\mathbf{S}_e = \mathbf{E}_a \mathbf{I}_a^*$, are both of interest. Proceed to evaluate $\mathbf{S}_a$, $\mathbf{S}_e$ by using Equation (4.54) to replace $\mathbf{I}_a$ and employing (4.82) and (4.83). To simplify matters, the resistance is neglected in the following expressions. The phase a real and reactive power at the terminal bus are

$$P_a = \frac{E_a V_a}{\omega_s L_d} \sin\theta + \frac{V_a^2}{2}\left(\frac{1}{\omega_s L_q} - \frac{1}{\omega_s L_d}\right) \sin 2\theta \tag{4.84}$$

$$Q_a = \frac{V_a^2}{2}\left(\frac{1}{\omega_s L_q} + \frac{1}{\omega_s L_d}\right) - \frac{E_a V_a}{\omega_s L_d} \cos\theta + \frac{V_a^2}{2}\left(\frac{1}{\omega_s L_d} - \frac{1}{\omega_s L_q}\right) \cos 2\theta \tag{4.85}$$

and at the internal bus

$$P_e = \frac{E_a V_a}{\omega_s L_q} \sin\theta, \quad Q_e = \frac{E_a V_a \cos\theta - E_a^2}{\omega_s L_d} \tag{4.86}$$

Now, consider Equation (4.76). Premultiply by $\omega_s M_f / R_f$ and reorganize to obtain

$$\frac{L_f}{R_f}\frac{d}{dt}\left(\frac{\omega_s M_f L_{fd}}{L_f} i_d + \omega_s M_f i_f\right) = -\omega_s M_f i_f + \frac{\omega_s M_f}{R_f} v_f \tag{4.87}$$

Define

$$\mathbf{E}'_a = \frac{1}{\sqrt{2}}\left(\frac{\omega_s M_f L_{fd}}{L_f} i_d + \omega_s M_f i_f\right) e^{j\delta} = E'_a e^{j\delta} \tag{4.88}$$

$$\mathbf{E}_a = \frac{\omega_s M_f i_f}{\sqrt{2}} e^{j\delta} = E_a e^{j\delta}, \quad E_{fd} = \frac{\omega_s M_f}{\sqrt{2} R_f} v_f \tag{4.89}$$

to obtain

$$\frac{L_f}{R_f}\frac{dE'_a}{dt} = -E_a + E_{fd} \tag{4.90}$$

From the definitions of $\mathbf{E}_a$, $\mathbf{E}'_a$, it can be seen that

$$E_a = E'_a - \frac{\omega_s M_f L_{fd}}{L_f} I_d \tag{4.91}$$

Resolving (4.92) along with the previous solutions for I_d, I_q, Equations (4.82), (4.83) and the definition of E'_a, (4.88), it is possible to obtain I_d, I_q, I_f *and* E_a as functions of E'_a. Ignoring the resistance, R, these relations are

$$E_a = \frac{E'_a + \sigma V_a \cos\theta}{1+\sigma} \tag{4.92}$$

$$I_d = \frac{E'_a/(\omega_s L_d) - V_a \cos\theta/(\omega_s L_d)}{1+\sigma} \tag{4.93}$$

$$I_q = \frac{V_a \sin\theta}{\omega_s L_q} \tag{4.94}$$

$$I_f = \frac{E'_a + \sigma V_a \cos\theta}{\omega_s M_f (1+\sigma)} \tag{4.95}$$

where

$$\sigma = \sqrt{\frac{3}{2}\frac{M_f^2}{L_f L_d}}$$

Finally, (4.90) can be written

$$\tau_f \frac{dE'_a}{dt} = -E'_a - \sigma V_a \cos\theta + (1+\sigma) E_{fd} \tag{4.96}$$

with

$$\tau_f = \frac{(1+\sigma) L_f}{R_f}$$

Recall that θ in (4.96) represents the relative angle between the voltage at the terminal bus and the machine rotor. If both the bus angle and the rotor angle are measured relative to some other common reference, then $\theta = \theta_a - \delta$, where θ_a is the phase angle of $\mathbf{V}_a$ and δ is the rotor angle. Thus, (4.96) becomes

$$\tau_f \frac{dE'_a}{dt} = -E'_a - \sigma V_a \cos(\theta - \delta) + (1+\sigma) E_{fd} \tag{4.97}$$

where δ is obtained from the mechanical equations, rewritten here

$$\frac{d\delta}{dt} = \upsilon$$

$$J\frac{d\upsilon}{dt} = L_{fd} i_f i_q + T$$

Premultiply the second of these equations by ω_2, and notice that $\omega_s T = P_m$, the mechanical power delivered to the rotating shaft, and $\omega_s L_{fd} i_f i_q = 3P_e$, the electrical power delivered from the shaft. Then, the mechanical equations are

$$\frac{d\delta}{dt} = \upsilon, \quad \omega_s J\frac{d\upsilon}{dt} = P_m - 3P_e \tag{4.98}$$

Equations (4.84), (4.85), (4.97), and (4.98) constitute a complete set of equations for the synchronous machine acting as a motor. The generator equations are obtained by reversing the positive direction of the stator currents, equivalent to changing $(\theta - \delta)$ to $(\delta - \theta)$.

A summary of the generator equations in per-unit form is given below:

$$\tau_f \frac{dE'_a}{dt} = -E'_a - \sigma V_a \cos(\delta - \theta) + (1 + \sigma) E_{fd} \tag{4.99}$$

$$\frac{d\delta}{dt} = \upsilon, \quad M\frac{d\upsilon}{dt} + D\upsilon = P_m - P_E, \quad M = \frac{\omega_s J}{S_G} \tag{4.100}$$

$$E_a = \frac{E'_a}{1 + \sigma} + \frac{\sigma}{1 + \sigma} V_a \cos(\delta - \theta) \tag{4.101}$$

$$I_d = \frac{E'_a / (\omega_s L_d) - V_a \cos(\delta - \theta) / (\omega_s L_d)}{1 + \sigma} \tag{4.102}$$

$$I_q = \frac{V_a \sin(\delta - \theta)}{\omega_s L_q}$$

$$I_f = \frac{E'_a + \sigma V_a \cos(\delta - \theta)}{\omega_s M_f (1 + \sigma)} \tag{4.103}$$

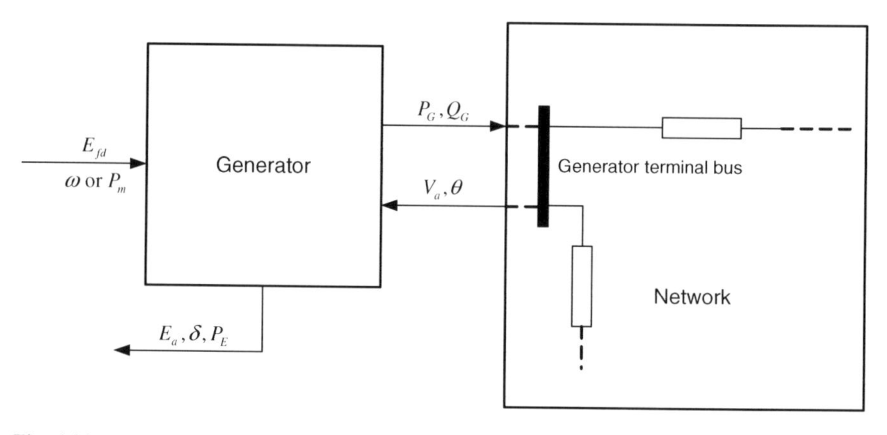

Fig. 4.23 Input-output summary of the simplified generator dynamic model with network interconnection

$$P_E = 3\frac{E_a V_a}{X_q}\sin(\delta-\theta) \tag{4.104}$$

$$Q_E = 3\frac{E_a V_a \cos(\delta-\theta) - E_a^2}{\omega_s L_d} \tag{4.105}$$

$$P_G = 3\left(\frac{E_a V_a}{X_d}\sin(\delta-\theta) + \frac{V_a^2}{2}\left(\frac{1}{X_q}-\frac{1}{X_d}\right)\sin 2(\delta-\theta)\right) \tag{4.106}$$

$$Q_G = 3\left(\frac{V_a^2}{2}\left(\frac{1}{X_q}+\frac{1}{X_d}\right) - \frac{E_a V_a}{X_d}\cos(\delta-\theta) + \frac{V_a^2}{2}\left(\frac{1}{X_d}-\frac{1}{X_q}\right)\cos 2(\delta-\theta)\right) \tag{4.107}$$

Here, a mechanical friction term with small parameter D has been added in (4.100). All voltages, powers, and impedance are in per-unit values. A summary of the model with its interconnection to a network is shown in Figure 4.23.

Remark 4.18 Voltage behind reactance model. In some instances, the time constant τ_f is substantially larger than the time period of interest. In such circumstances, it is appropriate to assume that E'_a is constant. Thus, Equation (4.101) can be dropped. This is the classical and frequently used *voltage behind reactance model.*

4.4.3 Power Flow Equations

4.4.3.1 Bus Admittance Matrix

Consider the ith bus within an N bus network. Let I_i denotes the *bus current* injected into the network at bus i. Then, KVC applies in the form

$$I_i = \sum_{k=1}^{N} I_{ik}, \tag{4.108}$$

where I_{ik} is the current flowing out of bus i over the transmission link connecting bus i with bus k (if any). The current I_{ii} is the flow out of bus i through any constant admittance load to ground connected to bus i.

Define the vectors of bus currents and bus voltages

$$\mathbf{I} = \begin{bmatrix} I_1 \\ \vdots \\ I_N \end{bmatrix}, \quad \mathbf{V} = \begin{bmatrix} V_1 \\ \vdots \\ V_N \end{bmatrix} \tag{4.109}$$

If each transmission element is modeled by an admittance, then the relation between $\mathbf{I}$ and $\mathbf{V}$ can be expressed

$$\mathbf{I} = \mathbf{Y}_{bus}\mathbf{V} \tag{4.110}$$

where the $N \times N$ bus admittance matrix $\mathbf{Y}_{bus}$ is

$$\mathbf{Y}_{bus} = \begin{bmatrix} Y_{11} & \cdots & Y_{1N} \\ \vdots & \ddots & \vdots \\ Y_{N1} & \cdots & Y_{NN} \end{bmatrix} \tag{4.111}$$

Each of the elements Y_{ij} can be defined in terms of the element admittances:

$$Y_{ij} = \begin{cases} y_{ii} + \sum_{k \neq i} y_{ik} & j = i \\ -y_{ij} & j \neq i \end{cases} \tag{4.112}$$

where $y_{ij} = g_{ij} + j\, b_{ij}$ denotes the admittance between bus i and bus j.

Example 4.19 Consider, as a simple example, the three-bus system of Figure 4.24. Note that transmission links are modeled as ideal inductors and the shunt elements as ideal capacitors. The bus admittance matrix is easily computed using (4.112) to be:

$$\mathbf{Y} = \begin{bmatrix} -j19.98 & j10 & j10 \\ j10 & -j19.98 & j10 \\ j10 & j10 & -j19.98 \end{bmatrix}$$

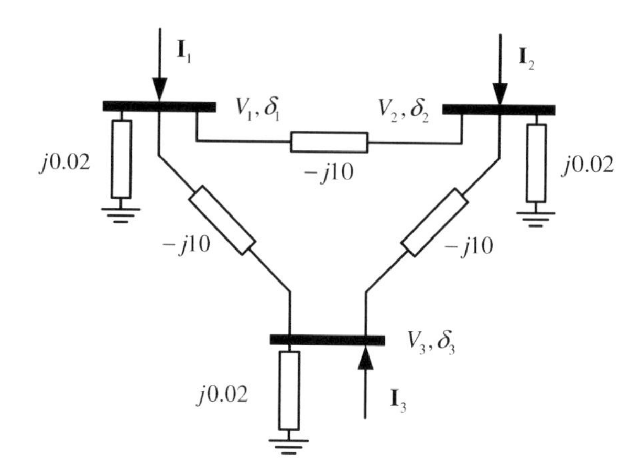

Fig. 4.24 Three-bus example from [27]

4.4.3.2 Power Flow Equations

Using (4.112) the bus power injection at bus i can be obtained in the form

$$S_i = V_i I_i^* = \sum_{k=1}^{N} Y_{ik}^* V_i V_k^* = \sum_{k=1}^{n} (G_{ik} - jB_{ik}) |V_i| \left|V_k\right| e^{j(\delta_i - \delta_k)}$$

Replacing the exponential by equivalent trigonometric functions leads to

$$S_i = \sum_{k=1}^{N} (G_{ik} - jB_{ik}) |V_i| \left|V_k\right| (\cos(\delta_i - \delta_k) + j \sin(\delta_i - \delta_k))$$

Separating this complex expression leads to expressions for real and reactive power injections for bus i.

$$P_i = \sum_{k=1}^{N} |V_i| \left|V_k\right| (B_{ik} \sin(\delta_i - \delta_k) + G_{ik} \cos(\delta_i - \delta_k)) \tag{4.113}$$

$$Q_i = \sum_{k=1}^{N} |V_i| \left|V_k\right| (G_{ik} \sin(\delta_i - \delta_k) - B_{ik} \cos(\delta_i - \delta_k)) \tag{4.114}$$

Equations (4.113) and (4.114), with $i = 1, \ldots, N$, are called the *power flow equations*, represent a complete description of the network. Note that the left-hand side of the equations, P_i, Q_i, represent the real and reactive power injections into the bus, whereas the right-hand expressions represent the total electric power flow out of the bus into the transmission system. These expressions are functions of the network bus voltage vector. Below, the right-hand expressions will be represented, respectively, by $p_i(\mathbf{V})$ and $q_i(\mathbf{V})$.

Example 4.20 Once again, consider the system of Figure 4.24. The power flow equations are

$$
\begin{aligned}
&P_1 - 10\,|V_1|\,|V_2|\sin(\delta_1-\delta_2) - 10\,|V_1|\,|V_3|\sin(\delta_1-\delta_3) = 0\\
&P_2 + 10\,|V_1|\,|V_2|\sin(\delta_1-\delta_2) - 10\,|V_1|\,|V_3|\sin(\delta_2-\delta_3) = 0\\
&P_3 + 10\,|V_1|\,|V_2|\sin(\delta_1-\delta_3) + 10\,|V_1|\,|V_3|\sin(\delta_2-\delta_3) = 0\\
&Q_1 - 19.98\,|V_1|^2 + 10\,|V_1|\,|V_2|\cos(\delta_1-\delta_2) + 10\,|V_1|\,|V_3|_3\cos(\delta_1-\delta_3) = 0\\
&Q_2 - 19.98\,|V_2|^2 + 10\,|V_1|\,|V_2|\cos(\delta_1-\delta_2) + 10\,|V_1|\,|V_3|\cos(\delta_2-\delta_3) = 0\\
&Q_3 - 19.98\,|V_3|^2 + 10\,|V_1|\,|V_3|\cos(\delta_1-\delta_3) + 10\,|V_2|\,|V_3|\cos(\delta_2-\delta_3) = 0
\end{aligned}
$$

Notice that in an N bus network there are N pairs of equations, one pair for each bus. Thus, there are in all $2N$ equations. Each bus has associated with it four independent variables, $P_i, Q_i, |V_i|, \delta_i$. Thus, there are $4N$ independent variables if it is assume that the admittance values are all given. It is to be anticipated that with specification of $2N$ of the bus variable, the $2N$ equations could be solved for the remaining $2N$ bus variables. To be precise, divide the $4N$ bus variables into two sets: $x \in R^{2N}$ and $y \in R^{2N}$. The $2N$ power flow equations, (4.113), (4.114), $i = 1, \ldots, N$, can be rewritten in the form $0 = f(x, y), f \in R^{2N}$. For a given solution pair, $y = y^*, x = x^*$, for there to exist a local solution $y = \phi(x)$ with $y^* = \phi(x^*)$, the Implicit Function Theorem 5.2 requires

$$\det \frac{\partial f(x^*, y^*)}{\partial y} \neq 0$$

Consequently, the selection of a set of independent variables is not arbitrary. In fact, notice that there is a translational symmetry in the power flow equations with respect to the angle variables δ_i. In fact, if all N angles are included in y then it is easily shown that

$$\det \frac{\partial f(x^*, y^*)}{\partial x} \equiv 0$$

As a result, it is standard practice to choose one bus called a *slack bus* or *reference bus* at which the angle is specified.

The appropriate causality at each bus, that is the specification of the two inputs and two outputs, depends on the device(s) attached to the bus. The network shown in Figure 4.25 connects to external devices via three types of buses:

1. Generator bus. What is termed a generator bus refers to a classical generator model, in which the bus is the generator internal bus, see Figure 4.11. This means that the inductance and resistance of the equivalent circuit is included as a branch within the network. At this type of bus, the bus (or port) inputs are voltage magnitude and angle $|V|, \delta$ and the outputs are real and reactive power, P, Q. One such generator could be a reference generator in which case δ is set to zero. Or the attached device might be an "infinite bus" reference in which $|V|, \delta$ are specified constants.

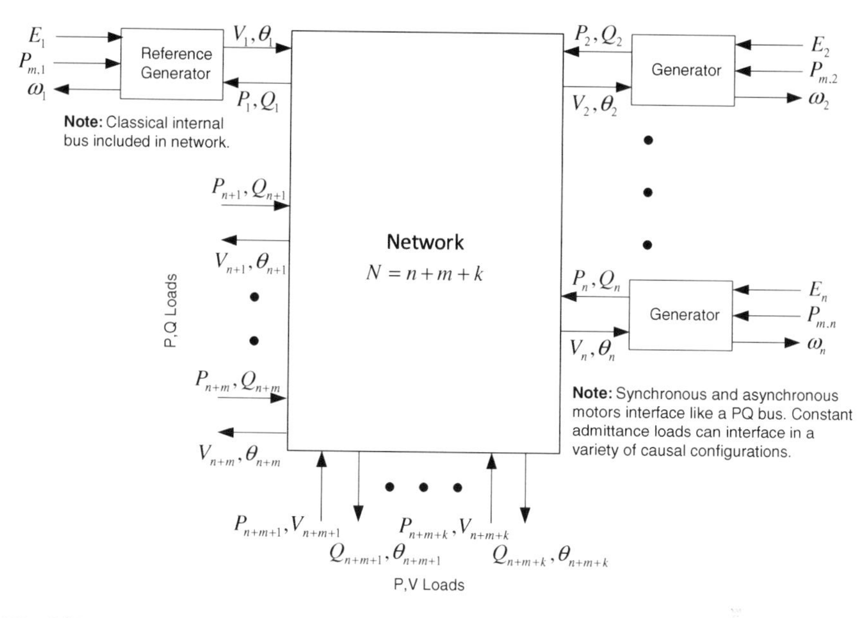

Fig. 4.25 Network interconnection diagram showing causality structure

2. PV bus. The causality for this bus is such that the bus inputs are P and V and the outputs are δ and Q. A constant admittance load might be the attached device.
3. PQ bus. The inputs in this case are the real and reactive power P, Q and the bus outputs are voltage magnitude and angle $|V|$, δ. Devices that could attach to this bus include a generator (terminal bus), synchronous or asynchronous motors, or constant P and Q loads.

4.4.3.3 The Classical Model

The classical model of a power system is defined as a system with n generator buses, m PV buses, and k PQ bus. The voltage behind reactance generator model is assumed valid. Each generator bus is taken to be the generator internal bus. Consequently, the internal bus voltage magnitude is constant and θ is determined by the mechanical swing equations (4.100). Assume the $N = n + m + k$ buses are labeled in the order: generator buses, PV buses, and PQ buses. Define the network input and out vectors, respectively, $U \in R^{2N}$ and $y \in R^{2N}$. First, note that

- All generator buses, in the classical model, take inputs V, θ and provide outputs P, Q.
- All PV buses take inputs P, V and provide outputs θ, Q.
- All PQ buses take inputs P, Q and provide outputs V, θ.

Thus,

$$\begin{aligned} U &= (\theta_1, \ldots, \theta_n, V_1, \ldots, V_{n+m}, P_{n+1}, \ldots, P_N, Q_{n+m+1}, \ldots Q_N) \\ y &= (\theta_{n+1}, \ldots, \theta_N, V_{n+m+1}, \ldots, V_n, P_1, \ldots, P_n, Q_1, \ldots, Q_{n+m}) \end{aligned} \tag{4.115}$$

Now, divide the input vector into two parts, $U = (\theta, u)$ where

$$\theta = (\theta_1, \ldots, \theta_n), u = (V_1, \ldots, V_{n+m}, P_{n+1}, \ldots, P_N, Q_{n+m+1}, \ldots Q_N)$$

The vector u is a vector of exogenous system inputs and the vector θ will become part of the state vector, $x \in R^{2n}$:

$$x = (\theta_1, \ldots, \theta_n, \upsilon_1, \ldots, \upsilon_n), \quad \dot{\theta}_i = \upsilon_i \tag{4.116}$$

Define the functions $f : R^n \times R^{2N} \times R^{2N-n} \to R^n$ and $g : R^n \times R^{2N} \times R^{2N-n} \to R^{2N-n}$:

$$f(\theta, y, u) = \left(P_{M,1} - p_1, \ldots, P_{M,n} - p_n\right) \tag{4.117}$$

$$g(\theta, y, u) = \begin{pmatrix} P_{L,n+1} - p_{n+1}, \ldots, P_{L,N} - p_N, Q_{M,1} - q_1, \ldots, Q_{M,n} - q_n, \\ - q_{n+1}, \ldots, -q_{n+m}, Q_{L,n+m+1} - q_{n+m+1}, \ldots, Q_{L,N} - q_N \end{pmatrix} \tag{4.118}$$

The dynamics of the system can be written

$$\begin{aligned} \dot{\theta} &= \upsilon \\ M\dot{\upsilon} + D\upsilon &= f(\theta, y, u) \\ 0 &= g(\theta, y, u) \end{aligned} \tag{4.119}$$

Remark 4.21 Reduced Network. If a network bus has nothing attached to it other than transmission lines and a constant admittance load, then it can be eliminated resulting in a smaller set of equations. This is easily accomplished as follows. Consider an N bus network with admittance matrix $\mathbf{Y}$. Suppose the network contains N_2 buses to be eliminated, and $N_1 = N - N_2$ that are to be retained. Suppose the buses are ordered such that the last N_2 buses are the constant admittance load buses. The network admittance matrix and the current and voltage vectors can be partitioned so that

$$\begin{bmatrix} \mathbf{I}_1 \\ \mathbf{I}_2 \end{bmatrix} = \begin{bmatrix} \mathbf{Y}_{11} & \mathbf{Y}_{12} \\ \mathbf{Y}_{21} & \mathbf{Y}_{22} \end{bmatrix} \begin{bmatrix} \mathbf{V}_1 \\ \mathbf{V}_2 \end{bmatrix}$$

but $\mathbf{I}_2 = 0$, so

$$\mathbf{I}_2 = -\mathbf{Y}_{21}\mathbf{V}_1 + \mathbf{Y}_{22}\mathbf{V}_2 = 0 \Rightarrow \mathbf{V}_2 = \mathbf{Y}_{22}^{-1}\mathbf{Y}_{21}\mathbf{V}_1$$

Now,

$$\mathbf{I}_1 = \mathbf{Y}_{11}\mathbf{V}_1 + \mathbf{Y}_{21}\mathbf{V}_2 = \left(\mathbf{Y}_{11} - \mathbf{Y}_{21}\mathbf{Y}_{22}^{-1}\mathbf{Y}_{21}\right)\mathbf{V}$$

or

$$\mathbf{I}_1 = \mathbf{Y}_{red}\mathbf{V}_1, \quad \mathbf{Y}_{red} = \left(\mathbf{Y}_{11} - \mathbf{Y}_{21}\mathbf{Y}_{22}^{-1}\mathbf{Y}_{21}\right) \tag{4.120}$$

It follows that if a system has n generators and all loads are constant admittance, then all network buses other than the n generator internal buses can be eliminated by this process. Thus, the dynamical equations (4.117) reduce to

$$\begin{aligned} \dot{\theta} &= \upsilon \\ M\dot{\upsilon} + D\upsilon &= f(\theta, u) \end{aligned} \tag{4.121}$$

with $\theta = (\theta_1, \ldots, \theta_n)$, $\upsilon = (\omega_1 - \omega_s, \ldots, \omega_n - \omega_s)$, and $u = (V_1, \ldots, V_n)$, where all of the n voltage magnitudes, V_i, are known.

More generally, suppose there are n generators, m constant admittance load buses, and k PQ buses, for a total of $N = n + k + m$ buses including the n generator internal buses. The m constant admittance buses can be eliminated to give a reduced network with $N_r = n + k$ buses. As above, assume the buses to be eliminated are the last m, and assume the generator internal buses are the first n. The classical model now reduces to the form

$$\begin{aligned} \dot{\delta} &= \upsilon \\ M\dot{\upsilon} + D\upsilon &= f_1(\delta, y, u) \\ 0 &= f_2(\delta, y, u) \end{aligned} \tag{4.122}$$

with

$$\begin{aligned} \delta &= (\theta_1, \ldots, \theta_n) \\ u &= \left(V_1, \ldots, V_n, P_{n+1}, \ldots, P_{N_r}, Q_{n+1}, \ldots, Q_{N_r}\right) \\ y &= \left(\theta_{n+1}, \ldots, \theta_{N_r}, V_{n+1}, \ldots, V_{N_r}\right) \end{aligned}$$

Notice that $f_1 : R^{4N_r} \to R^n$ and $f_2 : R^{4N_r} \to R^{2N_r - n}$.

Chapter 5
Power System Dynamics: Foundations

> *"My methods are really methods of working and thinking; this is why they have crept in everywhere anonymously."*
>
> —Emmy Noether

5.1 Introduction

In this chapter, we briefly review basic material about nonlinear ordinary differential equations that is important background for later chapters. After a preliminary discussion of the basic properties of differential equations including the existence and uniqueness of solutions, we turn to a short discussion of stability in the sense of Lyapunov. In addition to stating the most important theorems on stability and instability, we provide a number of illustrative examples. As part of this discussion, we introduce Lagrangian systems—a topic to be treated at great length later. This chapter is concerned exclusively with dynamical systems with smooth systems. It is presumed that the material discussed is not new to the reader, and we provide only a short summary of those elements considered immediately relevant. For a more complete discussion, many excellent textbooks are available. We reference a number of them in the sequel.

5.2 Preliminaries

A *linear vector space*, $\mathcal{V}$- over the field R, is a set of elements called vectors such that:

1. For each pair $x, y \in \mathcal{V}$, the sum $x + y$ is defined, $x + y \in \mathcal{V}$ and $x + y = y + x$.
2. There is an element "0" in $\mathcal{V}$ such that for every $x \in \mathcal{V}$, $x + 0 = x$.

H.G. Kwatny and K. Miu-Miller, *Power System Dynamics and Control*,
Control Engineering, DOI 10.1007/978-0-8176-4674-5_5

3. For any number $a \in R$ and vector $x \in \mathcal{V}$, scalar multiplication is defined and $ax \in \mathcal{V}$.
4. For any pair of numbers $a, b \in R$ and vectors $x, y \in \mathcal{V}$: $1 \cdot x = x$, $(ab)x = a(bx)$, $(a+b)x = ax + bx$.

A linear vector space is a *normed linear space* if for each vector $x \in \mathcal{V}$, there corresponds a real number $\|x\|$ called the *norm* of x which satisfies:

1. $\|x\| > 0,\ x \neq 0,\ \|0\| = 0$
2. $\|x + y\| \leq \|x\| + \|y\|$ (triangle inequality)
3. $\|ax\| = |a|\ \|x\|\ \forall a \in R,\ x \in \mathcal{V}$

When confusion can arise as to which space a norm is defined in we replace $\|\cdot\|$ by $\|\cdot\|_{\mathcal{V}}$.

A sequence $\{x_k\} \subset \mathcal{V}$, $\mathcal{V}$ a normed linear space, *converges* to $x \in \mathcal{V}$ if

$$\lim_{k \to \infty} \|x_k - x\| = 0$$

A sequence $\{x_k\} \subset \mathcal{V}$ is a *Cauchy sequence* if for every $\varepsilon > 0$, there is an integer, $N(\varepsilon) > 0$ such that $\|x_n - x_m\| < \varepsilon$ if $n, m > N(\varepsilon)$. Every convergent sequence is a Cauchy sequence but not vice versa. The space is *complete* if every Cauchy sequence is a convergent sequence. A complete normed linear space is called a *Banach* space.

The most basic Banach space of interest herein is n-dimensional *Euclidean* space, the set of all n-tuples of real numbers, denoted R^n. The most common types of norms applied to R^n are the p-norms, defined by

$$\|x\|_p = \left(|x_1|^p + \cdots + |x_n|^p\right)^{1/p}, \quad 1 \leq p < \infty$$

and

$$\|x\|_\infty = \max_{i \in \{1,\ldots,n\}} |x_i|$$

An ε-*neighborhood* of an element x of the normed linear space $\mathcal{V}$ is the set $S(x, \varepsilon) = \{y \in V \mid \|y - x\| < \varepsilon\}$. A set A in $\mathcal{V}$ is *open* if for every $x \in A$ there exists an ε-neighborhood of x also contained in A. An element x is a *limit point* of a set $A \subset \mathcal{V}$ if each ε-neighborhood of x contains points in A. A set A is *closed* if it contains all of its limit points. The *closure* of a set A, denoted $\bar{A}$, is the union of A and its limit points. A set A is *dense* in $\mathcal{V}$ if the closure of A is $\mathcal{V}$.

If B is a subset of $\mathcal{V}$, A is a subset of R, and $\{V_a, a \in A\}$ is a collection of open subsets of $\mathcal{V}$ such that $\cup_{a \in A} V_a \supset B$, then the collection V_a is called an *open covering* of B. A set B is *compact* if every open covering of B contains a finite number of subsets which is also an open covering of B. For a Banach space, this is equivalent to the property that every sequence $\{x_n\}$, $x_n \in B$, contains a subsequence which converges to an element of B. A set B is *bounded* if there exists a number $r > 0$ such that $B \subset \{x \in \mathcal{V} \mid \|x\| < r\}$. A set B in R^n is compact if and only if it is closed and bounded.

A function f taking a set A of a space $\mathcal{X}$ into a set B of a space $\mathcal{Y}$ is called a *mapping* of A into B, and we write $f : A \to B$. A is the *domain* of the mapping, and B is the *range* or *image*. The image of f is denoted $f(A)$. f is *continuous* if, given $\varepsilon > 0$, there exists $\delta > 0$ such that

$$\|x - y\| < \delta \Rightarrow \|f(x) - f(y)\| < \varepsilon$$

A function f defined on a set A is said to be *one-to-one* on A if and only if for every $x, y \in A$, $f(x) = f(y) \Rightarrow x = y$. If f is one-to-one, it has an inverse denoted f^{-1}. If the one-to-one mapping f and its inverse f^{-1} are continuous, f is called a *homeomorphism* of A onto B.

Suppose $\mathcal{X}$ and $\mathcal{Y}$ are Banach spaces, and $f : \mathcal{X} \to \mathcal{Y}$. f is a *linear map* if $f(a_1x_1 + a_2x_2) = a_1 f(x_1) + a_2 f(x_2)$ for all $x_1, x_2 \in \mathcal{X}$ and $a_1, a_2 \in R$ (or C). In general, we can write a linear mapping in the form $y = Lx$, where L is an appropriately defined "linear operator." A linear map f is said to be *bounded* if there is a constant K such that $\|f(x)\|_{\mathcal{Y}} \leq K \|x\|_{\mathcal{X}}$ for all $x \in \mathcal{X}$. A linear map $f : \mathcal{X} \to \mathcal{Y}$ is bounded if and only if it is continuous. A linear map from $R^n \to R^m$ is characterized by an $m \times n$ matrix of real elements, example $y = Ax$. The "size" of the matrix A can be measured by the *induced p-norm* (or gain) of A

$$\|A\|_p = \sup_{x \neq 0} \frac{\|Ax\|_p}{\|x\|_p}$$

for which we write the following special cases

$$\|A\|_1 = \max_{1 \leq j \leq n} \sum_{i=1}^{m} |a_{ij}|$$

$$\|A\|_2 = \sqrt{\lambda_{\max}(A^T A)}$$

$$\|A\|_\infty = \max_{1 \leq i \leq m} \sum_{j=1}^{n} |a_{ij}|$$

Here, $\lambda_{\max}(A^T A)$ denotes the largest eigenvalue of the nonnegative matrix $A^T A$.

f is said to be (*Frechet*) *differentiable* at a point $x \in A$ if there exists a bounded linear operator $L(x)$ mapping $\mathcal{X} \to \mathcal{Y}$ such that for every $h \in \mathcal{X}$ with $x + h \in A$

$$\|f(x + h) - f(x) - L(x)h\| \,/\, \|h\| \to 0$$

as $\|h\| \to 0$. $L(x)$ is called the derivative of f at x. If $f : R^n \to R^m$ is differentiable at x, then $L(x) = \partial f(x) / \partial x$, the Jacobian of f with respect to x. If f and f^{-1} have continuous first derivatives, f is a *diffeomorphism*.

A function $f : A \to B$ is said to belong to the class C^k of functions if it has continuous derivatives up to order k. It belongs to the class C^∞ if it has continuous derivatives of any order. C^∞ functions are sometimes called *smooth*. A function f is said to be *analytic* if for each $x_0 \in A$ there is a neighborhood U of x_0 such that the Taylor series expansion of f at x_0 converges to $f(x)$ for all $x \in U$.

Consider a transformation $T : \mathcal{X} \to \mathcal{X}$, where $\mathcal{X}$ is a Banach space. $x \in \mathcal{X}$ is a *fixed point* of T if $x = T(x)$. Suppose A is a subset of Banach space, $\mathcal{X}$ and T is a mapping of A into a Banach space $\mathcal{B}$. The transformation T is a *contraction* on A if there exists a number $0 \leq \lambda < 1$ such that

$$\|T(x) - T(y)\| \leq \lambda \|x - y\|, \quad \forall x, y \in A$$

Proposition 5.1 (Contraction Mapping Theorem) *Suppose A is a closed subset of a Banach space $\mathcal{X}$ and $T : A \to A$ is a contraction on A. Then*

1. *T has a unique fixed point $\bar{x} \in A$*
2. *If $x_0 \in A$ is arbitrary, then the sequence $\{x_{n+1} = T(x_n),\ n = 0, 1, \ldots\}$ converges to $\bar{x}$.*
3. *$\|x_n - \bar{x}\| \leq \lambda^n \|x_1 - x_0\| / (1 - \lambda)$, where $\lambda < 1$ is the contraction constant for T on A.*

Proof [84], page 5.

We will make use of the following important theorem.

Proposition 5.2 (Implicit Function Theorem) *Suppose $F : R^n \times R^m \to R^n$ has continuous first partial derivatives and $F(0, 0) = 0$. If the Jacobian matrix $\partial F(x, y)/\partial x$ is nonsingular, then there exists neigborhoods U, V of the origin in R^n, R^m, respectively, such that for each $y \in V$ the equation $F(x, y) = 0$ has a unique solution $x \in U$. Furthermore, this solution can be given as $x = g(y)$, i.e., $F(g(y), y) = 0$ on V, where g has continuous first derivatives and $g(0) = 0$.*

Proof [84], page 8.

5.3 Ordinary Differential Equations

5.3.1 *Existence and Uniqueness*

Let $t \in R, x \in R^n$, D an open subset of R^{n+1}, $f : D \to R^n$ a map and let $\dot{x} = dx/dt$. We will consider differential equations of the type

$$\dot{x} = f(x, t), \quad x \in R^n, \ t \in R \tag{5.1}$$

When t is explicitly present in the right-hand side of (5.1), then the system is said to be *nonautonomous*. Otherwise, it is *autonomous*. A solution of (5.1) on a time interval $t \in [t_0, t_1]$ is a function $x(t) : [t_0, t_1] \to R^n$, such that $dx(t)/dt = f(x, t(t))$ for each $t \in [t_0, t_1]$. We can visualize an individual solution as a graph $x(t) : t \to R^n$. For autonomous systems, it is convenient to think of $f(x)$ as a "vector field" on the space R^n. $f(x)$ assigns a vector to each point $x \in R^n$. As t varies, a solution $x(t)$ traces a path through R^n. These curves are often called *trajectories* or *orbits*. At each point $x \in R^n$, the trajectory $x(t)$ is tangent to the vector $f(x)$. The collection of all trajectories in R^n is called the *flow* of the vector field $f(x)$. This point of view can be extended to nonautonomous differential equations in which case the vector field $f(x, t)$ and its flow vary with time.

Example 5.3 (*Phase portraits*) For two-dimensional systems, the trajectories can be plotted in a plane. We will consider two systems, the Van der Pol system

$$\begin{bmatrix} \dot{x}_1 \\ \dot{x}_2 \end{bmatrix} = \begin{bmatrix} x_2 \\ -0.8(1 - x_1^2)x_2 - x_1 \end{bmatrix}$$

and the damped pendulum

$$\begin{bmatrix} \dot{x}_1 \\ \dot{x}_2 \end{bmatrix} = \begin{bmatrix} x_2 \\ -x_2/2 - \sin x_1 \end{bmatrix}$$

Both of these systems are in so-called phase variable form (the first equation, $\dot{x}_1 = x_2$, defines velocity) so the trajectory plots are called phase portraits. These are shown in Figures 5.1 and 5.2.

The above-mentioned examples illustrate several important properties of nonlinear dynamical systems. In both cases, the flow directions are to the right in the upper half plane and to the left in the lower half plane (recall $\dot{x}_1 = x_2$).

Fig. 5.1 Phase portrait for the Van der Pol equation

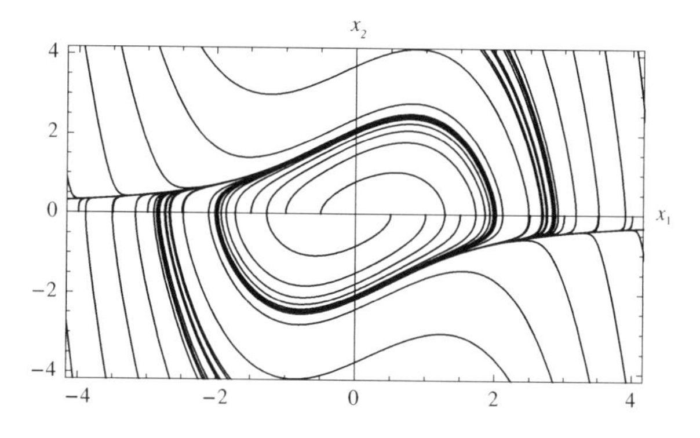

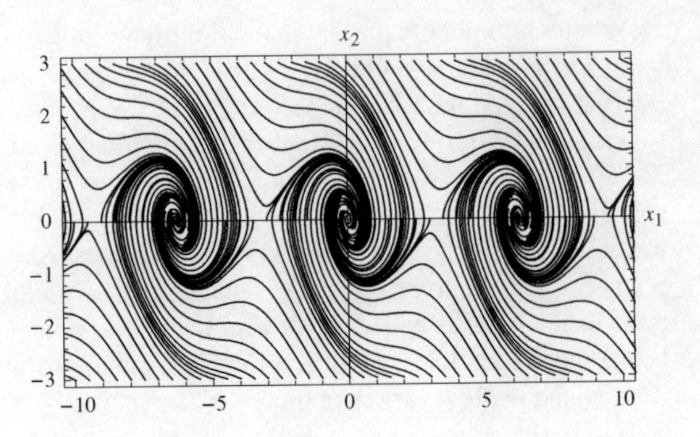

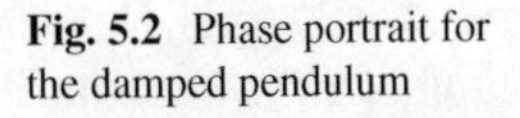

Fig. 5.2 Phase portrait for the damped pendulum

Thus, it is easily seen that trajectories of the pendulum ultimately converge to rest points corresponding to the pendulum hanging straight down. These have the property that $f(x) = 0$. Any point $x \in R^n$ satisfying the condition $f(x) = 0$ is called an *equilibrium point*. The pendulum has an infinite number of equilibria spaced π radians apart. Some of these are attracting (the pendulum points straight down) and some repelling (straight up).

In contrast, all trajectories of the Van der Pol equation approach a periodic trajectory. Such an isolated periodic trajectory is called a *limit cycle*. Some systems can exhibit multiple limit cycles, and they can be repelling as well as attracting. Equilibria and limit cycles are two types of "limit sets" that are associated with differential equations. We will define limit sets precisely below. As a matter of fact, these are the only type of limit sets exhibited by two-dimensional systems. More exotic ones, like "strange attractors," require at least three-dimensional state spaces.

The existence and uniqueness of solutions to (5.1) depend on the properties of the function f. In many applications, $f(x, t)$ is continuous in the variables t and x. We will impose a somewhat less restrictive characterization of f. We say that a function $f : R^n \rightarrow R^n$ is *locally Lipschitz* on an open and connected subset $D \subset R^n$, if each point $x_0 \in D$ has a neighborhood U_0 such that

$$\|f(x) - f(x_0)\| \leq L \, \|x - x_0\| \tag{5.2}$$

for some constant L and all $x \in U_0$. The function $f(x)$ is said to be *Lipschitz* on the set D if it satisfies the (local) Lipschitz condition uniformly (with the same constant L) at all points $x_0 \in D$. It is *globally Lipschitz* if it is Lipschitz on $D = R^n$. We apply the terminology "Lipschitz in x" to functions $f(x, t)$, provided the Lipschitz condition holds uniformly for each t in a given interval of R.

Note that C^0 functions need not be Lipschitz but C^1 functions always are. The following theorems relate the notion of Lipshitz with the property of continuity.

Lemma 5.4 *Let $f(x, t)$ be continuous on $D \times [a, b]$, for some domain $D \subset R^n$. If $\partial f / \partial x$ exists and is continuous on $D \times [a, b]$, then f is locally Lipschitz in x on $D \times [a, b]$.*

Proof (following Khalil [106], p. 77) For $x_0 \in D$ there is an r sufficiently small that

$$D_0 = \left\{x \in R^n \mid \|x - x_0\| < r\right\} \subset D$$

The set D_0 is convex and compact. Since f is C^1, $\partial f/\partial x$ is bounded on $[a, b] \times D_0$. Let L_0 denote such a bound. If $x, y \in D_0$, then by the mean value theorem, there is a point z on the line segment joining x, y such that

$$\|f(x, t) - f(t, y)\| = \left\|\frac{\partial f(t, z)}{\partial x}(x - y)\right\| \leq L_0 \|x - y\|$$

■

The proof of this lemma is easily adapted to prove the following.

Proposition 5.5 *Let $f(x, t)$ be continuous on $[a, b] \times R^n$. If f is C^1 in $x \in R^n$ for all $t \in [a, b]$ then f is globally Lipschitz in x if and only if $\partial f/\partial x$ is uniformly bounded on $[a, b] \times R^n$.*

Let us state the key existence result.

Proposition 5.6 (Local Existence and Uniqueness) *Let $f(x, t)$ be piecewise continuous in t and satisfy the Lipschitz condition*

$$\|f(x, t) - f(t, y)\| \leq L \|x - y\|$$

for all $x, y \in B_r = \{x \in R^n \mid \|x - x_0\| < r\}$ and all $t \in [t_0, t_1]$. Then there exists a $\delta > 0$ such that the differential equation with initial condition

$$\dot{x} = f(x, t), \quad x(t_0) = x_0 \in B_r$$

has a unique solution over $[t_0, t_0 + \delta]$.

Proof ([106], p. 74) A continuation argument leads to the following global extension.

Proposition 5.7 (Global Existence and Uniqueness) *Suppose $f(x, t)$ is piecewise continuous in t and satisfies*

$$\|f(x, t) - f(t, y)\| \leq L \|x - y\|$$

$$\|f(x, t_0)\| < h$$

for all $x, y \in R^n$ and all $t \in [t_0, t_1]$. Then the equation

$$\dot{x} = f(x, t), \quad x(t_0) = x_0$$

has a unique solution over $[t_0, t_1]$.

5.3.1.1 Continuous Dependence on Parameters and Initial Data

Let $\mu \in R^k$ and consider the parameter-dependent differential equation

$$\dot{x} = f(x, t, \mu), \quad x(t_0) = x_0 \tag{5.3}$$

We will show that a solution $x(t; t_0, x_0, \mu)$ defined on a finite time interval $[t_0, t_1]$ is continuously dependent on the parameter μ and the initial data t_0, x_0.

Definition 5.8 Let $x(t; t_0, \xi_0, \mu_0)$ denote a solution of (5.3) defined on the finite interval $t \in [t_0, t_1]$ with $\mu = \mu_0$ and $x(t_0; t_0, \xi_0, \mu_0) = \xi_0$. The solution is said to *depend continuously on μ at μ_0* if for any $\varepsilon > 0$ there is a $\delta > 0$ such that such that for all μ in the neighborhood $U = \{\mu \in R^k \mid \|\mu - \mu_0\| < \delta\}$, (5.3) has a solution $x(t; t_0, \xi_0, \mu)$ such that

$$\|x(t; t_0, \xi_0, \mu) - x(t; t_0, \xi_0, \mu_0)\| < \varepsilon$$

for all $t \in [t_0, t_1]$. Similarly, the solution is said to *depend continuously on ξ at ξ_0* if for any $\varepsilon > 0$ there is a $\delta > 0$ such that for all ξ in the neighborhood $X = \{\xi \in R^k \mid \|\xi - \xi_0\| < \delta\}$, (5.3) has a solution $x(t; t_0, \xi, \mu_0)$ such that

$$\|x(t; t_0, \xi, \mu_0) - x(t; t_0, \xi_0, \mu_0)\| < \varepsilon$$

for all $t \in [t_0, t_1]$.

The following result establishes the basic continuity properties of (5.3) on finite time intervals.

Proposition 5.9 *Suppose $f(x, t, \mu)$ is continuous in (x, t, μ) and locally Lipschitz in x (uniformly in t and μ) on $[t_0, t_1] \times D \times \{\|\mu - \mu_0\| < c\}$ where $D \subset R^n$ is an open and connected set. Let $x(t; t_0, \xi_0, \mu_0)$ denote a solution of* (5.3) *that belongs to D for all $[t_0, t_1]$. Then given $\varepsilon > 0$ there is $\delta > 0$ such that*

$$\|\xi - \xi_0\| < \delta, \; \|\mu - \mu_0\| < \delta$$

implies that there is a unique solution $x(t; t_0, \xi, \mu)$ of (5.3) *defined on $t \in [t_0, t_1]$ and such that*

$$\|x(t; t_0, \xi, \mu) - x(t; t_0, \xi_0, \mu_0)\| < \varepsilon, \; \forall t \in [t_0, t_1]$$

Proof ([106], p. 86) We emphasize that the results on existence and continuity of solutions hold on finite time intervals $[t_0, t_1]$. Stability, as we shall see below, requires us to consider solutions defined on infinite intervals. We will often tacitly assume that they are so defined. Continuity issues with respect to both initial conditions and parameters for solutions on infinite time intervals are quite subtle.

5.3.2 *Invariant Sets*

In the following paragraphs, we shall restrict attention to autonomous systems

$$\dot{x} = f(x), \quad x(t_0) = x_0 \tag{5.4}$$

In many instances, the results can be extended to nonautonomous systems by extending the nonautonomous differential equation with the addition of a new state $\dot{x}_{n+1} = 1$ to replace t in the right-hand side of the differential equation.

Let us denote by $\Psi(t, x)$ the flow of the vector field f on R^n defined by (5.4), i.e., $\Psi(t, x)$ is the solution of (5.4) with $\Psi(0, x) = x$:

$$\frac{\partial \Psi(x,t)}{\partial t} = f(\Psi(x,t)), \quad \Psi(0,x) = x$$

Definition 5.10 A set of points $S \subset R^n$ is *invariant* with respect to f if trajectories beginning in S remain in S both forward and backward in time, i.e., if $s \in S$, then $\Psi(t, s) \in S, \forall t \in R$.

Obviously, any entire trajectory of (5.4) is an invariant set. Such an invariant set is minimal in the sense that it does not contain any proper subset which is itself an invariant set.

A set S is invariant if and only if $\Psi(t, S) \mapsto S$ for each $t \in R$.

5.3.2.1 Nonwandering Sets

Definition 5.11 A point $p \in R^n$ is a *nonwandering point* with respect to the flow Ψ if for every neighborhood U of p and $T > 0$, there is a $t > T$ such that $\Psi(t, U) \cap U \neq \emptyset$. The set of nonwandering points is called the *nonwandering set* and denoted Ω. Points that are not nonwandering are called *wandering points*.

The nonwandering set is a closed, invariant set. For proofs and other details see, for example, Guckenheimer and Holmes [83], Arrowsmith and Place [12], or Sibirsky [177]. The detailed structure of the nonwandering set is an important aspect of the analysis of strange attractors.

Obviously, fixed points and periodic trajectories belong to Ω.

5.3.2.2 Limit Sets

Definition 5.12 A point $q \in R^n$ is said to be an *ω-limit point* of the trajectory $\Psi(t, p)$ if there exists a sequence of time values $t_k \to +\infty$ such that

$$\lim_{t_k \to \infty} \Psi(t_k, p) = q$$

q is said to be an *α-limit point* of $\Psi(t, p)$ if there exists a sequence of time values $t_k \to -\infty$ such that

$$\lim_{t_k \to -\infty} \Psi(t_k, p) = q$$

The set of all ω-limit points of the trajectory through p is the *ω-limit set*, $\Lambda_\omega(p)$, and the set of all α-limit points of the trajectory through p is the *α-limit set*, $\Lambda_\alpha(p)$.

Hirsch and Smale [88] remind us that α and ω are the first and last letters of the Greek alphabet and, hence, the terminology.

Proposition 5.13 *The α-, ω-limit sets of any trajectory are closed invariant sets and they are subsets of the nonwandering set Ω.*

Proof Hirsch and Smale [88] or Sibirsky [177] for closed, invariant sets. That they are subsets of Ω is obvious.

We can make some simple observations:

1. If $r \in \Psi(t, p)$, then $\Lambda_\omega(r) = \Lambda_\omega(p)$ and $\Lambda_\alpha(r) = \Lambda_\alpha(p)$, i.e., any two points on a given trajectory have the same limit points.
2. If p is an equilibrium point, i.e., $f(p) = 0$ or $p = \Psi(t, p)$, then $\Lambda_\omega(p) = \Lambda_\alpha(p) = p$.
3. If $\Psi(t, p)$ is a periodic trajectory $\Lambda_\omega(p) = \Lambda_\alpha(p) = \Psi(R, p)$, i.e., the α and ω limit sets are the entire trajectory.

Finally, let us state the following important result.

Proposition 5.14 *A homeomorphism of a dynamical system maps ω-, α-limit sets into ω-, α-limit sets.*

Proof [177].

5.4 Lyapunov Stability

5.4.1 Autonomous Systems

In the following paragraphs, we consider autonomous differential equations and assume that the origin is an equilibrium point:

$$\dot{x} = f(x), \quad f(0) = 0 \tag{5.5}$$

with $f : D \to R^n$, locally Lipschitz in the domain D.

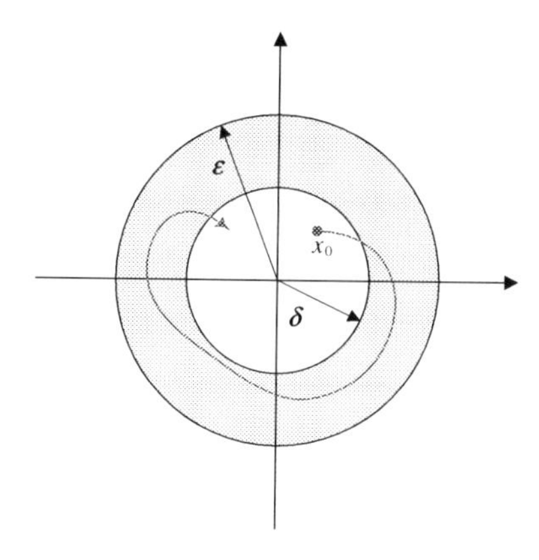

Fig. 5.3 Definition of Lyapunov stability

Definition 5.15 The origin of (5.5) is

1. a *stable equilibrium point* if for each $\varepsilon > 0$, there is a $\delta(\varepsilon) > 0$ such that

$$\|x(0)\| < \delta \Rightarrow \|x(t)\| < \varepsilon \ \forall t > 0$$

2. *unstable* if it is not stable, and
3. *asymptotically stable* if δ can be chosen such that

$$\|x(0)\| < \delta \Rightarrow \lim_{t\to\infty} x(t) = 0$$

The concept of Lyapunov stability is depicted in Figure 5.3.

The next seemingly trivial observation is nonetheless useful. Among other things, it highlights the distinction between stability and asymptotic stability.

Lemma 5.16 (Necessary condition for asymptotic stability) Consider the dynamical system $\dot{x} = f(x)$ and suppose $x = 0$ is an equilibrium point, i.e., $f(0) = 0$. Then $x = 0$ is asymptotically stable only if it is an isolated equilibrium point.

Proof If $x = 0$ is not an isolated equilibrium point, then in every neighborhood U of 0 there is at least one other equilibrium point. Thus, that not all trajectories beginning in U tend to 0 as $t \to \infty$. ■

For linear systems, the following result is easily obtained.

Proposition 5.17 *The origin of the linear system $\dot{x} = Ax$ is a stable equilibrium point if and only if*

$$\left\|e^{At}\right\| \leq N < \infty \ \forall t > 0$$

It is asymptotically stable if and only if, in addition, $\left\|e^{At}\right\| \to 0, \ t \to \infty$

Proof Exercise (choose $\delta = \varepsilon/N$)

5.4.1.1 Positive Definite Functions

Definition 5.18 A function $V : R^n \to R^n$ is said to be

1. *positive definite* if $V(0) = 0$ and $V(x) > 0,\ x \neq 0$,
2. *positive semidefinite* if $V(0) = 0$ and $V(x) \geq 0,\ x \neq 0$,
3. *negative definite* (*negative semidefinite*) if $-V(x)$ is positive definite (positive semidefinite)

For a quadratic form $V(x) = x^T Qx,\ Q = Q^T$, the following statements are equivalent:

1. $V(x)$ is positive definite
2. The eigenvalues of Q are positive real numbers
3. All of the principal minors of Q are positive

$$|q_{11}| > 0, \quad \begin{vmatrix} q_{11} & q_{12} \\ q_{21} & q_{22} \end{vmatrix} > 0, \ldots, |Q| > 0$$

Definition 5.19 A C^1 function $V(x)$ defined on a neighborhood D of the origin is called a *Lyapunov function* relative to the flow defined by $\dot{x} = f(x)$ if it is positive definite and it is nonincreasing along trajectories of the flow, i.e.,

$$V(0) = 0,\ V(x) > 0,\ x \in D - \{0\}$$

$$\dot{V} = \frac{\partial V(x)}{\partial x} f(x) \leq 0$$

5.4.2 *Basic Stability Theorems*

Stability of a dynamical system may be determined directly from an examination of the trajectories of the system or from a study of Lyapunov functions. The basic idea of the Lyapunov method derives from the idea of energy exchange in physical systems. A general physical conception is that stable systems dissipate energy so that the stored energy of a stable system decreases or at least does not increase as time evolves. The notion of a Lyapunov function is thereby an attempt to formulate a precise, energy-like theory of stability.

Proposition 5.20 (Lyapunov Stability Theorem) *If there exists a Lyapunov function $V(x)$ on some neighborhood D of the origin, then the origin is stable. Furthermore, if $\dot{V}$ is negative definite on D then the origin is asymptotically stable.*

Proof Given $\varepsilon > 0$ choose $r \in (0, \varepsilon]$ such that

$$B_r = \left\{ x \in R^n \,|\, \|x\| < r \right\} \subset D$$

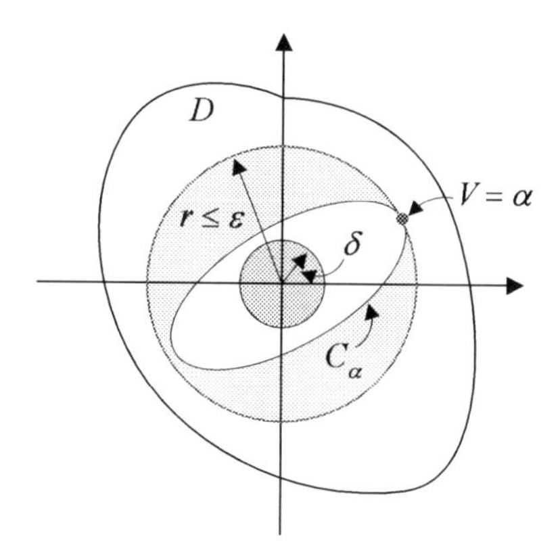

Fig. 5.4 Sets used in proof of the Lyapunov stability theorem, Proposition 5.20

Now, we can find a level set $C_\alpha = \{x \in R^n \,|V(x) = \alpha\}$ which lies entirely within B_r. Refer to Figure 5.4. The existence of such a set follows from the fact that since V is positive and continuous on B_r, it has a positive minimum, α, on ∂B_r. The level set C_α defined by $V(x) = \alpha$ must lie entire in B_r.

Now, since V is continuous and vanishes at the origin, there exists a $\delta > 0$ such that B_δ lies entirely within the set bounded by C_α, i.e.,

$$\Omega_\alpha = \left\{x \in R^n \,|V(x) \le \alpha\right\}$$

Since V is nonincreasing along trajectories, trajectories which begin in B_δ must remain in Ω_α, $\forall t > 0$. Hence, they remain in B_ε. In the event that $\dot{V}$ is negative definite, V decreases steadily along trajectories. For any $0 < r_1 < r$, there is a $\beta < \alpha$ such that B_β lies entirely within B_{r_1}. Since $\dot{V}$ has a strictly negative maximum in the annular region $B_r - B_{r_1}$, any trajectory beginning in the annular region must eventually enter B_{r_1}. Thus, all trajectories must tend to the origin as $t \to \infty$. ■

Unlike linear systems, an asymptotically stable equilibrium point of a nonlinear system may not attract trajectories from all possible initial states. It is more likely that trajectories beginning at states in a restricted vicinity of the equilibrium point will actually tend to the equilibrium point as $t \to \infty$. The above theorem can be used to establish stability and also to provide estimates of the *domain of attraction* using level sets of the Lyapunove function $V(x)$.

The following theorem due to LaSalle allows us to more easily characterize the domain of attraction of a stable equilibrium point and is a more powerful result than the basic Lyapunov stability theorem because the conditions for asymptotic stability do not require $\dot{V}$ to be negative definite.

Proposition 5.21 (LaSalle Invariance Theorem) *Consider the system defined by equation* (5.5). *Suppose $V(x) : R^n \to R$ is C^1 and let Ω_c designate a component of the region $\{x \in R^n \,|V(x) < c\}$. Suppose Ω_c is bounded and that within Ω_c $\dot{V}(x) \le 0$. Let E be the set of points within Ω_c where $\dot{V} = 0$, and let M be the largest* *invariant set of* (5.5) *contained in* E. *Then every solution* $x(t)$ *of* (5.5) *beginning in* Ω_c *tends to* M *as* $t \to \infty$.

Proof (following [127]) $\dot{V}(x) \leq 0$ implies that $x(t)$ starting in Ω_c remains in Ω_c. $V(x(t))$ nonincreasing and bounded implies that $V(x(t))$ has a limit c_0 as $t \to \infty$ and $c_0 < c$. By continuity of $V(x)$, $V(x) = c_0$ on the positive limit set $\Lambda_\omega(x_0)$ of $x(t)$ beginning at $x_0 \in \Omega_c$. Thus, $\Lambda_\omega(x_0)$ is in Ω_c and $\dot{V}(x) = 0$ on $\Lambda_\omega(x_0)$. Consequently, $\Lambda_\omega(x_0)$ is in E, and since it is an invariant set, it is in M. ■

Note that the theorem does not specify that $V(x)$ should be positive definite, only that it has continuous first derivatives and that there exist a bounded region on which $V(x) < c$ for some constant c. A number of useful results follow directly from this one.

Corollary 5.22 *Let* $x = 0$ *be an equilibrium point of* (5.5). *Suppose* D *is a neighborhood of* $x = 0$ *and* $V : D \to R$ *is* C^1 *and positive definite on* D *such that* $\dot{V}(x) \leq 0$ *on* D. *Let* $E = \left\{x \in D \,\middle|\, \dot{V}(x) = 0\right\}$ *and suppose that the only entire solution contained in* E *is the trivial solution. Then the origin is asymptotically stable.*

Corollary 5.23 *Let* $x = 0$ *be an equilibrium point of* (5.5). *Suppose*

1. $V(x)$ *is* C^1
2. $V(x)$ *is radially unbounded (Barbashin-Krasovskii condition), i.e.,*

$$\|x\| \to \infty \Rightarrow V(x) \to \infty$$

3. $\dot{V}(x) \leq 0, \ \forall x \in R^n$
4. *the only entire trajectory contained in the set* $E = \left\{x \in D \,\middle|\, \dot{V}(x) = 0\right\}$ *is the trivial solution.*

Then the origin is globally asymptotically stable.

The stability theorems provide only sufficient conditions for stability, and construction of a suitable Lyapunov function may require a fair amount of ingenuity. The event that attempts to establish stability does not bear fruit, and it may be useful to try to confirm instability.

Proposition 5.24 (Chetaev Instability Theorem) *Consider equation* (5.5) *and suppose* $x = 0$ *is an equilibrium point. Let* D *be a neighborhood of the origin. Suppose there is a function* $V(x) : D \to R$ *and a set* $D_1 \subset D$ *such that*

1. $V(x)$ *is* C^1 *on* D,
2. *the origin belongs to the boundary of* D_1, ∂D_1,
3. $V(x) > 0$ *and* $\dot{V}(x) > 0$ *on* D_1,
4. *On the boundary of* D_1 *inside* D, *i.e., on* $\partial D_1 \cap D$, $V(x) = 0$

Then the origin is unstable

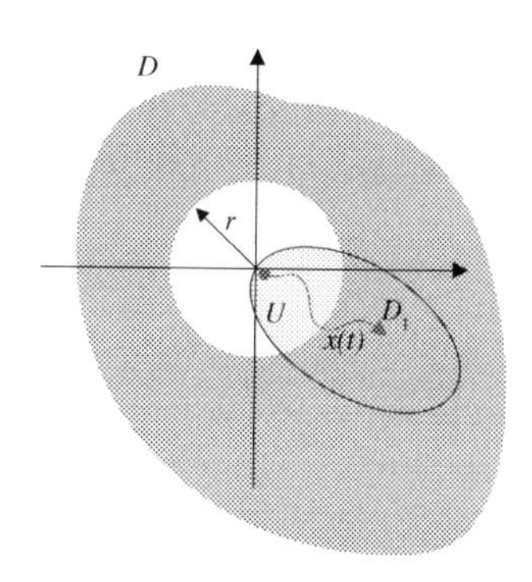

Fig. 5.5 Sets used in proof of the Chetaev instability theorem, Proposition 5.24

Proof Choose an r such that $B_r = \{x \in R^n \,|\|x\| \leq r\}$ is in D. Refer to Figure 5.5. For any trajectory beginning inside $U = D_1 \cap B_r$ at $x_0 \neq 0$, $V(x(t))$ increases indefinitely from $V(x_0) > 0$. But by continuity, $V(x)$ is bounded on U. Hence, $x(t)$ must leave U. It cannot do so across its boundary interior to B_r so it must leave B_r. ■

5.4.2.1 Stability of Linear Systems

Consider the linear system

$$\dot{x} = Ax \tag{5.6}$$

Proposition 5.25 *Consider the Lyapunov equation*

$$A^T P + PA = -Q \tag{5.7}$$

(a) *If there exists a positive definite pair of symmetric matrices P, Q satisfying the Lyapunov equation then the origin of the system (5.6) is asymptotically stable.*
(b) *If there exists a pair of symmetric matrices P, Q such that P has at least one negative eigenvalue and Q is positive definite, then the origin is unstable.*

Proof Consider (a) first. Choose $V(x) = x^T Px$ and compute $\dot{V} = x^T(A^T P + PA)x = -x^T Qx$. The assumptions and the LaSalle stability theorem lead to the conclusion that all trajectories tend to the orgin as $t \to \infty$. Case (b) requires application of Chetaev's instability theorem. In this case, consider $V(x) = -x^T Px$. Recall that for symmetric P the eigenvalues of P are real, they may be positive, negative, or zero. On the positive eigenspace, $V < 0$, on the negative eigenspace, $V > 0$, and on the zero eigenspace, $V = 0$. Since P has at least one negative eigenvalue, the negative eigenspace is nontrivial and there is a set of points, D, for which $V > 0$. Let B_ε be an open sphere of small radius ε centered at the origin. Since V is continuous, the boundary of D in B_ε, $\partial D \cap B_r$, consists of points of points at which $V = 0$. It includes the origin and is never nonempty (even if all eigenvalues of P are negative). For $V(x) = -x^T Px$, $\dot{V} = x^T Qx$, and it is always positive since Q is assumed positive definite. Thus, the conditions of Proposition 5.24 are satisfied. ■

Suppose $Q > 0$ and the P has a zero eigenvalue. If the matrix P has a zero eigenvalue, then there are points $x \neq 0$ such that $V(x) = x^T Px = 0$. But at such points $\dot{V}(x) = -x^T Qx < 0$. Since V(x) is continuous, this means that there must be points at which V assumes negative values. Thus, P must also have a negative eigenvalue. Thus, we have the following corollary to Proposition 5.25.

Corollary 5.26 *The linear system* (5.6) *is asymptotically stable if and only if for every positive definite symmetric Q there exists a positive definite symmetric P that satisfies the Lyapunov equation* (5.7).

5.4.2.2 Lagrangian Systems

The Lyapunov analysis of the stability of nonlinear dynamical systems evolved from a tradition of stability analysis via energy functions that goes back at least to Lagrange and Hamilton. We will consider a number of examples which are physically motivated and for which there are energy functions that serve as natural Lyapunov function candidates. Consider the class of Lagrangian systems characterized by the set of second-order differential equations

$$\frac{d}{dt}\frac{\partial L(q,v)}{\partial v} - \frac{\partial L(q,v)}{\partial q} = Q^T \tag{5.8}$$

where

1. $q \in R^n$ denotes a vector of *generalized coordinates* and $v = dq/dt$ the vector *generalized velocities.*
2. $L : R^{2n} \to R$ is the *Lagrangian*. It is constructed from the kinetic energy function $T(q,v)$ and the potential energy function $U(q)$, via $L(q,v) = T(q,v) - U(q)$.
3. The kinetic energy has the form

$$T(q,v) = \tfrac{1}{2} v^T M(q) v$$

   where for each fixed q, the matrix $M(q)$ is positive definite.
4. The potential energy is related to a force vector $f(q)$ via

$$U(q) = \int f(q)dq$$

5. $Q(q,v,t)$ is a vector of generalized forces.

Equation (5.8) evaluates to

$$M(q)\dot{v} + \left[\frac{\partial}{\partial q}(M(q)v) - \frac{1}{2}\left(\frac{\partial}{\partial q}(M(q)v)\right)^T\right]v + \frac{\partial U(q)}{\partial q^T} = Q \tag{5.9}$$

which can be written

$$M(q)\dot{v} + C(q,v)v + \frac{\partial U(q)}{\partial q^T} = Q \tag{5.10}$$

Remark 5.27 The total energy of the Lagrangian system (5.9) or (5.11) is

$$E(q,v) = \frac{1}{2}v^T M(q)v + U(q)$$

We can readily compute the time rate of change of energy

$$\dot{E} = \frac{1}{2}v^T\left(\frac{\partial M(q)v}{\partial q} - 2C(q,v)\right)v + v^T Q$$

In general, if $E(0,0) = 0$, $E(q,v) > 0$ otherwise, and $\dot{E} \leq 0$, then E is a Lyapunov function.

Suppose that $Q = 0$ so that the system is conservative. Consequently, we expect

$$\dot{E} = \frac{1}{2}v^T\left(\frac{\partial M(x)v}{\partial x} - 2C(x,v)\right)v \equiv 0$$

A useful first-order form derivable from (5.9) is *Hamilton's equations* obtained as follows. Define the *generalized momentum* as

$$p^T = \frac{\partial L}{\partial \dot{q}} = \dot{q}^T M(q) \Rightarrow \dot{q} = M^{-1}(q)p \tag{5.11}$$

Define the *Hamiltonian* $H : R^{2n} \to R$

$$H(q,p) = \left[p^T\dot{q} - L(q,\dot{q})\right]_{\dot{q}\to M^{-1}p} = \tfrac{1}{2}p^T M^{-1}(q)p + U(q) \tag{5.12}$$

The Hamiltonian is the total energy expressed in momentum rather than velocity coordinates. Notice that Lagrange's equation can be written

$$\dot{p}^T - \frac{\partial L}{\partial q} = Q^T \tag{5.13}$$

Now, using the definition of H, (5.12), write

$$dH = \frac{\partial H}{\partial q}dq + \frac{\partial H}{\partial p}dp = dp^T\dot{q} + p^T d\dot{q} - \frac{\partial L}{\partial q}dq - \frac{\partial L}{\partial \dot{q}}d\dot{q} = dp^T\dot{q} - \frac{\partial L}{\partial q}dq$$

Using (5.13), we have

$$\frac{\partial H}{\partial q}dq + \frac{\partial H}{\partial p}dp = \dot{q}^T dp - (\dot{p} - Q)^T dq$$

Comparing coefficients of dp and dq, we have Hamilton's equations.

$$\dot{q} = \frac{\partial H(q,p)}{\partial p^T}, \quad \dot{p} = -\frac{\partial H(q,p)}{\partial q^T} + Q \tag{5.14}$$

Example 5.28 (*Soft Spring*) Consider a system of with kinetic and potential energy functions

$$T = \frac{x_2^2}{2}, \quad U = \frac{x_1^2}{1+x_1^2}$$

Lagrange's equations in first-order form ($\dot{x}_1 = x_2$) with viscous damping are

$$\begin{bmatrix} \dot{x}_1 \\ \dot{x}_2 \end{bmatrix} = \begin{bmatrix} x_1 \\ -2\frac{x_1}{(1+x_1^2)^2} - cx_2 \end{bmatrix}$$

If we take the total energy as a candidate Lyapunov function,

$$V(x_1, x_2) = \tfrac{1}{2}x_2^2 + \frac{x_1^2}{1+x_1^2}$$

an easy calculation shows that $\dot{V} = -cx_2 \leq 0$ for $c > 0$. Furthermore, the set $\dot{V} = 0$ consists of the x_1-axis, and the only entire solution contained therein is the trivial solution. We conclude that the origin is asymptotically stable. We cannot, however, conclude global asymptotic stability because the Lyapunov function is not radially unbounded. Let us look at the level sets of V, shown in Figure 5.6. The state trajectories are shown in Figure 5.7.

Example 5.29 (*Variable Mass*) Consider a system with variable inertia, typical of a crankshaft. The kinetic and potential energy functions are

$$T = (2 - \cos 2x_1)\,x_2^2, \quad U = x_1^2 + \tfrac{1}{4}x_1^4$$

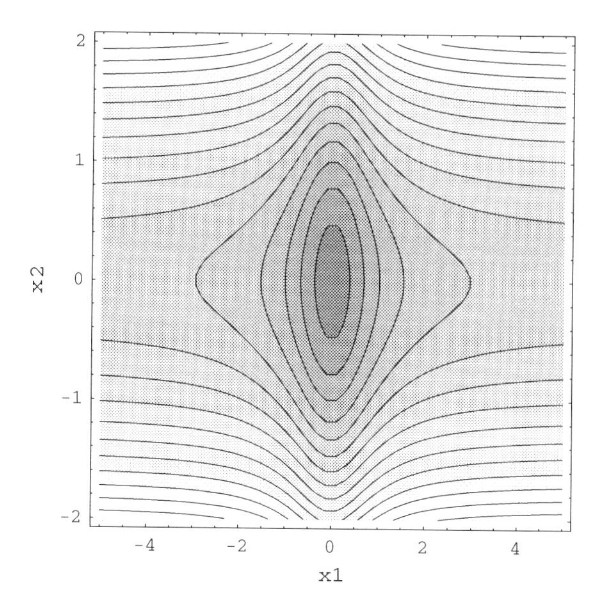

Fig. 5.6 Level sets soft spring total energy

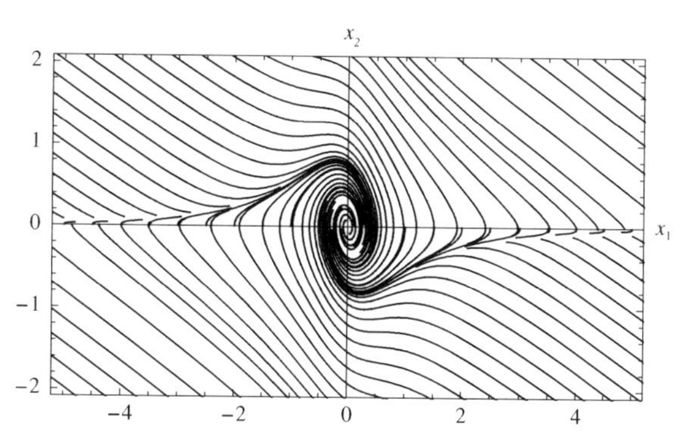

Fig. 5.7 Phase portrait for soft spring system

Systems with variable mass are much easier to analyze using Hamilton's equations, so we define the generalized momentum $p = (2 - \cos 2x_1)\, x_2$ and the Hamiltonian

$$H(x_1, p) = \frac{p^2}{2(2 - \cos 2x_1)} + x_1^2 + \tfrac{1}{4}x_1^4$$

Again with viscous damping, Hamilton's equations are

$$\begin{bmatrix} \dot{x}_1 \\ \dot{p} \end{bmatrix} = \begin{bmatrix} \frac{p}{2-\cos 2x_1} \\ -2x_1 - x_1^3 - \frac{p^2 \sin 2x_1}{(2-\cos 2x_1)^2} + \frac{2cp}{(2-\cos 2x_1)^2} \end{bmatrix}$$

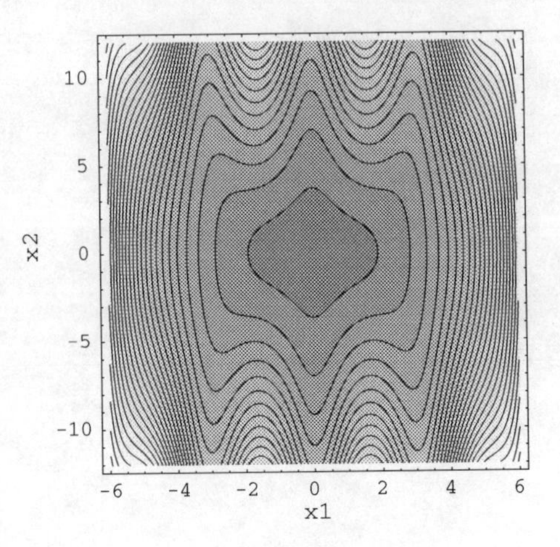

Fig. 5.8 Level curves for variable mass total energy

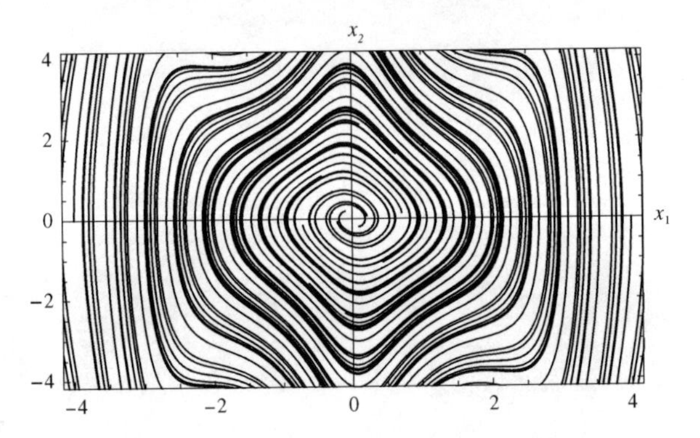

Fig. 5.9 Phase portrait for variable mass system

It is not difficult to compute $\dot{H}$, indeed,

$$\dot{H} = \frac{\partial H}{\partial x_1}\dot{x}_1 + \frac{\partial H}{\partial p}\dot{p} = \frac{2cp^2}{(\cos 2x_1 - 2)^3}$$

We conclude that $\dot{H} \leq 0$ for $c > 0$. Moreover, the only entire trajectory in the set $\dot{H} = 0$ is the trivial solution, and since H is radially unbounded, we can conclude that the origin is globally asymptotically stable. Let us look at the level curves of H, Figure 5.8, and at the state space trajectories, Figure 5.9.

Example 5.30 (*Multiple Equilibria*) Consider the system

$$\ddot{x} + \left|x^2 - 1\right| \dot{x}^3 - x + \sin\left(\frac{\pi x}{2}\right) = 0$$

Notice that the system has three equilibria $(x, \dot{x}) = (0, 0), (-1, 0), (1, 0)$. We can determine their stability by examining the system phase portraits or using a Lyapunov

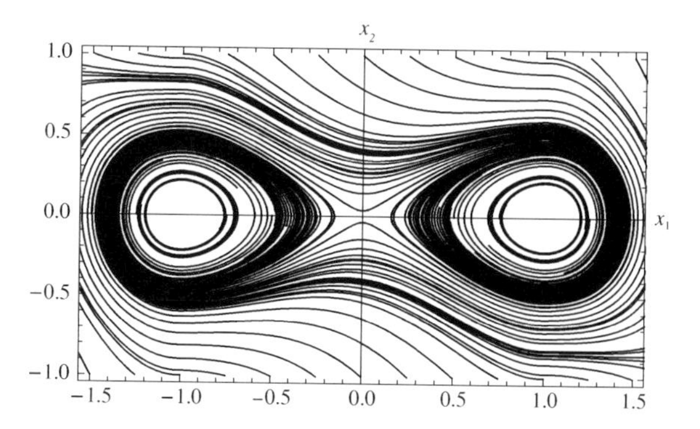

Fig. 5.10 Phase portrait for the multiple equilibria system

analysis based on total energy as the candidate Lyapunov function. First, let us examine the phase portraits shown in Figure 5.10. We see that the equilibrium point $(0, 0)$ is unstable and that the other two, $(\pm 1, 0)$, are asymptotically stable.

Now, let us consider the Lyapunov viewpoint. The total energy is

$$V = \frac{\dot{x}^2}{2} + \frac{x^2}{2} - \frac{2}{\pi}\left(1 - \cos\left(\frac{\pi x}{2}\right)\right)$$

A straightforward calculation leads to

$$\dot{V} = -\left|x^2 - 1\right| \dot{x}^2$$

The LaSalle theorem 5.21 can now be applied. Let us view the level surfaces shown in Figure 5.11.

Notice that there are level surfaces that bound compact sets that include the equilibrium point $(1, 0)$. Pick one and designate it Ω_{c_1}. Moreover, $\dot{V} \leq 0$ everywhere,

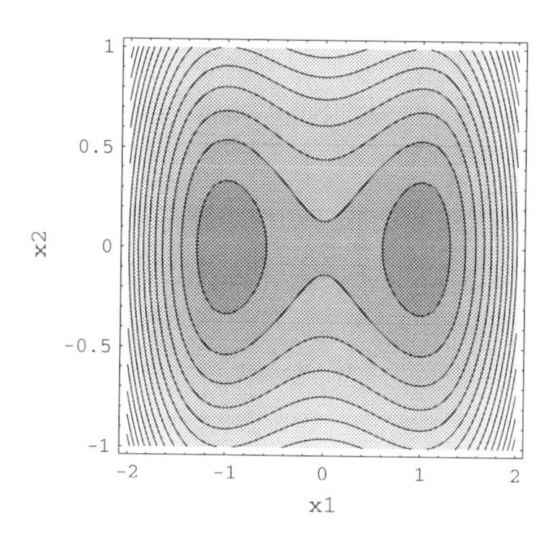

Fig. 5.11 Level sets for the multiple equilibria system

hence specifically in Ω_{c_1}, and the maximum invariant set contained in Ω_{c_1} is the equilibrium point. Consequently, all trajectories beginning in Ω_{c_1} tend to $(1, 0)$ so it is an asymptotically stable equilibrium point. A similar conclusion can be reached for the equilibrium point $(-1, 0)$.

5.4.3 *First Integrals and Chetaev's Method*

Identifying an appropriate Lyapunov function is not always a simple matter. Even if a system can be associated with an energy function, the energy function itself may not be a suitable Lyapunov function. There exist many approaches to systematize the search for a Lyapunov function. One of these involves building a Lyapunov function from first integrals.

Definition 5.31 (*First Integral*) A *first integral* of the dynamical system

$$\dot{x} = f(x, t)$$

is a scalar function $\phi(x, t)$ that is constant along trajectories, i.e., it satisfies

$$\dot{\phi}(x, t) = \frac{\partial \phi(x, t)}{\partial x} f(x, t) + \frac{\partial \phi(x, t)}{\partial t} \equiv 0$$

Remark 5.32 For simplicity, consider the system $\dot{x} = f(x)$. Suppose the n functions $\phi_1(x), \ldots, \phi_n(x)$ are locally independent around the point x_0, i.e.,

$$\det \frac{\partial}{\partial x} \begin{bmatrix} \phi_1(x) \\ \vdots \\ \phi_n(x) \end{bmatrix}_{x=x_0} \neq 0$$

Then, we can define a coordinate transformation $x \to z$ via $z = \phi(x)$. Suppose further that $\phi_1(x)$ is a first integral. Then, we can compute

$$\dot{z} = \left[\frac{\partial \phi(x)}{\partial x} f(x) \right]_{x=\phi^{-1}(z)}$$

Thus, we have $\dot{z}_1 = 0 \Rightarrow z_1 \equiv$ constant. Consequently, knowledge of one first integral reduces the problem to solving a system of $n - 1$ differential equations. Knowledge of n first integrals is tantamount to solving the original system of equations.

Now, consider the system

$$\dot{x} = f(x, t), \quad f(0, t) = 0$$

which has an equilibrium point at $x = 0$. We wish to consider the stability of the origin. If $\phi(x,t)$ is a first integral and it is also positive definite (in x), then it is an obvious candidate for a Lyapunov function. Suppose, however, that ϕ is not positive definite. Chetaev suggested the following approach. Suppose k first integrals $\phi_1(x,t), \ldots, \phi_k(x,t)$ are known, with $\phi_i(0,t) = 0$. Then, construct a candidate Lyapunov function of the form

$$V(x,t) = \sum_{i=1}^{k} \alpha_i \phi_i(x,t) + \sum_{i=1}^{k} \beta_i \phi_i^2(x,t)$$

where the $\alpha' s$ and $\beta' s$ are constants.

Example 5.33 (*Rigid Body*) Consider a rigid body with a body-fixed coordinate system x, y, z aligned along the principal axes. Suppose that the coordinates are arranged so that the principal inertias satisfy $I_x > I_y > I_z > 0$. Let $\omega_x, \omega_y, \omega_z$ denote the angular velocities in body coordinates. Euler's equations take the form

$$\begin{aligned}
\dot{\omega}_x &= -\left(\frac{I_z - I_y}{I_x}\right)\omega_z\omega_y = a\,\omega_z\omega_y \\
\dot{\omega}_y &= -\left(\frac{I_x - I_z}{I_y}\right)\omega_x\omega_z = -b\,\omega_x\omega_z \\
\dot{\omega}_z &= -\left(\frac{I_y - I_x}{I_z}\right)\omega_y\omega_x = c\,\omega_y\omega_x
\end{aligned}$$

with $a, b, c > 0$.

Note that any state $(\bar{\omega}_x, \bar{\omega}_y, \bar{\omega}_z)$ is an equilibrium state if any two of the angular velocity components are zero. In other words, any constant velocity rotation aligned with one of the body axis is an equilibrium point. Consider rotation about the x-axis $(\bar{\omega}_x, 0, 0)$ with $\bar{\omega}_x > 0$. Shift the coordinates $\omega_x \to \omega_x + \bar{\omega}_x$ so that the equilibrium point is at he origin of the new equations

$$\begin{aligned}
\dot{\omega}_x &= a\,\omega_z\omega_y \\
\dot{\omega}_y &= -b\,(\omega_x + \bar{\omega}_x)\,\omega_z \\
\dot{\omega}_z &= c\,(\omega_x + \bar{\omega}_x)\,\omega_y
\end{aligned}$$

Energy does not work as a Lyapunov function (obvious?). Consider what we wish to achieve. We seek $V(\omega_x, \omega_y, \omega_z)$ with the properties:

i. $V(0,0,0) = 0$
ii. $V(\omega_x, \omega_y, \omega_z) > 0$ if $(\omega_x, \omega_y, \omega_z) \neq (0,0,0)$
iii. $\dot{V} \leq 0$

Let us look at all functions that satisfy $\dot{V} = 0$, i.e., that satisfy the partial differential equation

$$\frac{\partial V}{\partial \omega_x} a\,\omega_z\omega_y + \frac{\partial V}{\partial \omega_y}\left(-b\,(\omega_x + \bar{\omega}_x)\,\omega_z\right) + \frac{\partial V}{\partial \omega_z} c\,(\omega_x + \bar{\omega}_x)\,\omega_y = 0$$

It is not difficult to confirm that all solutions take the form

$$f\left(\frac{b\omega_x^2 + 2b\omega_x\bar{\omega}_x + a\omega_y^2}{2a}, \frac{-c\omega_x^2 + 2c\omega_x\bar{\omega}_x + a\omega_z^2}{2a}\right)$$

In other words, we have identified two first integrals

$$\phi_1 = \frac{b\omega_x^2 + 2b\omega_x\bar{\omega}_x + a\omega_y^2}{2a}$$

$$\phi_2 = \frac{-c\omega_x^2 + 2c\omega_x\bar{\omega}_x + a\omega_z^2}{2a}$$

In accordance with Chetaev's method take

$$V\left(\omega_x, \omega_y, \omega_z\right) = c\phi_1 + b\phi_2 + (c\phi_1 - b\phi_2)^2 = \frac{1}{2}\left(\frac{8b^2c^2\bar{\omega}_x^2}{a^2}\omega_x^2 + c\omega_y^2 + b\omega_z^2\right) + h.o.t.$$

Thus, we have $V(0) = 0$ and on a neighborhood of the origin, $V > 0$ and $\dot{V} = 0$. Using Lyapunov's theorem, we conclude that spin about the x-axis is stable.

5.4.4 Remarks on Noether's Theorem

Recall the Lagrangian systems of Section 5.4.2.2. Emmy Noether proved[1] that for every smooth symmetry possessed by a conservative Lagrangian, there is a corresponding first integral. The Lagrange equations are ordinarily derived from the *principle of least action* which states that the path taken by the system between times t_1 and t_2 is one for which the *action integral*, S, is stationary, where

$$S(q(t)) = \int_{t_1}^{t_2} L(q(t), \dot{q}(t), t)\, dt \tag{5.15}$$

Thus, the *contemporaneous* variation must satisfy $\delta S = 0$. The contemporaneous variation of $q(t)$ is simply an arbitrary, small perturbation $\delta q(t),\ t \in [t_1, t_2]$. Noether's theorem can be obtained by applying a *noncontemporaneous* variation to S. A noncontemporaneous variation allows a perturbation Δt of t, so that [195, 19]

$$\Delta q(t) = \delta q(t) + \dot{q}(t)\Delta t \tag{5.16}$$

[1] See the translation of her 1918 paper, [158].

This leads to the requirement that

$$\frac{d}{dt}\left[\frac{\partial L}{\partial \dot{q}}\Delta q + \left(L - \frac{\partial L}{\partial \dot{q}}\dot{q}\right)\Delta t\right] = \Delta L \tag{5.17}$$

Thus, if $\Delta L = 0$, i.e., the Lagrangian is invariant with respect to the perturbation $(\Delta q, \Delta t)$, it must be true that

$$\frac{\partial L}{\partial \dot{q}}\Delta q + \left(L - \frac{\partial L}{\partial \dot{q}}\dot{q}\right)\Delta t = \text{const.} \tag{5.18}$$

To grasp the significance of this result, consider the following transformation, $\dot{q} \to p$, from Section 5.4.2.2, defined by

$$p^T = \frac{\partial L(\dot{q}, q, t)}{\partial \dot{q}}$$

and define the Hamiltonian

$$H(q, p) = \left[p^T \dot{q} - L(q, \dot{q})\right]_{\dot{q} \to M^{-1}(q)p}$$

so that

$$p^T \Delta q - H(q, p)\Delta t = \text{const.} \tag{5.19}$$

Now, consider two special cases as examples:

1. Suppose L is invariant with respect to time, i.e., $\Delta L = 0$ with $\Delta q = 0$ and $\Delta t = 1$, then Equation (5.19) asserts $H = \text{const.}$ implying conservation of energy.
2. Suppose L is invariant with respect to a coordinate q_i, i.e., $\Delta L = 0$ with $\Delta q_i = 1$, $\Delta q_j = 0$, $j \neq i$, and $\Delta t = 0$, then, again from (5.19), $p_i = \text{const.}$ implying conservation of momentum.

These results for conservative systems have been extended to nonconservative systems and systems with constraints [195, 19].

5.4.5 Stable, Unstable, and Center Manifolds

Consider the autonomous system (5.5) and suppose $x = 0$ is an equilibrium point so that $f(0) = 0$. Let $A := \partial f(0)/\partial x$. Define three subspaces of R^n:

1. The *stable subspace*, E^s: the eigenspace of eigenvalues with negative real parts;
2. The *unstable subspace*, E^u: the eigenspace of eigenvalues with positive real parts;
3. The *center subspace*, E^c: the eigenspace of eigenvalues with zero real parts.

An equilibrium point is called *hyperbolic* if A has no eigenvalues with zero real part; i.e., there is no center subspace, E^c. In the absence of center subspace, the linearization is a reliable predictor of important qualitative features of the nonlinear system. The basic result is given by the following theorem. First, some definitions.

Let f, g be C^r vector fields on R^n with $f(0) = 0$, $g(0) = 0$. M is an open subset of the origin in R^n.

Definition 5.34 Two vector fields f and g are said to be *C^k-equivalent* on M if there exists a C^k diffeomorphism h on M, which takes orbits of the flow generated by f on M, $\Phi(x, t)$, into orbits of the flow generated by g on M, $\Psi(x, t)$, preserving orientation but not necessarily parameterization by time. C^0-equivalence is referred to as *topological equivalence*. If there is such an h which does preserve parameterization by time, then $f and g$ are said to be *C^k-conjugate*. C^0-conjugacy is referred to as *topological-conjugacy*.

Proposition 5.35 (Hartman-Grobman Theorem) *Let $f(x)$ be a C^k vector field on R^n with $f(0) = 0$ and $A := \partial f(0)/\partial x$. If A is hyperbolic then there is a neighborhood U of the origin in R^n on which the nonlinear flow of $\dot{x} = f(x)$ and the linear flow of $\dot{x} = Ax$ are topologically conjugate.*

Proof (Chow & Hale [51], p. 108)

Definition 5.36 Let U be a neighborhood of the origin. We define the *local stable manifold* and *local unstable manifold* of the equilibrium point $x = 0$ as, respectively,

$$W^s_{loc} = \{x \in U \mid \Psi(x, t) \to 0 \; as \; t \to \infty \wedge \Psi(x, t) \in U \; \forall t \geq 0\}$$

$$W^u_{loc} = \{x \in U \mid \Psi(x, t) \to 0 \; as \; t \to -\infty \wedge \Psi(x, t) \in U \; \forall t \leq 0\}$$

Proposition 5.37 (Center Manifold Theorem) *Let $f(x)$ be a C^r vector field on R^n with $f(0) = 0$ and $A := \partial f(0)/\partial x$. Let the spectrum of A be divided into three sets $\sigma_s, \sigma_c, \sigma_u$ with*

$$\operatorname{Re} \lambda = \begin{cases} < 0 & \lambda \in \sigma_s \\ = 0 & \lambda \in \sigma_c \\ > 0 & \lambda \in \sigma_u \end{cases}$$

Let the (generalized) eigenspaces of $\sigma_s, \sigma_c, \sigma_u$ be E^s, E^c, E^u, respectively. Then there exist C^r stable and unstable manifolds W^s and W^u tangent to E^s and E^u, respectively, at $x = 0$ and a C^{r-1} center manifold W^c tangent to E^c at $x = 0$. The manifolds W^s, W^c, W^u are all invariant with respect to the flow of $f(x)$. The stable and unstable manifolds are unique, but the center manifold need not be.

Proof [141].

Example 5.38 (*Center Manifold*) Consider the system

$$\dot{x} = x^2, \quad \dot{y} = -y$$

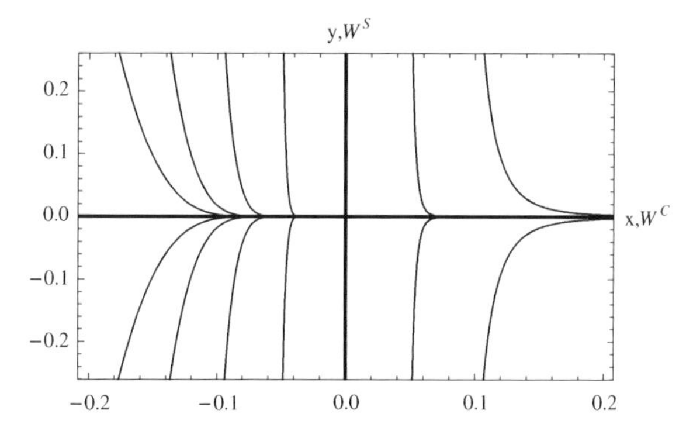

Fig. 5.12 Center manifold

from which it is a simple matter to compute

$$x(t) = x_0/(1 - tx_0), \; y(t) = y_0 e^{-t} \Rightarrow y(x) = \left[y_0 e^{-1/x_0} \right] e^{1/x}$$

The phase portrait is shown below. Observe that (0, 0) is an equilibrium point with:

$$A = \begin{bmatrix} 0 & 0 \\ 0 & -1 \end{bmatrix} \Rightarrow E^s = span \begin{Bmatrix} 0 \\ 1 \end{Bmatrix}, \quad E^c = span \begin{Bmatrix} 1 \\ 0 \end{Bmatrix}$$

Notice that the center manifold can be defined using any trajectory beginning with $x < 0$ and joining with it the positive x-axis. Also, the center manifold can be chosen to be the entire x-axis. This is the only choice which yields an analytic center manifold (Figure 5.12).

There are some important properties of these manifolds that will not be examined here. See, for example, [83] and [12]. Let us note, however, that existence and uniqueness of solutions insure that two stable (or unstable) manifolds cannot intersect or self-intersect. However, a stable and an unstable manifold can intersect. The global stable and unstable manifolds need not be simple submanifolds of R^n, since they may wind around in a complex manner, approaching themselves arbitrarily closely.

5.4.5.1 Motion on the Center Manifold

Consider the system of differential equations

$$\begin{aligned} \dot{x} &= Bx + f(x, y) \\ \dot{y} &= Cy + g(x, y) \end{aligned} \tag{5.20}$$

where $(x, y) \in R^{n+m}$, f, g, and their gradients vanish at the origin, and the eigenvalues of B have zero real parts, those of C negative real parts. The center manifold is tangent to E^c:

$$E^c = \text{span}\begin{bmatrix} I_n \\ 0_{m\times n} \end{bmatrix}$$

It has a local graph

$$W^c = \left\{(x, y) \in R^{n+m} \mid y = h(x)\right\}, \quad h(0) = 0, \quad \frac{\partial h(0)}{\partial x} = 0$$

Once h is determined, the vector field on the center manifold W^c (i.e., the surface defined by $y = h(x)$) can be projected onto the Euclidean space E^c as

$$\dot{x} = Bx + f(x, h(x)) \tag{5.21}$$

These calculations lead to the following result (see [83]).

Proposition 5.39 (Center Manifold Stability Theorem) *If the origin of* (5.21) *is asymptotically stable (resp. unstable) then the origin of* (5.20) *is asymptotically stable (resp. unstable).*

To compute h, we use the fact that on W^c it is required that $y = h(x)$ so that

$$\dot{y} = \frac{\partial h(x)}{\partial x}\dot{x} = \frac{\partial h(x)}{\partial x}[Bx + f(x, h(x)]$$

But $\dot{y}$ is also governed by (5.20) so we have the partial differential equation

$$\frac{\partial h(x)}{\partial x}[Bx + f(x, h(x)] = Ch(x) + g(x, h(x) \tag{5.22}$$

that needs to be solved along with the boundary conditions $h(0) = 0$, $\frac{\partial h(0)}{\partial x} = 0$.

Example 5.40 Consider the following two-dimensional system from Isidori [100].

$$\dot{x} = cyx - x^3$$

$$\dot{y} = -y + ayx + bx^2$$

where $a, b, andc$ are real numbers. It is easy to see that the origin is an equilibrium point and that it is in the form of (5.20) with $B = 0$ and $C = -1$. To compute h, we need to solve the partial differential equation

$$\frac{\partial h}{\partial x}\left[cxh(x)-x^3\right]+h(x)-ah(x)x-bx^2=0 \tag{5.23}$$

with boundary conditions $h(0)=0,\ \frac{\partial h(0)}{\partial x}=0$.

Assume a polynomial solution of the form

$$h(x)=a_0+a_1x+a_2x^2+a_3x^3+O(x^4)$$

In view of the boundary conditions, we must have $a_0=0$ and $a_1=0$. Substituting h into (5.23) leads to

$$(a_2-b)\,x^2+(a_3-aa_2)\,x^3+O\left(x^4\right)=0$$

From which it follows $a_2=b,\ a_3=a_2$, so that

$$h=bx^2+abx^3+O\left(x^4\right)$$

Thus, the motion on the center manifold is given by

$$\dot{x}=(-1+bc)x^3+abcx^4+O(x^5)$$

Thus, we have the following results:

(a) If $bc<1$, the motion on the center manifold is asymptotically stable,
(b) If $bc>1$, it is unstable,
(c) If $bc=1$ and $a\neq 0$, it is unstable
(d) If $bc=1$ and $a=0$, the above calculations are inconclusive. But in this special case, it is easy to verify that $h(x)=x^2$ and the center manifold dynamics are $\dot{x}=0$. So the motion is stable, but not asymptotically stable.

5.5 Analysis of Power System Stability

Lyapunov methods of stability assessment have been an important tool in power systems stability analysis for many decades [140, 14, 163, 191]. The important problem is to identify appropriate candidate Lyapunov functions. The total energy of the system of interest is often a good starting point, if it is known. As noted in Section 5.4.3, total energy may not have the desired properties, a problem that may be resolved by combining energy with other first integrals. An essential advantage of *natural* functions is the direct link between stability and the underlying physics of the system, providing an understanding of why stability breaks down in a given situation.

The Lagrange constructs of Chapters 3 and 4 provide insight into the energy functions associated with circuits and machines. In complex interconnected power systems, various manipulations, simplifications, and approximations lead to models in which the connection to natural energy functions or other first integrals may be lost. Thus, the notion of *energy-like* Lyapunov functions has been widely used in power systems (example [14, 149, 154, 112, 123, 47, 178]). In this approach, energy concepts motivate the form of the function. This formulation need not be entirely ad hoc as there is an extensive area of thought regarding the *inverse problem of analytical mechanics*, example [173, 9, 195, 19]. This perspective is taken in [112, 123].

In power system direct stability analysis, the importance of an energy function is clearly evident. Beginning with the work of Magnusson [140], energy analysis has been a recurrent theme over a span of many decades. Nevertheless, there remain difficulties with the application of energy functions to systems with loads. Much of the discussion in the literature centers on the issue of transfer conductances in the formulation of energy-like Lyapunov functions. The essential difficulty is the same whether constant admittance load models are employed and load buses eliminated, or constant power load models are employed and load buses retained.

Intrinsically more fundamental than the question of the existence of energy-like Lyapunov functions are questions about the energy function itself. If an energy function exists, then it is often convenient to use it as the basis for construction of a candidate Lyapunov function. Typically, such a Lyapunov function leads to sharp estimates of the domain of attraction of a stable equilibrium. Indeed, the energy function can often attribute a useful physical interpretation to the stability boundary and thereby suggest means of evaluating stability margins, example the potential energy boundary surface (PEBS) method [50]. Moreover, such a Lyapunov function can be used to study the affects of system parameter variations on the geometric properties of the domain of attraction. Should the system lose stability under parameter variations, the energy function, although no longer a Lyapunov function, may provide useful information about the mechanism of instability.

Attempts to construct exact global energy functions for power systems with loads have not yet proved satisfactory, but recent proposals show promise [33, 6]. On the other hand, local energy functions are easier to identify and they also can provide useful information about the nature of impending instability. In fact, when operating near stability limits, they may yield satisfactory estimates of the stability boundary and, if not, they can be useful first approximations for refined Lyapunov functions which provide improved estimates.

The following paragraphs discuss in more detail the use of energy functions in power system stability analysis.

5.5.1 Properties of Classical Power System Models

Consider the classical power system model composed of $n + m + k$ buses where buses $i = 1, \ldots, n$ are the internal buses of n generators, buses i = n + 1,..., $n + m$ are m PV load buses, and buses $i = n + m + 1, \ldots, n + m + k$ are k PQ load buses.

The interconnecting network is considered to be the equivalent reduced network resulting from the elimination of the constant admittance loads and network internal buses. It is convenient to write the dynamical equations of motion in the form

$$\begin{aligned} M\ddot{\delta}_g + D\dot{\delta}_g + f_1\left(\delta_g, \delta_l, V, \mu\right) &= 0 \\ f_2\left(\delta_g, \delta_l, V, \mu\right) &= 0 \end{aligned} \tag{5.24}$$

where M denotes the diagonal matrix of generator rotor inertias, D is the diagonal damping matrix, the vector of network bus angles is $\delta^T = \left(\delta_g^T, \delta_l^T\right)$, where δ_g is the n-vector of generator internal bus angles, δ_l is the $(m+k)$-vector of load bus angles, V is the k-vector of PQ load bus vectors, and μ is a p-dimensional parameter vector. The functions f_1 *and* f_2 are the classical load flow functions as defined in (4.122).

5.5.1.1 Translational Symmetry

The function $f^T = \left(f_1^T, f_2^T\right)$ has a *translational symmetry* in the angle variables, i.e., $f(\delta + \alpha \mathbf{1}, V, \mu) = f(\delta, V, \mu)$ for any real number α. The usual remedy for this situation is to define a swing bus and refer all other bus angles to it. For example, choose the first generator bus as the swing bus and define the transformation of angle variables:

$$\theta_1 = \delta_1, \ \theta_i = \delta_i - \delta_1 \ for \ i = 2, .., n$$

When this change of angle coordinates is made, the resultant equations are independent of θ_1. In the load flow problem, it is typical to make the change of variables, then drop the first equation and solve the remaining equations for the $n-1$ angle variables $\theta_2 \ldots \theta_n$.

The dynamic problem is somewhat more delicate. There are two cases, either the swing bus is an infinite bus or it is not. If the swing bus can be treated as an infinite bus, the transformation of variables can be applied and then set $\theta_1 \equiv 0$. This effectively results in dropping the first dynamic equation, precisely analogous to the load flow case. If the swing bus cannot be treated as an infinite bus, the first equation cannot, in general, be eliminated. Nevertheless, it is still useful to make a transformation of the angle coordinates. Two transformations are commonly used: the one described above or the so-called *center of angle* transformation which uses and average angle reference, example [58]. To justify the above remarks, consider the transformation $\delta = T\theta$ in matrix form:

$$T = \begin{bmatrix} 1\,0\,0 \cdots 0 \\ 1\,1\,0 \cdots 0 \\ 1\,0\,1 \quad\;\; 0 \\ \vdots\;\vdots \quad \ddots \\ 1\,0 \qquad\;\; 1 \end{bmatrix}, \quad T^{-1} = \begin{bmatrix} 1\;\;0\,0 \cdots 0 \\ -1\,1\,0 \cdots 0 \\ -1\,0\,1 \quad\;\; 0 \\ \vdots\;\;\vdots \quad \ddots \\ -1\,0 \qquad\;\; 1 \end{bmatrix}$$

The governing equations transform to

$$\begin{aligned} T_{11}^T M T_{11}\ddot{\theta}_g + T_{11}^T D T_{11}\dot{\theta}_g + T_{11}^T f_1(T_{11}\theta_g + T_{12}\theta_l, T_{12}\theta_g + T_{22}\theta_l, V, \mu) &= 0 \\ f_2(T_{11}\theta_g + T_{12}\theta_l, T_{12}\theta_g + T_{22}\theta_l, V, \mu) &= 0 \end{aligned} \tag{5.25}$$

where T is partitioned:

$$T = \left[\begin{array}{c|c} T_{11} & T_{12} \\ \hline T_{21} & T_{22} \end{array}\right] \sim \left[\begin{array}{c|c} n \times n & n \times (m+k) \\ \hline (m+k) \times n & (m+k) \times (m+k) \end{array}\right]$$

It is easy to confirm that setting $\theta_{g,1} = 0$ in (5.25) is equivalent to setting $\delta_1 = 0$ in (5.24). But this is only appropriate when the swing bus can be treated as an infinite bus. Otherwise, equation (5.25) can be used. By modifying the transformation T, one can obtain the equivalent to (5.25) in center of angle coordinates.

It is easy to overlook the significance of the translational symmetry because it has such an obvious physical interpretation. It means simply that only the relative motions of the angular displacements are unique. Thus, any equilibrium point of interest is actually a point in a one-dimensional manifold of equilibria in the 2n-dimensional state space, and it only makes sense to discuss the stability of the entire manifold. The usual remedy is to measure displacement relative to an arbitrary selected swing bus as above. In any case, the state space is reduced to dimension $2n - 1$, and the equilibrium manifold is collapsed to a point.

When the system is conservative, i.e., in the absence of damping and transfer conductances, the translational symmetry is directly associated with a conservation law or first integral: Total angular momentum is constant. Thus, a second reduction is obtainable so that a reduction of the state space to dimension $2n - 2$ can be achieved [126, 173].

As it turns out, this reduction is not restricted to purely conservative power systems. It is known [198] that it works when uniform damping is present (in the absence of transfer conductances), an often used approximation. It does not work, however, when arbitrary damping is present. The fact that various dimensions for the state space have been employed in the literature has sometimes made comparison between methods difficult. Willems [198]) describes the situation very well and builds a case for conducting the analysis in the $2n$-dimensional state space.

5.5.1.2 Stability of the Equilibrium Manifold

In the event that the network does not contain any load buses, the variables δ_l and V are absent as is the second equation of (5.24), i.e., f collapses to f_1. As a result, the governing equation (5.24) reduces to

$$M\ddot{\delta} + D\delta + f(\delta, \mu) = 0 \tag{5.26}$$

An equilibrium point of (5.26) is any point $\delta^* \in R^n$ such that $\dot{\delta}^* = 0$. Thus, equilibria are roots of the equation

$$f(\delta^*, \mu) = 0 \tag{5.27}$$

Let δ^* be a solution of (5.27). Then because of the translational symmetry of $f(\delta, \mu)$, all points of the type

$$\delta = \delta^* + c\mathbf{1}$$

with c an arbitrary real constant are also equilibria. Consequently, any equilibrium point belongs to a one-dimensional manifold of equilibrium points. Consider the equilibrium manifold $\mathcal{M} \subset R^n$ associated with the equilibrium point δ^*.

$$\mathcal{M} = \delta^* + c\mathbf{1}, \quad c \in R \tag{5.28}$$

In the $2n$-dimensional state space composed of points (ω, δ), the one-dimensional equilibrium manifold is the set of points

$$\tilde{\mathcal{M}} = \{(\omega, \delta) \,|\omega = 0, \delta \in \mathcal{M}\} \tag{5.29}$$

Our interest is the stability of $\tilde{\mathcal{M}}$. For any set $\tilde{\mathcal{M}} \subset R^{2n}$, an η-neighborhood $U_\eta\left(\tilde{\mathcal{M}}\right)$ is the set of points $x \in R^{2n}$ such that dist $\left(x, \tilde{\mathcal{M}}\right) < \eta$, [84].

Definition 5.41 An invariant set $\tilde{\mathcal{M}}$ of (5.26) is stable if for any $\epsilon > 0$ there is an $\eta > 0$ such that for any initial $\left(\omega^0, \delta^0\right)$ in $U_\eta\left(\tilde{\mathcal{M}}\right)$, the corresponding $(\omega(t), \delta(t))$ is in $U_\epsilon\left(\tilde{\mathcal{M}}\right)$ for all $t \geq 0$. $\tilde{\mathcal{M}}$ is asymptotically stable if it is stable and in addition each solution with initial state in $U_\eta\left(\tilde{\mathcal{M}}\right)$ approaches $\tilde{\mathcal{M}}$ as $t \to \infty$.

In the $2n$ dimensional state space, stability of a power system equilibrium point corresponds to the study of stability of a one-dimensional invariant set. Lyapunov methods are easily modified for this situation. Consider an autonomous system on an m-dimensional state space, and suppose Ω is a p-dimensional invariant set. Furthermore, suppose that Ω can be characterized in the following way. There exists a continuous function $g : R^m \to R^{m-p}$ such that

$$\Omega = \left\{y \in R^m \,|g(y) = 0\right\} \tag{5.30}$$

This is the situation for the power system classical model with $m = 2n$ and $p = 1$, and the map g is linear.

Definition 5.42 A scalar valued function $V(y)$ is said to be positive definite with respect to Ω in an open region $U \supset \Omega$ if

1. $V(y)$ and its first partial derivatives are continuous on U.
2. $V(y) = 0$ for $y \in \Omega$
3. $V(y) \geq W(g(y))$, where $W(g(y))$ is an ordinary positive definite function on Image of U under g, $g(U)$

If, in addition, $\dot{V} \leq 0$ on U, V is called a Lyapunov function (with respect to Ω).

A straightforward extension of Lyapunov's stability theorem is as follows:

Proposition 5.43 *If a Lyapunov function exists in some open neighborhood U of an invariant set Ω, then Ω is stable.*

Willems [198] provides a variant of the LaSalle Invariance Theorem, 5.21.

Proposition 5.44 *Suppose there exists a Lyapunov function $V(y)$ on a open region $U \supset \Omega$ such that*

1. *$V(y) = a$ on the boundary of U and $V(y) < a$ in U.*
2. *$g(y)$ is bounded in U.*
3. *$\dot{V}$ does not vanish identically on any trajectory in U that does not lie entirely in Ω.*

Then Ω is asymptotically stable and every trajectory in U tends to Ω as $t \to \infty$.

5.5.1.3 Conservative Power Systems

In the event that the network does not contain any load buses, then the variable ϕ and V are absent as is the second equation of (5.24). If, in addition $D = 0$, the equations reduce to

$$M\ddot{\delta} + f(\delta, \mu) = 0 \tag{5.31}$$

Note, if the reduced network does contain any transfer conductances, the system is conservative (lossless). Moreover, it can be easily shown that the Jacobian of f with respect to δ, i.e.,

$$\frac{\partial f(\delta, \mu)}{\partial \delta} = \left[\frac{\partial f(\delta, \mu)}{\partial \delta}\right]^T$$

which implies that there is a scalar function $U(\delta, \mu)$, called a *potential function*, such that

$$f(\delta, \mu) = -\frac{\partial U(\delta, \mu)}{\partial \delta} \tag{5.32}$$

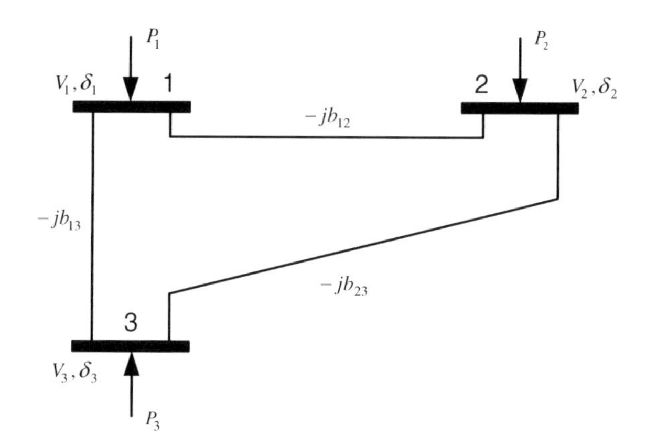

Fig. 5.13 Three-bus lossless network

It is common to refer to $U(\delta, \mu)$ as the potential energy function and to define the energy function as

$$E\left(\dot{\delta}, \delta, \mu\right) = \frac{1}{2}\dot{\delta}^T M \dot{\delta} + U(\delta, \mu) \tag{5.33}$$

It is also possible to define the Lagrangian

$$L\left(\dot{\delta}, \delta, \mu\right) = \frac{1}{2}\dot{\delta}^T M \dot{\delta} - U(\delta, \mu) \tag{5.34}$$

Note that (5.31) may be derived via Langrange's equations using the Lagrangian (5.29) (using an appropriate dissipation function if $D \neq 0$). The energy function (5.28) is the *Jacobi first integral* of the Lagrangian system associated with the Lagrangian (5.29).

Example 5.45 Three Bus Lossless Network. Consider the three-bus system shown in Figure 5.13. The three buses are generator internal buses. If uniform damping is assumed and bus 1 is taken as the reference bus, the governing equations are

$$\begin{aligned} \ddot{\theta}_1 + \gamma\dot{\theta}_1 &= \Delta P_1 - b_{13}\sin(\theta_1) - b_{12}\sin(\theta_1 - \theta_2) \\ \ddot{\theta}_2 + \gamma\dot{\theta}_2 &= \Delta P_2 + b_{12}\sin(\theta_1 - \theta_2) - b_{23}\sin(\theta_2) \end{aligned} \tag{5.35}$$

where

$$\theta_1 = \delta_2 - \delta_1, \theta_2 = \delta_3 - \delta_1, \Delta P_1 = P_2 - P_1, \Delta P_2 = P_3 - P_1$$

This system was studied by Aronovich and Neimark in 1961 as a Lagrangian system, as reported by Aronovich and Kartvelishvili [11]. This is a Lagrangian system with potential energy

$$U(\theta_1, \theta_2) = -\Delta P_1\theta_1 - \Delta P_2\theta_2 - b_{13}\cos(\theta_1) - b_{12}\cos(\theta_1 - \theta_2) - b_{23}\cos(\theta_2) \tag{5.36}$$

kinetic energy

$$T(\omega_1, \omega_2) = \frac{1}{2}\left(\omega_1^2 + \omega_2^2\right) \tag{5.37}$$

and generalized forces due to rotor friction

$$Q_d = \begin{bmatrix} -\gamma\omega_1 & -\gamma\omega_2 \end{bmatrix} \tag{5.38}$$

To study stability, choose a candidate Lyapunov function

$$V = T(\omega_1, \omega_2) + U(\theta_1, \theta_2) \tag{5.39}$$

and compute

$$\dot{V} = -\gamma\omega_1^2 - \gamma\omega_2^2 \le 0 \tag{5.40}$$

The key to stability analysis in this case is the Lasalle Invariance Theorem, Proposition 5.21. Accordingly, note that $T(0,0) = 0$ and $T(\omega_1, \omega_2) > 0\ \forall(\omega_1, \omega_2) \neq 0$ imply that equilibria corresponding to minimal extremal points of $U(\theta_1, \theta_2)$ are stable.

Clearly an understanding of the behavior of the potential energy is central to the stability characteristics of the system. The potential energy as expressed in (5.35) is a function of the angle coordinates θ_1 *and* θ_2 which are the configuration coordinates of this Lagrangian system. The configuration is 2π periodic in each of the two angles. Consequently, the configuration space can be envisioned as a torus. Consider the domain $[-\pi, \pi) \times [-\pi, \pi) \subset R^2$. Bend around the θ_2-axis so that the domain edges $\theta_1 = \pm\pi$ meet to form a cylinder. Now bend the cylinder around the θ_1-axis to form a torus. Because configuration motion along any trajectory evolves along the surface of the cylinder, the angular velocity at any point (θ_1, θ_2) belongs to the tangent plane to the cylinder at (θ_1, θ_2). The collection of all tangent planes (called the *tangent bundle*) is in fact the state space for this system.

The potential energy will be viewed as a function on the domain $[-\pi, \pi)\times[-\pi, \pi)$. In the following illustrations, in Figures 5.14–5.16, the parameters $b_{ij} = 1$ will be fixed and the parameters $\Delta P_1, \Delta P_1$ varied. Three different cases are considered, and for each case, the potential energy surface and a potential energy contour plot are shown.

Consider the case in Figure 5.14. It is easy to identify the six distinct equilibrium points:

$$(0,0), (0,-\pi), (-\pi,0), (-\pi,-\pi), (-2\pi/3, 2\pi/3), (2\pi/3, -2\pi/3)$$

Only one of these, the equilibrium at $(0,0)$ is stable. The level sets provide a clear and precise indication of the domain of stability.

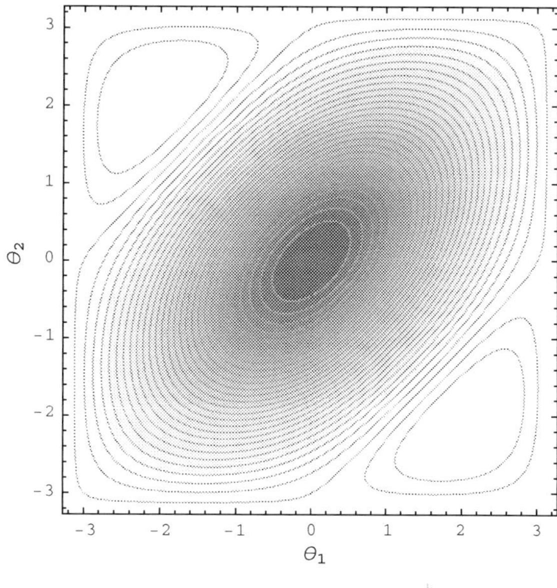

Fig. 5.14 Contour plot with $\Delta P_1 = 0, \Delta P_2 = 0, b_{12} = 1, b_{13} = 1, b_{23} = 1$

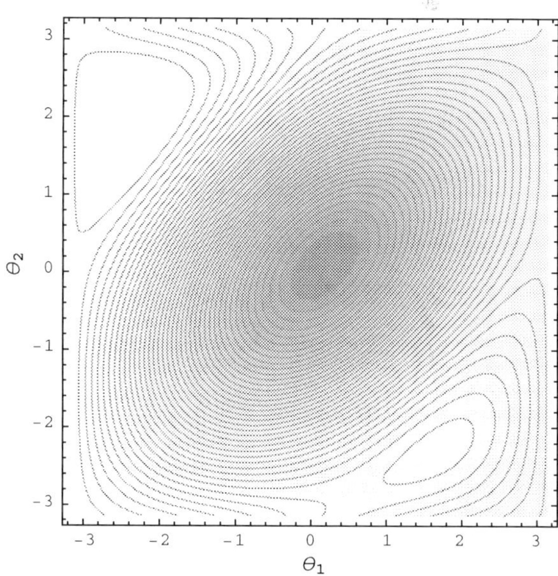

Fig. 5.15 Contour plot with $\Delta P_1 = .25, \Delta P_2 = 0, b_{12} = 1, b_{13} = 1, b_{23} = 1$

Figure 5.15 also shows six equilibria, with locations somewhat perturbed from those noted above:

$$(0.167251, 0.0836254)\,, (0.546935, -2.86813)\,, (-3.14159, 0.25268)\,,$$
$$(-3.14159, 2.88891)\,, (-2.39283, 1.94518)\,, (1.67864, -2.30227)$$

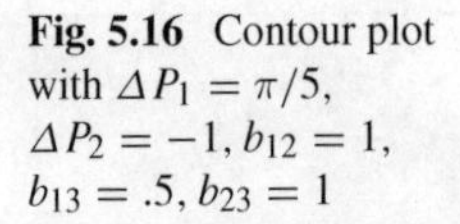

Fig. 5.16 Contour plot with $\Delta P_1 = \pi/5$, $\Delta P_2 = -1$, $b_{12} = 1$, $b_{13} = .5$, $b_{23} = 1$

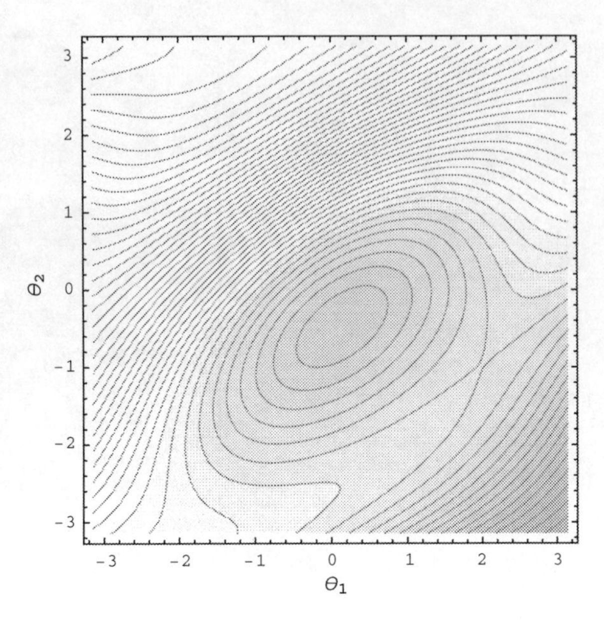

Once again, the single stable equilibrium at (0.167251, 0.0836254) is clearly evident. The contour lines create a more complex picture. Recall the torus picture, which implies that the counter lines leaving the upper boundary of Figure 5.15 re-emerge by entering the lower boundary. The contour lines may be complex—not necessarily forming a closed path. Even so, the there is a level set that defines the boundary of the stability domain.

Figure 5.16 corresponds to a larger change in parameters which results in only two equilibrium points with locations

$$(0.137867, -0.45604)\,, (1.99344, -0.974969)$$

The single stable equilibrium point, (0.137867, −0.45604), is again evident and the boundary of the domain of stability can be easily envisioned as the limit of the closed contour curves as they are expanded outward until the unstable equilibrium is reached. It is interesting to note that this limit will not be smooth although the energy function is. In general, the boundary of the domain of attraction of a stable equilibrium point is a combination of the stable manifolds of neighboring unstable equilibria.

5.5.2 Systems with Transfer Conductances

Power systems are nonconservative when energy is extracted from or added to the system. In classical power systems, specifically (5.24), important sources of nonconservative behavior include the presence of transfer conductances, $G_{ij} \neq 0$, typically attributed to transmission line losses or constant admittance loads; machine damping, $D \neq 0$; and other types of loads or power sources. The main consideration in the following discussion is the role of transfer conductances as well as the interaction of transfer conductances with machine damping.

The arguments leading to the energy function (5.28) for the system (5.26) may be extended to the more general case of the system (5.24) as described by Tsolas et al [188]. In [188], load buses are included but transfer conductances are not. The retention of load buses is intended to circumvent the introduction of transfer conductances, an approach suggested by Bergen and Hill [26]. The discussion below will be concerned with the local characterization of energy functions to better understand the role of transfer conductances. The matrix parameters of the linearized representation involve the same complexity (notably asymmetry) regardless of the load model employed. Therefore, the following analysis is constructed to admit any mix of constant impedance, voltage controlled, and constant power loads.

5.5.2.1 Transfer Conductances

Consider, first, the essential issue raised by transfer conductances. In the classical model, suppose the system is composed of n generator buses and no load buses so that the dynamics are governed by (5.26) and f_i takes the form

$$f_{1,i}(\delta,\mu) = \sum_{k=1}^{N} V_i V_k \left(B_{ik}\sin(\delta_i-\delta_k) + G_{ik}\cos(\delta_i-\delta_k)\right) - P_i, \quad i = 1,\ldots,n \tag{5.41}$$

and note that that the matrices B_{ij} and G_{ij} are symmetric (in the absence of phase-shifting transformers). Now compute the Jacobian of f_1 with respect to δ, denoted J,

$$J_{ij} = \frac{\partial f_{1,i}}{\partial \delta_j} = \left\{ \begin{array}{ll} \sum_{k=1}^{N} V_j V_k \left(B_{jk}\cos\left(\delta_j-\delta_k\right) - G_{jk}\sin\left(\delta_j-\delta_k\right)\right) & i=j \\ -V_i V_j B_{ij}\cos\left(\delta_i-\delta_j\right) + V_i V_j G_{ij}\sin\left(\delta_j-\delta_j\right) & i\neq j \end{array} \right\} \tag{5.42}$$

J is easily divided into a symmetric part $J_s = J_s^T$ and an antisymmetric part $J_a = -J_a^T$:

$$[J_s]_{ij} = \left\{ \begin{array}{ll} \sum_{k=1}^{N} V_j V_k \left(B_{jk}\cos\left(\delta_j-\delta_k\right) - G_{jk}\sin\left(\delta_j-\delta_k\right)\right) & i=j \\ -V_i V_j B_{ij}\cos\left(\delta_i-\delta_j\right) & i\neq j \end{array} \right\} \tag{5.43}$$

$$[J_a]_{ij} = \left\{ \begin{array}{ll} 0 & i=j \\ V_i V_j G_{ij}\sin\left(\delta_j-\delta_j\right) & i\neq j \end{array} \right\} \tag{5.44}$$

Clearly, if there are no transfer conductances, then $G_{ij} = 0\ i \neq j$, and the Jacobian J is symmetric. Thus, f_1 is derivable from a potential function, U, as in (5.27). The fact that in the presence of transfer conductances f_1 is not directly integrable is the essential difficulty in analyzing such systems. Notice that the self-conductances G_{ii} are not the issue. Transfer conductances come from losses in transmission lines, which are typically small, and also from self-conductances associated with constant admittance load buses that are removed using the network reduction process.

The asymmetry of J when transfer conductances are present means that $f_1(\delta, \mu)$ includes nonconservative forces of a type called circulatory forces in mechanics. Circulatory forces introduce effects that are quite different from dissipative forces and interact with dissipative forces in ways which are not intuitively obvious. The stability of mechanical systems under the influences of circulatory forces has been studied, most notably by Huseyin [93] and Leipholz [128]. The search for energy-like Lyapunov functions is closely related to questions of the existence of variational principles which produce a given set of differential equations. For an investigation of such issues see [111] and [18].

5.5.2.2 Transfer Conductances and no Damping

Consider the lossless system (5.31), and suppose it has an equilibrium point at $(\delta, \mu) = (\delta^*, \mu^*)$. This system is readily linearized at this point and takes the form

$$M\ddot{x} + Kx = 0 \tag{5.45}$$

where $x = \delta - \delta^*$, $K = \left[\partial f/\partial\delta\right]^*$, $M = M^T > 0$, and $K = K^T$. Furthermore, assume the a reference bus has not been defined so that f and hence Kx have a translational symmetry. A simple conclusion from the application of Proposition 5.44 is the following:

Proposition 5.46 *Consider Equation (5.45) with $M = M^T > 0$, $K = K^T$ and suppose Kx has a translational symmetry. The equilibrium manifold of (5.45) is stable if and only if $K \geq 0$ with precisely one zero eigenvalue.*

We wish to discuss the effect of small perturbations of K. Let $\mathcal{F}$ denote the set of real $n \times n$ matrices having the property of translational symmetry and $\|F\| \leq 1$.

Definition 5.47 The equilibrium manifold of (5.45) is *strongly stable* if there exists an $\epsilon_0 > 0$ such that the perturbed system

$$M\ddot{x} + (K + \varepsilon F)x = 0 \tag{5.46}$$

is stable for each ϵ, $|\varepsilon| < \varepsilon_0$ and each $F \in \mathcal{F}$

The following theorem from [112] provides a characterization of strongly stability of the equilibrium manifold of (5.45).

Proposition 5.48 *The equilibrium manifold of system (5.45) is strongly stable iff $K \geq 0$ with precisely one zero eigenvalue and in addition, the eigenvalues of K are distinct.*

Note that the requirements on K for stability of (5.45) admit repeated roots. Strong stability does not allow repeated roots. This can be viewed as a very simple type of resonance exclusion. Our notion of strong stability, tailored for the power system stability problem, may be viewed as a special case of strong stability for linear reciprocal systems as defined by Hale [84]. The functions belonging to the perturbation class used in [84] are periodic in time. As may be expected, the resulting resonance conditions are considerably more complex. In the context of power system stability, the use of time-varying (perhaps even stochastic) perturbations may be appropriate in view of the fact that the reduced bus admittance parameters change with load perturbations. Destabilization of the swing equations under stochastic perturbations has been observed by Loparo and Blankenship [136].

The subsequent analysis depends on the notion of a symmetrizable matrix introduced by Olga Taussky [182].

Definition 5.49 A real matrix A is symmetrizable if it be comes symmetric upon multiplication by a real, symmetric, positive definite matrix, S.

The following theorem of Taussky [182] is stated without proof.

Proposition 5.50 *The following properties are equivalent:*

- $A^T = SAS^{-1}$, *with* $S = S^T > 0$,
- *A is similar to a symmetric matrix,*
- *A is the product of two symmetric matrices, one of which is positive definite,*
- *A is symmetrizable,*
- *A has real characteristic roots and a full set of eigenvectors.*

The following corollary will prove useful.

Corollary 5.51 *If A is symmetrizable by a matrix S such that $SA = Q$, $S = S^T > 0$, then A has real characteristic roots and these roots have the same sign as those of Q.*

Proof That A has real roots follows from Proposition 5.50. Note that since $S > 0$, we can write $A = S^{-1}Q$ and S^{-1} can be factored $S^{-1} = BB^T$. Thus, $A = BB^TQ$ and $B^{-1}AB = B^TQB$ so that A is similar to the symmetric matrix B^TQT whose eigenvalues obviously have the same signs as those of Q. ■

The following proposition extends Proposition 5.46.

Proposition 5.52 *Consider the system (5.45)*

$$M\ddot{x} + Kx = 0$$

with $M = M^T > 0$ and suppose K is real and Kx has the translational symmetry property. The equilibrium manifold of (5.45) is stable if and only if there exists there exists a symmetric, positive definite matrix, S, such that $SM^{-1}K$ and satisfies the conditions of Proposition 5.46.

Proof Clearly (5.45) is unstable if there exists an nontrivial (eigen)vector, u, satisfying

$$\left[M\lambda^2 + K\right]u = 0$$

with corresponding eigenvalue λ having positive real part, or a repeated eigenvalue with zero real part and without a complete set of eigenvectors (we except the double root at the origin associated with the translational symmetry). It is easy to see that the roots of $\det\left[M\lambda^2 + K\right]$ are distributed symmetrically with respect to both the real and imaginary axes. It follows that the eigenvalues of a stable system must have zero real part and they must be associated with a complete set of eigenvectors (translational symmetry excepted). Moreover, this implies that $M^{-1}K$ must have positive real eigenvalues except for the single zero eigen value corresponding to the translational symmetry and a complete set of eigenvectors. It follows from Proposition 5.50 that $M^{-1}K$ is symmetrizable by a matrix S, and from Corollary 5.51 that $SM^{-1}K$ is nonnegative with the only zero eigenvalue corresponding to the translational symmetry. Thus, $SM^{-1}K$ satisfies the conditions of Proposition 5.46. ■

5.5.2.3 Transfer Conductances and Damping

Consider the system

$$M\ddot{x} + D\dot{x} + Kx = 0 \tag{5.47}$$

with $M^T = M > 0$, $D^T = D > 0$ and K has positive real eigenvalues except for precisely one zero eigenvalue corresponding to a translational symmetry of Kx. It is convenient to write (5.47) in the form

$$\ddot{x} + M^{-1}D\dot{x} + M^{-1}Kx = 0 \tag{5.48}$$

Definition 5.53 The system (5.47) or (5.48) is similar to a symmetric system if there exists a real transformation of coordinates $x \to y$, $y = Wx$, W real and nonsingular, such that the equations of motion have real symmetric coefficients in the new coordinate system.

The following proposition was given by Inman [98].

Proposition 5.54 *The system (5.47) is similar to a symmetric system if and only if $M^{-1}D$ and $M^{-1}K$ have a common symmetrizing matrix.*

For a system similar to a symmetric system, many well-known results apply. One is noted here.

Proposition 5.55 *If (5.47) is similar to a symmetric system then the equilibrium manifold* $\mathcal{M} = \{(\dot{x}, x) \,|\, \dot{x} = 0, x \in \text{span}\ (\mathbf{1})\}$ *is asymptotically stable.*

Proof The proof is a straightforward application of Proposition 5.44. Let S be the common symmetrizing matrix of Proposition 5.54 and define the candidate Lyapunov function

$$V(\dot{x}, x) = \tfrac{1}{2}\dot{x}^T S\dot{x} + \tfrac{1}{2}x^T\left(SM^{-1}K\right)x \tag{5.49}$$

Direct calculation leads to

$$\dot{V} = -\dot{x}^T\left(SM^{-1}D\right)\dot{x} \tag{5.50}$$

Clearly, V satisfies the positive definiteness requirements of Proposition 5.44, where $\mathcal{M}$ is the invariant set. Moreover, $\dot{V} \leq 0$ and the equality holds only for $\dot{x} = 0$. But all solutions satisfying $\dot{x} = 0$ lie entirely in $\mathcal{M}$. ■

One special case of interest is when $M^{-1}D$ and $M^{-1}K$ commute. This includes the case of uniform damping, that is, $M^{-1}D = \gamma I$, where γ is a positive scalar. When $M^{-1}D$ and $M^{-1}K$ commute, they have a common set of eigenvectors (Gantmacher, [76]) so that

$$\begin{aligned} M^{-1}D &= W^{-1}\Sigma W, \quad \Sigma = \text{diag}\,(\sigma_1, \ldots, \sigma_n) \\ M^{-1}S &= W^{-1}\Lambda W, \quad \Lambda = \text{diag}\,(\lambda_1, \ldots, \lambda_n) \end{aligned}$$

and thus, the matrix $S = W^T W$ is a common symmetrizing matrix. It follows that the conclusions of Proposition 5.55 apply. Thus, a power system with transfer conductances which is stable in the absence of damping is asymptotically stable in the presence of commutative (specially uniform) damping.

Suppose, however, that (5.48) is not similar to a symmetric system. We can still utilize the symmetrizing matrix, S, for $M^{-1}K$ and rewrite (5.48) as

$$\ddot{x} + (C + G)\dot{x} + SM^{-1}Kx = 0 \tag{5.51}$$

where

$$\begin{aligned} C &= \tfrac{1}{2}\left[\left(SM^{-1}D\right) + \left(SM^{-1}D\right)^T\right] \\ G &= \tfrac{1}{2}\left[\left(SM^{-1}D\right) - \left(SM^{-1}D\right)^T\right] \end{aligned} \tag{5.52}$$

so that C is symmetric and G is antisymmetric.

The stability of (5.51) can be characterized in terms of the matrix C, defined (5.52).

Proposition 5.56 *If $C \geq 0$, and the $n^2 \times n$ matrix*

$$\begin{bmatrix} C \\ C\left(SM^{-1}K\right) \\ \vdots \\ C\left(SM^{-1}K\right)^{n-1} \end{bmatrix}$$

has rank n then the equilibrium manifold of (5.51) is asymptotically stable.

Remark 5.57 This theorem extends a result of Walker and Schmitendorf [196] to the case $G \neq 0$.

Proof Once again, we use the Lyapunov function defined in (5.49). A simple computation shows that its time derivative along trajectories of (5.51) is

$$\dot{V} = -\dot{x}^T (C)\, \dot{x} \tag{5.53}$$

which is negative semidefinite. Thus, by Proposition 5.44, it is now necessary to show that any solution of (5.51) satisfying $x^T Cx = 0$ lies in the equilibrium manifold $\tilde{M}$. This will be accomplished by showing that under the hypothesis of the theorem only the trivial solution of (5.51) can satisfy this condition.

Assume that

$$\dot{x}^T C\dot{x} = 0 \tag{5.54}$$

on some nontrivial time interval (t_0, t_1). Premultiply (5.51) by $\dot{x}^T$ to obtain

$$\dot{x}^T S\ddot{x} + \dot{x}^T SM^{-1}Kx = 0 \tag{5.55}$$

on (t_0, t_1). Now choose t, $t_0 < t < t_1$ and integrate by parts over (t_0, t) to obtain

$$\dot{x}^T S\dot{x} + x^T SM^{-1}Kx = \int_{t_0}^{t} \ddot{x}^T S\ddot{x}\, dt + \int_{t_0}^{t} \dot{x}^T SM^{-1}K\dot{x}\, dt \tag{5.56}$$

The left-hand side is readily identifiable as V. Thus, a necessary condition that V is nondecreasing is

$$\dot{x}^T S\dot{x} + x^T SM^{-1}Kx = 0 \tag{5.57}$$

It is easy to prove that this condition is sufficient as well. Since $SM^{-1}K$ has a one-dimensional null space spanned by the vector $\mathbf{1}$, the only solutions of (5.51) satisfying (5.57) are of the form $\mathbf{1}\phi(t)$ where $\phi(t)$ is a scalar function of t. Direct substitution into (5.51) and premultiplication by $\mathbf{1}^T$ lead to the conclusion that $\ddot{\phi}(t) = 0$, or equivalently $\ddot{x}(t) = 0$. It follows that the right-hand side of (5.56) is constant. Thus, a solution of (5.51) satisfies (5.54) if and only if it satisfies (5.57).

Note that (5.54) and (5.57) can be satisfied simultaneously if and only if there exists a nontrivial vector q in the null spaces of both C and $SM^{-}K$. The conditions of the theorem preclude this as will be proved in the following. Assume that there exists a nontrivial q in the null space of $SM^{-}K$. It will be shown that it cannot lie in the null space of C. Write the sequence of relations

$$\begin{aligned} Cq &= Cq \\ C\left(SM^{-1}K\right) &= 0 \\ &\vdots \\ C\left(SM^{-1}K\right)^{n-1} &= 0 \end{aligned} \tag{5.58}$$

Since the coefficient matrix on the left has full rank by hypothesis, it has a left inverse. Let Σ denote the first n columns of the left inverse. Then

$$q = \Sigma C q \tag{5.59}$$

Clearly, $Cq \neq 0$. It follows that it does not possess a nontrivial solution satisfying (5.54). ■

If C is indefinite, then the equilibrium manifold may be unstable. The significance of this result arises from the fact that the definiteness properties of C do not directly follow from those of D. It is true that if D has positive real eigenvalues, then so does $C+G$. However, C may not have positive real eigenvalues, and it is C that determines the stability of (5.51). This observation was made by Huseyin and Hagedorn [94]. The important implication is that a power system, with transfer conductances, which is stable in the absence of dissipation may be destabilized by the addition of dissipation.

In the absence of transfer conductances, the energy function represents the perfect Lyapunov function in the sense that it globally characterizes the stability properties of the system. The energy function itself precisely determines the domain of stability of the stable equilibrium manifold. It is not known whether a global counterpart to the energy function exists in the presence of transfer conductances. In this regard, it is possible to give an interpretation of a Lyapunov function proposed by DiCaprio [62, 63]. In view of the remark following Proposition 5.56, it is reasonable to conjecture that if a global energy-like potential function exists for a system with transfer conductances, its local character will be that of (5.49). We can easily define a class of candidate Lyapunov functions which possess the following two properties: 1) They are locally equivalent to (5.49) and 2) they reduce globally to the conservative system energy function in the absence of transfer conductances.

First, extend the model of (5.31) to include damping,

$$M\ddot{\delta} + D\dot{\delta} + f(\delta) = 0 \tag{5.60}$$

where for convenience, the parameter μ has been suppressed. Denote an equilibrium point of interest by δ^* so that $f(\delta) = 0$. The function $f(\delta)$ can be nonuniquely separated

$$f(\delta) = f_a(\delta) + f_b(\delta)$$

such that f_a is integrable and f_b is not necessarily so $f_a(\delta^*) = f_b(\delta^*) = 0$. $f_a and f_b$ both have the translational symmetry property of f, and f reduces to f_a in the absence of transfer conductances. One simple choice for f_a is obtained from f by simply setting the transfer conductances to zero. There are many others. Let $U_a(\delta)$ represent a potential function from which f_a is derivable. As before, define the potential function V_a:

$$V_a\left(\dot{\delta}, \delta\right) = \tfrac{1}{2}\dot{\delta}^T M\dot{\delta} + \left(U_a(\delta) - U_a\left(\delta^*\right)\right)$$

or in local coordinates, $\delta = x + \delta^*$

$$V_a(\dot{x}, x) = \tfrac{1}{2}\dot{x}^T M\dot{x} + \tfrac{1}{2}x^T K_a x + G_a(x)$$

where

$$K_a = \left[\frac{\partial^2 U_a}{\partial x \partial x^T}\right]^* = \left[\frac{\partial f_a}{\partial x}\right]^*, \quad G_a(x) = U_a\left(x + \delta^*\right) - \left(U_a\left(\delta^*\right) + \tfrac{1}{2}x^T K_a x\right)$$

It follows directly from the construction of U_a that V_1 has the desired global property; i.e., it reduces to the conservative system energy function in the absence of transfer conductances. However, in general, it is not locally equivalent to (5.49).

An alternative is to evaluate the Jacobian, K, of f, at δ^* and compute the matrix S that symmetrizes $M^{-1}K$. Consider the system

$$\ddot{\delta} + SM^{-1}D\dot{\delta} + SM^{-1}f(\delta) = 0$$

This time divides $SM^{-1}f(\delta)$ into two parts

$$SM^{-1}f(\delta) = f_a(\delta) + f_b(\delta)$$

where f_a *and* f_b satisfy the same conditions as above. Now take

$$V(\dot{x}, x) = \tfrac{1}{2}\dot{x}^T S\dot{x} + \tfrac{1}{2}x^T\left[SM^{-1}K_a\right]x + \tilde{G}(x)$$

where

$$K_a = \left[\frac{\partial f}{\partial \delta}\right]^*, \quad \tilde{G}(x) = U_a\left(x + \delta^*\right) - \left(U_a\left(\delta^*\right) + \tfrac{1}{2}x^T\left[SM^{-1}K_a\right]x\right).$$

Remark 5.58 More General System Configurations. The discussion in this Section 5.5.2 has assumed that the system model was of the form (5.26), i.e., a n generator buses and no load buses. Of course, many more general systems can be reduced to this model by load bus elimination. But the local analysis discussed above applies to even more general systems. Consider the system of (5.24). Let $\left(\delta_g^*, \delta_l^*, V^*, \mu^*\right)$ be an equilibrium point. Suppose that it is *strictly causal*[2] in the sense that there exist unique functions $\delta_l\left(\delta_g, \mu\right), V\left(\delta_g, \mu\right)$ satisfying $f_2\left(\delta_g, \delta_l\left(\delta, \mu\right), V\left(\delta, \mu\right), \mu\right) = 0$ on a neighborhood of $\left(\delta_g^*, \delta_l^*, V^*, \mu^*\right)$ with $\delta_l\left(\delta_g^*, \mu^*\right) = \delta_l^*$ and $V\left(\delta_g^*, \mu^*\right) = V^*$. Under these circumstances, the linearized dynamics of (5.24) reduce to

$$M\ddot{x} + D\dot{x} + Kx = 0 \tag{5.61}$$

where

$$K = \left[D_{\delta_g} f_1 - \begin{bmatrix} D_{\delta_l} f_1 \\ D_V f_1 \end{bmatrix} \begin{bmatrix} D_{\delta_l} f_2 \\ D_V f_2 \end{bmatrix}^{-1} \right]^* \tag{5.62}$$

and $x = \delta_g - \delta_g^*$.

[2]More discussion about this concept will be found below, particularly in Section 6.2.

Chapter 6
Power System Dynamics: Bifurcation Behavior

> *".... the need is not so much for 'more mathematics' as for a better understanding of the potentialities of its application."*
> —Theodore v. Karman and Maurice Biot, "Mathematical methods in Engineering"

6.1 Introduction

This chapter begins with a summary of the basic properties of systems described by differential-algebraic equations (DAEs) and moves on to study singularities and bifurcations of DAEs. The study of local behavior around bifurcation points of the equilibrium equations is important as such points typically involve some sort of static or dynamic instability phenomenon. Computational methods for finding these *static bifurcation* points and generating models for examining local behavior are considered next. Locating Hopf (dynamic) bifurcation points are also examined.

Power system applications occupy the final section of the chapter. Issues discussed include static stability and voltage collapse, the relation between static bifurcation and dynamic stability margins, bus variable sensitivity near bifurcation points, and other. Several examples of static and Hopf bifurcation in power networks are given.

6.2 Systems Described by Differential-Algebraic Equations

As seen above in Chapter 4, a power system can often be described by the (semi-explicit) differential-algebraic equation (DAE) model (example see [87]):

$$\begin{aligned} \dot{x} &= f(x, y, \mu) \\ 0 &= g(x, y, \mu) \end{aligned} \tag{6.1}$$

H.G. Kwatny and K. Miu-Miller, *Power System Dynamics and Control*,
Control Engineering, DOI 10.1007/978-0-8176-4674-5_6

where $x \in R^n$, $y \in R^m$ and $\mu \in R^k$. x and y are independent variables and μ is a vector of system parameters. The functions f and g are assumed to be smooth, i.e., k times differentiable for some $k \geq 1$.

An example is the classical power system model composed of n generators, m PV load buses, and p PQ buses. As commonly expressed, these equations take the form:

$$\begin{aligned} \dot{\delta} &= \omega \\ M\dot{\omega} + D\omega + f_g(\delta, \theta, V, \mu) &= 0 \\ f_l(\delta, \theta, V, \mu) &= 0 \end{aligned} \tag{6.2}$$

where $\omega, \delta \in R^n$, $\theta \in R^{m+p}$, $V \in R^p$ and $f_l \in R^{m+2p}$. The generator inertia matrix, M, is nonsingular so (6.2) reduces to (6.1).

There are two essential features of the model (6.1) to be emphasized: 1) the explicit parameter dependence, and 2) the differential-algebraic structure. Consideration of the change in system behavior that occurs as a consequence of parameter variation is central theme of the discussion below. This perspective is the key to formulating concepts of stability that allow systematic examination of voltage collapse phenomenon among other things. In addition to the obvious computational issues, a differential-algebraic structure can produce behaviors not present in purely differential equations [121, 193]. Examples will be given below.

Systems described by DAEs are frequently encountered in engineering and are studied as dynamical systems that evolve on manifolds. An insightful introduction to this point of view is the discussion of nonlinear RLC circuits in [89]. Numerical methods for solving DAEs and other examples and references may be found in [32]. The conceptual framework within which (6.1) is to be considered will be discussed below along with power system examples that illustrates the main issues.

6.3 Basic Properties of DAEs

For each fixed μ trajectories evolve on the state space, $\mathcal{M}_\mu$,

$$\mathcal{M}_\mu = \{(x, y) \in R^{n+m} \mid g(x, y, \mu) = 0\}$$

Typically, $\mathcal{M}_\mu$ is composed of one or more disconnected manifolds called *components* [28]. In general, when referring to $\mathcal{M}_\mu$ we mean one of these components called the *principal component*.

$\mathcal{M}_\mu$ is a regular n-dimensional manifold in R^{n+m} if

$$rank \left[\frac{\partial g}{\partial x} \; \frac{\partial g}{\partial y} \right] = m \tag{6.3}$$

The structure of $\mathcal{M}_\mu$ depends on the parameter μ. Even for very simple power systems (6.3) y not be satisfied for some values of μ.

The manifold $\mathcal{M}_\mu$ is the state space for the dynamical system defined by (6.1) which induces a vector field on $\mathcal{M}_\mu$. If $\det[\partial g/\partial y] \neq 0$ at a point $(x, y) \in \mathcal{M}_\mu$, the vector field is locally well defined at (x, y) by the relations

$$\dot{x} = f(x, y, \mu) \tag{6.4}$$

and

$$\dot{y} = -\left[\frac{\partial g}{\partial y}\right]^{-1} \frac{\partial g}{\partial x} f(x, y, \mu) \tag{6.5}$$

As a matter of fact, the Implicit Function Theorem provides that it is possible to locally solve the algebraic equation in (6.1) to obtain $y = \varphi(x, \mu)$ so that the flow on $\mathcal{M}_\mu$ is locally defined by the ordinary differential equation

$$\dot{x} = f\left(x, \varphi\left(x, \mu\right), \mu\right) \tag{6.6}$$

If this is the case, then it is easy to show that the vector field defined by $\dot{x}$ and $\dot{y}$ lies in the tangent space to $\mathcal{M}_\mu$ at (x, y). On the other hand, if at the point $(x, y) \in \mathcal{M}_\mu$, $\det[\partial g/\partial y] = 0$, the vector field may not be well defined. Typically, such *singular points* lie on co-dimension-1 submanifolds of singular points in $\mathcal{M}_\mu$.[1]

In power systems, such points were encountered by DeMarco and Bergen [60] in connection with transient stability studies ("impasse points"), by Kwatny et al. [121] in connection with bifurcation analysis ("noncausal points") and others ("impasse surfaces" in [90]; "singularity" in [194]).

Definition 6.1 Suppose $\mathcal{M}_\mu$ is a regular manifold for all μ near μ^* and that $\det\left[\partial g/\partial y\right] \neq 0$ at the point $\mu = \mu^*$, $(x, y) = (x^*, y^*) \in \mathcal{M}$. Then, (x^*, y^*, μ^*) is said to be causal. Otherwise, it is noncausal.

If (x^*, y^*, μ^*) is causal, then the trajectories of the DAE (6.1) are locally defined by the ordinary differential equation (6.6).

6.4 Singularities and Bifurcations of DAEs

Definition 6.2 The point (x^*, y^*, μ^*) is an *equilibrium point* of (6.1) if

$$\begin{aligned} f\left(x^*, y^*, \mu^*\right) &= 0 \\ g\left(x^*, y^*, \mu^*\right) &= 0 \end{aligned} \tag{6.7}$$

Consider the set

$$\mathcal{M} = \left\{ (x, y, \mu) \in R^{n+m+k} \middle| \, g\left(x, y, \mu\right) = 0 \right\} \tag{6.8}$$

---

[1] The *co-dimension* of a k-dimensional submanifold of an n-dimensional manifold is $n - k$.

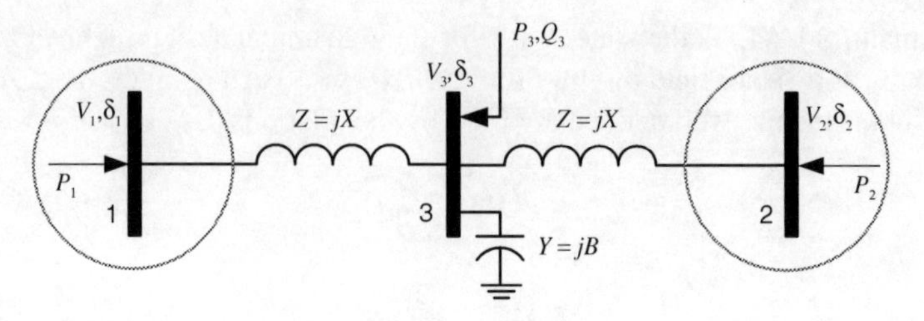

Fig. 6.1 A three-bus system with two-generator buses and a load bus

This set forms a regular manifold of dimension $n + k$ in R^{n+m+k} provided

$$\operatorname{rank}\left[\frac{\partial g}{\partial x}\ \frac{\partial g}{\partial y}\ \frac{\partial g}{\partial \mu}\right] = m \text{ on } \mathcal{M} \tag{6.9}$$

Similarly, the set

$$\mathcal{F} = \left\{(x, y, \mu) \in R^{n+m+k} \,\middle|\, f(x, y, \mu) = 0\right\} \tag{6.10}$$

forms a regular manifold of dimension $m + k$ in R^{n+m+k} provided

$$\operatorname{rank}\left[\frac{\partial f}{\partial x}\ \frac{\partial f}{\partial y}\ \frac{\partial f}{\partial \mu}\right] = n \text{ on } \mathcal{F} \tag{6.11}$$

Equilibria are the points in the intersection of these manifolds. It will be assumed that the intersection of $\mathcal{M}$ and $\mathcal{F}$ is *transversal.* A transversal intersection implies that either $\mathcal{M}$ and $\mathcal{F}$ do not intersect at all or that the intersection forms a regular k-dimensional submanifold of R^{n+m+k}. Note that if the intersection is not transversal, then an arbitrarily small perturbation of $\mathcal{M}$ or $\mathcal{F}$ will cause it to be. That is, a nontransversal intersection is nongeneric [8].

In what follows, it is assumed that $\mathcal{M}$ and $\mathcal{F}$ are regular manifolds, i.e., that (6.9) and (6.11) are true and that their intersection is transversal.

Example 6.3 Three-Bus Network. As an illustration of these concepts consider the network illustrated in Figure 6.1. This network was used in [121] to illustrate some of the properties of power systems described by DAEs. Although extremely simple, configurations like this involving one or two generators feeding a remote load have often been used in discussions of voltage stability and control [37, 59, 79, 91, 103].

P_1, P_2, P_3 are the bus real power injections. Equilibrium solutions exist only if $P_1 + P_2 + P_3 = 0$. Assume that this is the case, for convenience, fix some of the parameters as follows. The generator internal bus voltages are $V_1 = 1, V_2 = 1$, and the generator inertia constants are $M_1 = 1, M_2 = 1$. Eliminate the translational symmetry by using bus 1 as a reference bus and defining $\theta = \delta_2 - \delta_1$ and $\phi = \delta_3 - \delta_1$. Finally, define $\Delta P = P_2 - P_1$. The equations of motion can then be written

$$\begin{aligned}
\ddot{\theta} &= -V\,(\sin(\theta - \phi) + \sin\phi) + \Delta P \\
0 &= V\,(\sin(\phi - \theta) + \sin\phi) - P_3 \\
0 &= -V\,(\cos\phi + \cos(\phi - \theta)) + (2 - B)\,V^2 - Q_3
\end{aligned} \tag{6.12}$$

These three equations define the three independent variables (V, θ, ϕ) in terms of four parameters $(\Delta P, P_3, Q_3, B)$. Any motion is constrained by the two algebraic equations. Within the three-dimensional (3-D) space of independent variables, these equations define a one-dimensional manifold called a configuration manifold. Each point on this configuration manifold has a one-dimensional tangent space. At any point in the configuration manifold, the velocity vector belongs to the tangent space at that point. When these tangent spaces are collected together, they form a two-dimensional state space (called the *tangent bundle*). Clearly, the configuration manifold, and hence the state space, can change its shape as the system parameters vary. In [121], it is shown that for almost all values of the parameters the configuration space is a closed curve so that the state space is topologically equivalent to a cylinder. Moreover, the noncausal points form two one-dimensional submanifolds that divide the state space (cylinder) into two sheets. The parameter dependence of the system equilibria is investigated for various values ΔP while the remaining parameters are fixed: $P_3 = -1$, $Q_3 = 0$, $B = 1$.

As ΔP is decreased from 0 to -1, the system has two stable equilibria, one in each of the two sheets, which move as indicated in Figure 6.2. As ΔP decreases

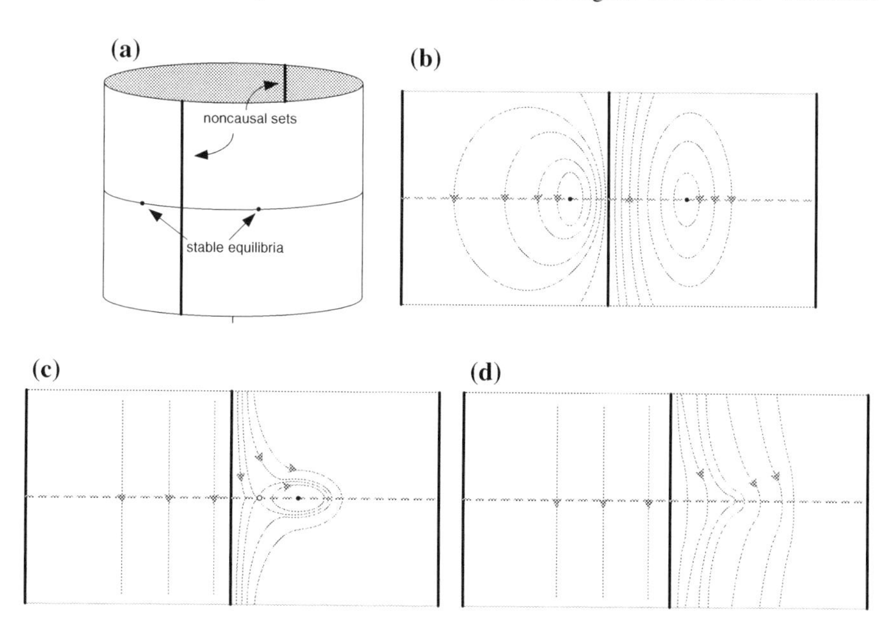

Fig. 6.2 (a) The cylindrical state space is divided into two sheets by the noncausal sets or "impasse" surfaces. (b) The cylindrical space is flattened out in order to illustrate the state trajectories on the two sheets. (c) As ΔP decreases below -1, the left equilibrium point migrates into the right sheet and changes from stable to unstable. (d) Further decreases in ΔP cause a bifurcation that leaves no equilibria. The post-bifurcation trajectories are illustrated here

below -1, the left equilibrium moves through the noncausal set into the adjacent sheet and becomes unstable. Further decreases in ΔP cause the two equilibria to meet and annihilate each other. As shown in [121], if the parameter B is reduced with all others held fixed, then the radius of the cylindrical state space decreases so that the state space shrinks to a line as $B \to 0$. The crossing of the left equilibrium point across the noncausal surface as ΔP is reduced is accompanied by an exchange of stability from stable to unstable. It is called a *singularity-induced bifurcation*. Further movement to the right of unstable equilibrium ultimately involves collision with the stable equilibrium point and disappearance of both. This is a classical *saddle-node bifurcation*.

Another, and quite different, example of a power system with DAE description is given in [193, 194]. Here again, two-dimensional state space is divided by a one-dimensional submanifold of noncausal points. The analysis in [193, 194] illustrates a one-parameter variation that causes an equilibrium point to pass through the noncausal surface accompanied by an exchange of stability of one eigenvalue that diverges through infinity.

6.5 Bifurcation of Flows Near Equilibria

An important factor contributing to improved understanding of nonlinear phenomenon is the simple notion that it is far more profitable to study families of nonlinear systems rather than individual nonlinear systems. It is the differences in behavior that exist between members of a family that is most revealing. For instance, the detailed investigation of a system containing a limit cycle is not nearly so informative as the study of a family that contains the system and that exhibits the birth and extinction of the limit cycle.

6.5.1 *Equivalence of Flows and Structural Stability*

A model of a power system, such as (4.121), is at best an approximation and there is always a concern that conclusions drawn from it may not be consistent with reality. One basic question that should be asked is how sensitive are the predictions of the model to small perturbations of it? We are particularly interested in qualitative properties of the system – namely stability – hence, it is essential to know whether qualitative features of the flow change under perturbations of the model. Thus, the classical definition of equivalence of flows is appropriate [12, 83, 89].

For now, ignore the parameter μ and consider the DAE (6.1) with independent variables $x \in R^n$ and $y \in R^m$. The DAE defines a flow on the state space $\mathcal{M} = \{(x, y) \,|g(x, y) = 0\}$. $\mathcal{M} \subset R^{n+m}$ is assumed to be a regular manifold of dimension n. The flow defined on $\mathcal{M}$ will be designated Φ_t.

Definition 6.4 Two flows Φ_t and Ψ_t that evolve on an n-dimensional state space are said to *topologically equivalent* if there exists a homeomorphism taking trajectories of Φ_t into trajectories of Ψ_t, preserving their time orientation.

Roughly speaking, this implies that one flow can be transformed to the other by a continuous deformation of the state space.

Let U be a bounded, open set in R^{n+m} and suppose $\mathbf{F}(U)$ denotes the set of all smooth (C^1) maps $F : U \to R^{n+m}$ defined on U. The magnitude of any map $F \in \mathbf{F}(U)$ is taken to be its C^1-norm, i.e.,

$$\|F\| = \sup_{\xi \in U} \left\{ \sum_{i=1}^{n+m} |F_i| + \sum_{i,j=1}^{n+m} \left| \frac{\partial F_i}{\partial \xi_j} \right| + \right\} \tag{6.13}$$

An ϵ-*neighborhood* of $F \in \mathbf{F}(U)$ is

$$\mathcal{N}_\varepsilon = \{G \in \mathbf{F}(U) \, | \, \|G - F\| < \varepsilon\} \tag{6.14}$$

The magnitude of a DAE or a neighborhood of a DAE can be characterized by identifying F with $\{f, g\}$.

Definition 6.5 Consider a DAE $\{f, g\} \in \mathbf{F}(U)$ with an equilibrium point $(x^*, y^*) \in \mathcal{M}$, then $\{f, g\}$ is *locally structurally stable* at (x^*, y^*) if there exists a neighborhood U in $\mathcal{M}$ of (x^*, y^*) and an $\epsilon > 0$ such that for every $\left\{\tilde{f}, \tilde{g}\right\} \in \mathcal{N}_\varepsilon(\{f, g\})$ there is a neighborhood $\tilde{U}$ of $(\tilde{x}^*, \tilde{y}^*) \in \mathcal{M}$ such that $\varphi_t \, |U$ and $\tilde{\varphi}_t \left| \tilde{U} \right.$ are locally topologically equivalent.

The main results of interest here regarding structural stability are summarized in the following proposition.

Theorem 6.6 *Suppose the DAE $\{f, g\}$ has an equilibrium point at $(x^*, y^*) \in \mathcal{M}$. Then, $\{f, g\}$ is locally structurally stable at (x^*, y^*) if and only if the equilibrium point is causal and hyperbolic.*

Remark 6.7 Remarks on Proof. Sufficiency is very straightforward because the fact that the system is causal at (x^*, y^*) implies that there is a local representation of the dynamics as an ordinary differential equation and hence standard results apply based on the Hartman–Grobman theorem [12, 83]. The only additional requirement to establish necessity is to verify that a noncausal equilibrium point is not structurally stable. But this is clearly true in view of Theorem 2 in [193].

6.5.2 *Bifurcation Points*

A k-parameter family of DAEs is a C^r, $r \geq 1$, map $\phi : \mathcal{P} \subset R^k \to \mathbf{F}(U)$. $\mathcal{P}$ represents a parameter space, and it is tacitly assumed that $\phi(\mathcal{P})$ is a k-dimensional submanifold of $\mathbf{F}(U)$. A family of DAEs will be denoted $\{f(\mu), g(\mu)\}$ where $\mu \in \mathcal{P}$ is the parameter. The immediate goal is to characterize the qualitative properties that occur in the family when its parameters are varied. But a seen above, a small change from a given parameter value μ_0 can induce a behavioral change only if the system $\{f(\mu_0), g(\mu_0)\}$ is structurally unstable. Hence, we have the following definition of a (local) bifurcation point [51, 83].

Definition 6.8 A value μ_0 for which the flow of $\{f(\mu), g(\mu)\}$ is not locally structurally stable near an equilibrium point $\left(x_0^*, y_0^*\right)$ of $\{f(\mu_0), g(\mu_0)\}$ is a bifurcation value of μ. The pair $(\mu_0, \{f(\mu_0), g(\mu_0)\})$ is called a bifurcation point.

This definition of a bifurcation point in the family $\{f(\mu), g(\mu)\}$ has a deficiency in that the family may include a structurally unstable member, but may not exhibit any distinctive behavior. For example, the trivial system $\{f(\mu), g(\mu)\} = \{0, y - \mu\}$, $x, y, \mu \in R$, defines a flow that is structurally unstable and unchanged for all values of μ.

An alternative definition of a bifurcation point is given in [12]: The family has a bifurcation point at $\mu = \mu_0$ if in every neighborhood of μ_0, there are family members that exhibit topologically different behaviors. While this definition ensures the existence of dissimilar behaviors near bifurcation points, it suffers annoying technicalities that arise from the fact that while all bifurcation points are structurally unstable, not every structurally unstable point is a bifurcation point. Moreover, the essential difficulty is not really eliminated, just postponed. The remedy is to introduce the concept of a generic family.

6.5.3 *Genericity*

Given a well-defined set $\mathcal{G}$ of mathematical objects, such as a set of algebraic equations, vector fields, or DAEs, it is useful to identify properties that are common to virtually all elements in the given set. Such properties are called generic properties. Formally, a property is said to be generic if it is shared by a residual subset (a countable intersection of open dense sets) [193] of the set $\mathcal{G}$. The elements in $\mathcal{G}$ which exhibit a generic property (the generic points) form an open set in $\mathcal{G}$, and typically the nongeneric points lie on submanifolds of $\mathcal{G}$ with co-dimension ≥ 1. In some contexts, structural stability is a generic property in which case structurally unstable elements are referred to as nongeneric. When examining an arbitrarily selected individual object from $\mathcal{G}$, one expects to observe only generic properties. However, in applications it is often necessary to consider a collection or family of objects within which individual members that do exhibit nongeneric properties are encountered. It is useful to distinguish between those nongeneric behaviors that are likely to be found in families and those that are not. A k-parameter family in $\mathcal{G}$ is a C^1 map $s : \mathcal{P} \subset R^k \to \mathcal{G}$. Here, it is again assumed that $s(\mathcal{P})$ is a k-dimensional submanifold of $\mathcal{G}$. The family $s(\mathcal{P})$ contains nongeneric points if it intersects a manifold $\mathcal{A}$ of nongeneric points. If the intersection of $s(\mathcal{P})$ and $\mathcal{A}$ is transverse then the intersection is stable in the sense that small changes in the family do not eliminate it. It can be said that these nongeneric points contained in $s(\mathcal{P})$ persist or are nonremovable under perturbations.

In many important cases, it is possible to prove that the set of k-parameter families that are transverse to a given submanifold $\mathcal{A}$ of $\mathcal{P}$ form a residual set (in the set of all k-parameter families). Theorems of this type, called *transversality theorems*, are important in bifurcation analysis. Transversality concepts are discussed at length in [8, 78, 80, 88]. The important implication is that it makes sense to speak of *generic families* relative to a generic property of interest. This is particularly useful when structural stability is a generic property. In this case, generic families that contain structurally unstable members (bifurcation points) have the property that such members remain even when the family is perturbed. By focusing on bifurcations contained in generic families, it is unnecessary to deal with special cases that can be eliminated by a small change in the family. Roughly speaking, the main result of significance here is that generic k-parameter families contain bifurcation points of co-dimension k or less.

6.5.4 Normal Forms

A normal form is a convenient way of representing a class of equivalent systems. It is a member of the class which is *simple* in some convenient and acceptable sense. Unlike a canonical form, the normal form is not chosen to meet any specific criteria; but the general idea is that it should clearly exhibit the essential features of the system. In view of the Hartman–Grobman theorem, the local behavior of a nonlinear vector field at a hyperbolic equilibrium point is completely described by its linearization. Indeed, there is a transformation of coordinates which establishes the equivalence. Hence, we need only look at the linearization to determine whether two such nonlinear systems have similar behavior. The linear dynamics (or perhaps its Jordan form) represent a normal form for comparing local dynamics. But behavior near a hyperbolic equilibrium is not particularly interesting because it is not sensitive to perturbations. It is necessary to establish normal forms of vector fields for nonhyperbolic equilibria.[2] The basic idea is to seek a transformation of coordinates that brings the given vector field into an *almost* linear form leaving only the nonlinear terms that are not removable by any smooth transformation. These nonlinear terms are essential to the complete characterization of the local behavior. Algorithms for reduction of a flow defined by a differential equation to normal form and examples can be found in [8, 12, 83] as well as many other sources. The key result is given below following some necessary definitions.

---

[2]The present discussion of normal forms is confined to vector fields, which for DAEs means normal forms near causal equilibria. However, there is an emerging theory for DAEs [168].

Let X, Y be smooth vector fields on R^n. The Lie bracket of the vector field Y with respect to the vector field X results in the map $ad_X : R^n \rightarrow R^n$ defined by

$$ad_X(Y) = [X, Y] = \frac{\partial Y}{\partial x} X - \frac{\partial X}{\partial x} Y$$

Following Arnold [8], consider the transformation of the linear differential equation $\dot{y} = Ay$ under the near-identity transformation $x = y + h(y)$ where $h(y)$ is a vector field of polynomials of order $r \geq 2$. Then direct computation yields

$$\dot{x} = \dot{y} + \frac{\partial h(y)}{\partial y} \dot{y} = Ay + \frac{\partial h(y)}{\partial y} Ay = \left(I + \frac{\partial h(y)}{\partial y} \right) A \left(x - h(x) + O\left(|x|^{r+1} \right) \right)$$

which can be reduced to

$$\dot{x} = Ax + ad_L(h(x)) + O\left(|x|^{r+1} \right) \tag{6.15}$$

This result can be inverted. Let $v(x)$ be a vector field whose elements are homogeneous polynomials of degree 2. Consider the nonlinear differential equation

$$\dot{x} = Ax + v(x) + O\left(|x|^{r+1} \right) \tag{6.16}$$

Then, (6.16) can be transformed by a near-identity transformation $y = x + h(x)$, $h(x)$ a polynomial vector field of degree $r \geq 2$, to

$$\dot{y} = Ay + O\left(|y|^{r+1} \right) \tag{6.17}$$

provided the exists an $h(x)$ that satisfies the *homological* equation

$$ad_{Ax}(h(x)) = v(x) \tag{6.18}$$

This approach can be extended to remove successively higher-order terms using the same procedure. If the a solution of (6.18) does not exist it is possible to reframe the question to eliminate as many of the polynomial terms as possible. This approach is used to prove the Poincaré and Poincaré–Dulac Theorems.

Now, we are in a position to state these results in the form of the following theorem from [83].

Theorem 6.9 *Normal Form Reduction Theorem. Let* $\dot{x} = \phi(x)$, $x \in R^n$, *be a smooth system of differential equations with* $\phi(0) = 0$ *and* $L = D_x\phi(0)\,x$. *Define* H_k *to be the linear space of vector fields whose components are homogeneous polynomials of degree k. Choose a complement* G_k *for* $ad_L(H_k)$ *in* H_k, *so that* $H_k = ad_L(H_k) + G_k$. *Then, there is an analytical change of coordinates,* $x \rightarrow y$, *in a neighborhood of the origin that transforms the system to*

Table 6.1 Examples of Normal Forms for Dynamic Bifurcations

Name	Normal Form
Saddle-node	$\dot{x} = -x^2 + O\left(\lvert x\rvert^3\right)$
Hopf	$\dot{x} = \begin{bmatrix} 0 & -1 \\ 1 & 0 \end{bmatrix} x + \left(x_1^2 + x_1^2\right)\left(a \begin{bmatrix} x_1 \\ x_2 \end{bmatrix} + b \begin{bmatrix} x_1 \\ x_2 \end{bmatrix}\right) + O\left(\lvert x\rvert^5\right), \quad a \neq 0$
Cusp, R	$\dot{x} = -x^3 + O\left(\lvert x\rvert^4\right)$
Cusp, R^2	$\dot{x} = \begin{bmatrix} 0 & 1 \\ 0 & 0 \end{bmatrix} x + \begin{bmatrix} a \\ b \end{bmatrix} x_1^2 + O\left(\lvert x\rvert^3\right), \quad a \neq 0, b \neq 0$
Generalized Hopf	$\dot{x} = \begin{bmatrix} 0 & -1 \\ 1 & 0 \end{bmatrix} x + a\left(x_1^2 + x_2^2\right)^2 x + O\left(\lvert x\rvert^7\right), \quad a \neq 0$

$$\dot{y} = Ay + g^2(y) + \cdots + g^r(y) + O\left(\lvert y\rvert^{r+1}\right)$$

with

$$A = D_x\phi(0), \quad g^k \in G_k, \ 2 \leq k \leq r$$

A constructive proof and examples may be found in [83]. Some examples of normal forms are given in Table 6.1.

6.5.5 *Deformations and Unfoldings*

Unfoldings provide an efficient characterization of all behaviors exhibited by systems in the vicinity of a system that is locally structurally unstable at an equilibrium point. The notion of a versal deformation or unfolding again goes back to Poincaré, [8, 12, 83].

Consider the smooth DAE $\{f, g\}$ defined in some neighborhood of $(x^*, y^*) \in R^n \times R^m$. For convenience, take $(x^*, y^*) = (0, 0)$. Any family $\{f(\mu), g(\mu)\}$, locally defined at $(x, y, \mu) = (0, 0, 0)$ in $R^n \times R^m \times R^p$, with $\{f(0), g(0)\} = \{f, g\}$, is said to be a *deformation* of $\{f, g\}$.

Two deformations of $\{f, g\}$, $\{f(\mu), g(\mu)\}$ and $\left\{\tilde{f}(\mu), \tilde{g}(\mu)\right\}$, are equivalent if there is a continuous transformation of coordinates $h : \mathcal{N} \subseteq R^n \times R^m \times R^p \to R^n \times R^m$, $\mathcal{N}$ a neighborhood of $(0, 0, 0)$, with $h(0, 0, 0) = 0$, such that for each μ, h is a homeomorphism that exhibits the topological equivalence of their flows.

A deformation $\left\{\tilde{f}(\mu), \tilde{g}(\mu)\right\}$ defined on $R^n \times R^m \times R^p$ is *induced* by a deformation $\{f(\gamma), g(\gamma)\}$ defined on $R^n \times R^m \times R^q$ if there is a continuous change of parameters $\phi : R^q \to R^p$, $\gamma = \phi(\mu)$, $0 = \phi(0)$, such that $\left\{\tilde{f}(\mu), \tilde{g}(\mu)\right\} = \{f(\phi(\mu)), g(\phi(\mu))\}$. A deformation $\{f(\gamma), g(\gamma)\}$ with q parameters is *versal* if

Table 6.2 Examples of Unfoldings for Dynamic Bifurcations

Name	Miniversal Unfolding
Saddle-node	$\dot{x} = \mu - x^2 + O\left(\lvert x\rvert^3\right)$
Hopf	$\dot{x} = \begin{bmatrix} \mu & -1 \\ 1 & \mu \end{bmatrix} x + \left(x_1^2 + x_1^2\right)\left(a\begin{bmatrix} x_1 \\ x_2 \end{bmatrix} + b\begin{bmatrix} x_1 \\ x_2 \end{bmatrix}\right) + O\left(\lvert x\rvert^5\right)$
Cusp, R	$\dot{x} = \mu_0 + \mu_1 x - x^3 + O\left(\lvert x\rvert^4\right)$
Cusp, R^2	$\dot{x} = \begin{bmatrix} 0 \\ \mu_0 \end{bmatrix} + \begin{bmatrix} \mu_1 & 1 \\ 0 & 0 \end{bmatrix} x + \begin{bmatrix} a \\ b \end{bmatrix} x_1^2 + O\left(\lvert x\rvert^3\right)$
Generalized Hopf	$\dot{x} = \begin{bmatrix} \mu_1 & -1 \\ 1 & \mu_1 \end{bmatrix} x + \mu_2\left(x_1^2 + x_2^2\right) x + a\left(x_1^2 + x_2^2\right)^2 x + O\left(\lvert x\rvert^7\right)$

every other deformation is equivalent to one induced by it, and *miniversal* (sometimes called *universal* [81]) if q is the smallest number of parameters needed to define a versal deformation.

If $\{f, g\}$ is structurally unstable, then a deformation of it is said to "unfold the singularity" and a deformation is often referred to as an *unfolding*. Versal deformations or unfoldings are important because they reveal all possible behaviors that might be observed in perturbations of $\{f, g\}$. Miniversal unfoldings are especially significant because they do this with a minimum number of parameters. The dimension of the γ-space (R^q) is a measure of the degeneracy of the singularity. It follows from analysis of the miniversal unfolding $\{f(\gamma), g(\gamma)\}$, that there exists a neighborhood of 0 in γ-space which is divided into open regions by surfaces of co-dimension 1 such that throughout each region $\{f(\gamma), g(\gamma)\}$ exhibits equivalent behavior. The surfaces across which the behavior changes are called bifurcation surfaces. These bifurcation surfaces can intersect, thereby defining (bifurcation) surfaces of higher co-dimension. The origin lies at an intersection of co-dimension q. This bifurcation is referred to as a singularity of co-dimension q. The miniversal unfoldings associated with the singularities of Table 6.1 are given in Table 6.2.

6.5.6 Deformations and Unfoldings in Other Contexts

The concept of deformation, described above for DAEs, is frequently applied to other mathematical objects. It is only necessary that it makes sense to speak of parameter-dependent families of those objects and to have an appropriate concept of equivalence. For example, in addition to DAEs and vector fields, we note the following:

Algebraic equations – the study of zeros of equations $f = 0$ under perturbations ([51, 78]). Two deformations of a map $f : R^n \times R^p \rightarrow R^n$, $f(x, \mu)$ and $\tilde{f}(x, \mu)$ are

Table 6.3 Bifurcations of Algebraic Equations up to Co-dimension 4

Name	Codim	Normal Form	Unfolding
Saddle-node [78, 81, 185]	1	x^2	$\mu_0 + x^2$
Cusp [78, 81, 185]	2	x^3	$\mu_0 + \mu_1 x + x^3$
Swallowtail [78, 81, 185]	3	x^4	$\mu_0 + \mu_1 x + \mu_2 x^2 + x^4$
Butterfly [78, 81, 185]	4	x^5	$\mu_0 + \mu_1 x + \mu_2 x^2 + \mu_3 x^3 + x^5$
Hilltop [81]	4	$\begin{bmatrix} x^2 - y^2 \\ 2xy \end{bmatrix}$	$\begin{bmatrix} \mu_0 + \mu_1 x + \mu_2 y + x^2 - y^2 \\ \mu_3 + 2xy \end{bmatrix}$
Hilltop [81]	4	$\begin{bmatrix} x^2 \\ y^2 \end{bmatrix}$	$\begin{bmatrix} \mu_0 + \mu_1 y + x^2 \\ \mu_2 + \mu_3 x + y^2 \end{bmatrix}$

locally equivalent at $(0, 0)$ if there exists a continuous, near-identity transformation of coordinates $x = h(y, \mu)$ defined on a neighborhood of $(0, 0)$, and a continuous, invertible matrix $S(y, \mu)$ such that $f(y, \mu) = S(y, \mu)\tilde{f}(h(y, \mu), \mu)$. Since S is invertible, the zeros of $f(y, \mu)$ correspond to those of $\tilde{f}(x, \mu)$ (in the domain of definition of h). Because of the relevance of this topic to the study of the solution structure of load flow equations, we will discuss the bifurcation of algebraic equations in some detail below.

Table 6.3 provides a summary of bifurcations of algebraic equations (called singularities) up to co-dimension 4. Bifurcations of algebraic equations are important to because it is often useful to study the equilibrium point structure and parametrically induced changes to it, separately from, or as an adjunct to, considering dynamical issues. Notice that the bifurcations of co-dimension less than 4 in Table 6.3 involve only one independent variable, whereas bifurcations of co-dimension 4 may involve two. Singularities involving two or more independent variables have not been completely classified – see the interesting discussions in [51], Chapter 7 and [81], Chapter 9. It is also useful to note that any singularity involving a single independent variable directly corresponds to a “catastrophe” which gives some additional interpretations [78, 81].

Matrices – the study of Jordan forms of matrices under perturbations [8]: Two deformations of an $n \times n$ matrix A, A_μ, and $\tilde{A}_\mu$ are equivalent if they are related by a near-identity similarity. transformation, itself dependent on the same parameters, $T(\mu)$ with $T(0) = I$.

Pencils – the study of the zero structure of linear systems under perturbations [25]: Two deformations of an $n \times m$ pencil $sA + B$, $sA_\mu + B_\mu$, and $s\tilde{A}_\mu + \tilde{B}_\mu$ are equivalent if they are related to a near-identity, strict equivalence transformation [76], i.e., $sA_\mu + B_\mu = P(\mu)\left[s\tilde{A}_\mu + \tilde{B}_\mu\right]Q^{-1}(\mu)$, with $P(0) = I$, $Q(0) = I$.

6.6 Numerical Computation

6.6.1 Static Bifurcation Points

Locating Bifurcation Points: Consider the load flow equations, rewritten here as

$$\begin{aligned} f(x, y, \mu) &= 0 \\ g(x, y, \mu) &= 0 \end{aligned} \tag{6.19}$$

It is convenient to consolidate these by joining x and y to create the N-vector X, $N = n+m$, and joining the functions f and g to create the function $F : R^N \times R^p \to R^N$, so that (6.19) becomes

$$F(X, \mu) = 0 \tag{6.20}$$

The goal now is to investigate the roots of F as a function of the parameters μ. Conventional methods for finding roots of algebraic equations, such as the *Newton–Raphson* (NR) method, can be modified for this purpose. The continuation (or homotopy) method [107] enables tracing a root from an initial solution pair X_0, μ_0. Another important tool, necessary when the solution structure (6.20) includes singular (bifurcation) points, is a modification of the Newton–Raphson method called the *Newton–Raphson–Seydel* (NRS) method [176]. Applications of these computational constructs have been widely applied in power systems, example continuation methods in [36, 48, 95, 105], and the NRS method in [3, 5, 17, 35, 39].

The basic idea of continuation is simple and has many applications. Consider a one-parameter family of mathematical problems $\mathcal{P}(\mu)$. Suppose solutions to $\mathcal{P}(\mu)$ are attainable by iteration, for any μ, provided a good initial estimate is available. Moreover, suppose that a solution to the problem $\mathcal{P}(\mu_0)$ is known, but it is desired to find a solution to $\mathcal{P}(\mu^*)$. The idea is to sequentially solve a sequence of problems $\mathcal{P}(\mu_0)$, $i = 0, \ldots, K$, terminating with $\mu_K = \mu^*$. Successive solutions are extrapolated to get a starting value for an iterative solution of the next problem.

Consider the application of this approach to (6.20). Suppose that a solution (X_0, μ_0) is known, i.e., $F(X_0, \mu_0) = 0$. We wish to find the solution X^* corresponding to the parameter value μ^*. To apply the continuation method divide the interval $[\mu_0, \mu^*]$ into a large number K of subintervals and generate successive solutions, X_i, $i = 1, \ldots, K$, as follows. To generate the first solution, X_1, take as a starting value $X_1^0 = X_0$ and then apply Newton's method (for example) to determine X_1. For subsequent starting values, X_i^0, $i > 1$, we could continue the zeroth-order extrapolation, i.e.,

$$X_i^0 = X_{i-1} \tag{6.21}$$

or use a linear extrapolation (called the secant method in [48])

$$X_i^0 = X_{i-1} + (X_{i-1} - X_{i-2}) \frac{\mu_i - \mu_{i-1}}{\mu_{i-1} - \mu_{i-2}} \tag{6.22}$$

followed by Newton's method.

Remark 6.10 Using a Differential Equation Solver. This method can be taken to the extreme so that $K \to \infty$, $\Delta\mu \to 0$, in which case the successive solutions are connected by the differential equation

$$F_X(X, \mu)\, dX + F_\mu(X, \mu)\, d\mu = 0$$

or

$$F_X(X, \mu) \frac{dX}{d\mu} + F_\mu(X, \mu) = 0 \tag{6.23}$$

In principle, an implicit ordinary differential equation solver can be used, with μ as the independent variable, to obtain the curve $X(\mu)$ for $\mu \in [\mu_0, \mu^*]$.

The computation method described, based on the NR iteration, breaks down at (static) bifurcation points, i.e., when F_X is singular (rank $(F_X) < N$). In generic one-parameter families, the dimension of ker (F_X) at a bifurcation point is precisely one (rank $(F_X) = N-1$). Thus to locate such a point, we seek values for $X \in R^N$, $\mu \in R^1$ and nontrivial $v \in R^N$ or $w \in R^N$ that satisfy

$$F(X, \mu) = 0 \tag{6.24}$$

$$F_X(X, \mu)\, v = 0 \quad \text{or} \quad w^T F_X(X, \mu) = 0 \tag{6.25}$$

along with the nontriviality requirement, that can be expressed

$$\|v\| = 1 \quad \text{or} \quad \|w\| = 1 \tag{6.26}$$

The NR approach can be applied to (6.24), (6.25) and (6.26) to locate bifurcation points. This is the NRS approach to locating static bifurcation points. Data that satisfies these equations will be denoted X_b, μ_b, v_b, w_b. Note that the vectors v_b, w_b have special significance. They are, respectively, the right and left eigenvectors corresponding to the zero eigenvalue of the Jacobian $J_b = F_X(X, \mu)$. The vector v_b spans the kernel of J_b. It identifies those dependent variables that play a role in the bifurcation. That is, only those elements of X that correspond to nonzero entries in v_b will exhibit the bifurcation behavior. w_b is orthogonal to all of the columns. It will be seen below to provide useful information in multi-parameter cases [64, 67].

Once a bifurcation point is located, it is feasible to modify it to locate points on the equilibrium surface that are near the bifurcation point by replacing (6.25) by

$$[F_X(X,\mu) - \lambda I]\, v = 0 \tag{6.27}$$

for values $\lambda \in [-\varepsilon_1, \varepsilon_2]$ with $\varepsilon_1, \varepsilon_2 > 0$. Thus, for a given λ, the $2N+1$ equations

$$\begin{aligned} F(X,\mu) &= 0 \\ [F_X(X,\mu) - \lambda I]\, v &= 0 \\ \|v\| &= 1 \end{aligned} \tag{6.28}$$

are to be solved for the $2N+1$ variables X, μ, v using the NR iteration. This approach is applied in the voltage stability toolbox [17].

Remark 6.11 The bifurcation point X_b, μ_b, v_b (corresponding to $\lambda = 0$ is a regular solution of (6.24), (6.25) and (6.26) so that the implicit function theorem guarantees existence of a unique solution $X(\lambda), \mu(\lambda), v(\lambda)$ of Equation (6.28) on a neighborhood of $\lambda = 0$.

Remark 6.12 There are variants of the continuation-NRS approach to improve computational speed for large power systems. See, for example [39].

Now consider the multi-parameter case. In the following discussion, it will be assumed that dim ker $[F_X(X,\mu)] = 1$. When the number of parameters is $p = 1, 2, 3$, this is the generic case (see the discussion [51], Chapter 7, and [81], Chapter 9). The strategy is to locate the bifurcation parameter value closest to a given parameter value μ_0 Since coordinates used to define μ may be arbitrarily shifted, let us take $\mu_0 = 0$. The problem of locating the closest static bifurcation point in multi-parameter power systems has been raised by several investigators including [66, 68, 102, 109, 137, 203].

A straightforward approach is minimized $\|\mu\|^2$ subject to the constraints (6.24), (6.25) and (6.26). Thus, to do this, combine the $2N+1$ constraints to form the constraint function

$$\psi(X, v, \mu) = \begin{bmatrix} F(X,\mu) \\ F_X(X,\mu)\, v \\ \|v\| - 1 \end{bmatrix} \tag{6.29}$$

and introduce the $2N+1$-dimensional vector of Lagrange multipliers, $\lambda \in R^{2N+1}$. Construct the Lagrangian

$$L(X, v, \mu, \lambda) = \|\mu\|^2 + \lambda^T \psi(X, v, \mu) \tag{6.30}$$

The necessary conditions for a minimum are obtained from setting the variation of L to zero, $\delta L(X, v, \mu, \lambda) = 0$:

$$\frac{\partial L}{\partial \lambda} = \psi(X, v, \mu) = 0 \tag{6.31}$$

$$\frac{\partial L}{\partial X} = \lambda^T \frac{\partial \psi}{\partial X} = 0 \tag{6.32}$$

$$\frac{\partial L}{\partial v} = \lambda^T \frac{\partial \psi}{\partial v} = 0 \tag{6.33}$$

$$\frac{\partial L}{\partial \mu} = 2\mu^T + \lambda^T \frac{\partial \psi}{\partial \mu} = 0 \tag{6.34}$$

Note that the last three equations can be written as

$$\Phi^T \lambda = b \tag{6.35}$$

where

$$\Phi = \left[\frac{\partial \psi}{\partial X} \; \frac{\partial \psi}{\partial v} \; \frac{\partial \psi}{\partial \mu} \right], \quad b = \left[0_{1\times N} \; 0_{1\times N} \; \mu^T \right]^T \tag{6.36}$$

Let Φ^* denote the right inverse of Φ[3] and obtain from (6.35) the relations

$$\lambda = \Phi^* b \tag{6.37}$$

$$\left[I - \Phi^* \Phi \right] b = 0 \tag{6.38}$$

Notice that $[I - \Phi^* \Phi]$ has only $k - 1$ linearly independent rows. Having eliminated λ, the problem is now completely characterized by (6.38) along with (6.31). Thus, the dimension of the system of equations to be solved has increased from $2N + 1$ in the one-parameter case to $2N + k$. Points (X, v, μ) that satisfy the enlarged system are the extremal bifurcation points. One way to compute them using numerical iteration is given by the following algorithm. Let the parameter vector μ be expressed $\mu = me$, where m is a scalar and e is a unit vector that specifies a direction in parameter space.

Static Bifurcation Search Algorithm

Step 1, Initialize: Choose an initial search direction $e_0 \in R^k$ and initial values $x_0 \in R^N$, $v_0 \in R^N$, $\mu_0 \in R^k$. Set $i = 1$.
Step 2, Iterate: While $\|e_i - e_{i-1}\| > \varepsilon$ and $i \leq i_{max}$

Step a, Solve: Recursively solve $\psi(X, v, me_{i-1}) = 0$ for (X_i, v_i, m_i) starting with $(X_{i-1}, v_{i-1}, m_{i-1})$ using the NRS method.
Step b, Update Direction: Update the search direction e_{i-1} to e_i in order to reconcile $[I - \Phi^* \Phi] b = 0$. There are various ways to do this. See for example [66, 68, 115, 137].
Step c, Update Index: Set $i \to i + 1$ and return to Step 2.

---

[3] For a generic function $\psi : R^{2N+k} \to R^{2N+1}$, the solution set of $\psi = 0$ is a smooth $k-1$-dimensional manifold and rank $\Phi = 2N + 1$, its maximum rank on the set [10], Section 31. Thus, for a generic family Φ^* exists on a neighborhood of any solution of $\psi = 0$.

Computing Unfoldings—The Lyapunov–Schmidt Reduction: When a bifurcation point is located, the next step is to determine the normal form and its unfolding. Once again, consider the map $F : R^N \times R^p \to R^N$. and suppose, for convenience, that $(0, 0) \in R^N \times R^p$ and it is a bifurcation point. Using the method of Lyapunov–Schmidt, the study of the zeros of $F(X, \mu)$ can be reduced to the study of the zeros of the so-called (reduced) bifurcation equation, which typically involves only a few dependent variables. The essentials of the method of Lyapunov–Schmidt will be briefly reviewed in a somewhat simplified version that is adequate for our immediate needs. A more general construction will be discussed below when addressing the Hopf bifurcation [51, 81].

Define $J = F_X(0, 0)$ so that

$$F_X(X, \mu) = JX + N(X, \mu) \tag{6.39}$$

where

$$N(0, 0) = 0, \quad N_X(0, 0) = 0 \tag{6.40}$$

Now, the goal is to study the solution set of

$$JX + N(X, \mu) = 0 \tag{6.41}$$

in a neighborhood of (0, 0).

If J has rank r, then there exist full-rank matrices L, R of dimension $N \times r$ and $r \times N$, respectively, such that $J = LR$. Furthermore, L has a left inverse L^* and R has a right inverse R^*. Now, the space R^N can be decomposed,

$$R^N = \ker\left(R^{*T}\right) \times \text{image}\left(R^*\right)$$

Let U be a matrix with $N - r$ columns that span $\ker\left(R^{*T}\right)$.[4] Then new coordinates $X \mapsto (u, v)$ can be defined via the transformation

$$X = Uu + R^*v \tag{6.42}$$

Proposition 6.13 *Let W be a matrix with $N - r$ columns that span* $\ker L^*$, *then (6.41) is equivalent to*

$$v + L^*N\left(Uu + R^*v, \mu\right) = 0 \tag{6.43}$$

$$W^T N\left(Uu + R^*v, \mu\right) = 0 \tag{6.44}$$

where $u \in R^{N-r}$ and $v \in R^r$ are new coordinates as defined by (6.42).

---

[4]Take $R^* = R^T\left(RR^T\right)^{-1}$, then $\ker R^{*T} \sim \ker R$. Similarly, if $L^* = \left(L^T L\right)^{-1} L^T$ then $\ker L^* \sim \ker L^T$.

Proof This is a well-known result that is easily obtained by substituting (6.42) into (6.41) and then premultiplying by the nonsingular matrix

$$T = \begin{bmatrix} L^* \\ W^T \end{bmatrix}$$

Note that nonsingularity follows from the fact that $R^N = \ker R^{*T} \times \text{image } R^*$. ∎

Applying the implicit function theorem to (6.43), it follows that there is a unique, smooth function $v^*(u, \mu)$ defined n a neighborhood of (0, 0) that satisfies it. Thus, on a neighborhood of (0, 0) (6.43) becomes

$$f(u, \mu) = W^T N \left(Uu + R^* v^*(u, \mu), \mu \right) = 0 \tag{6.45}$$

Equation (6.45) is referred to as the *reduced bifurcation equation.*

The number of independent equations represented by (6.44) is $\tilde{r} = n - r$. It is easily verified that

$$f(0, 0) = 0, \quad f_u(0, 0) = 0 \tag{6.46}$$

The zero structure of the family $F(X, \mu) = 0$ is completely characterized near the singular point $(X, \mu) = (0, 0)$ by the zero structure of $f(u, \mu)$ near the point $(u, \mu) = (0, 0)$. The goal now is to investigate the zero structure of $f(u, \mu)$ near its singular point (0, 0). One approach is to determine if the family $f(u, \mu)$ is equivalent to its universal unfolding. If so, all of the properties $f(u, \mu)$ can be inferred by from an established catalog of unfoldings. To make this determination, we seek an appropriate (near-identity) transformation $(u, \mu) \mapsto (z, \gamma)$, which reduces $f(u, \mu)$ to $\varphi(z, \gamma)$ such that on a neighborhood of (0, 0), the zeros of φ coincide with those of f, the parameter γ is of minimum dimension q, and φ is a polynomial of some degree in z. Thus, the characterization of the zeros of the function $F(X, \mu)$ is ultimately reduced to the study of the much simpler problem

$$\varphi(z, \gamma) = 0, \quad \varphi : R^{\tilde{r}} \times R^q \to R^{\tilde{r}} \tag{6.47}$$

In the case of $\tilde{r} = 1$, the following theorem is well known.

Proposition 6.14 *Suppose the reduced bifurcation equation (6.45) is smooth, and the first nonvanishing derivative with respect to u is $r \geq 2$;*

$$D_u^i f(0, 0) = 0, \ldots i = 0, \ldots, r-1 \quad \text{and} \quad D_u^r f(0, 0) \neq 0$$

Then

1. *there exists a smooth change of variables $z(u, \mu) : R \times R^p \to R$ and $\gamma(\mu) : R^p \to R^q$ so that (6.45) is reducible to the following polynomial on a neighborhood of $(u, \mu) = (0, 0)$.*

$$\gamma_0(\mu) + \gamma_1(\mu) z + \cdots + \gamma_{r-2}(\mu) z^{r-2} + z^r$$

2. *if in addition,* $\partial\gamma(0)/\partial\mu$*, is of full rank, then (6.45) is locally equivalent to the universal unfolding of* f *at* $(0, 0)$

$$\varphi(z, \gamma) = \gamma_0 + \gamma_1 z + \cdots + \gamma_{r-2} z^{r-2} + z^r \tag{6.48}$$

Proof The proposition is a variant of well-known results [51], Section 6.8, and [12], Section 4.1. Conclusion 1) follows from the Malgrange preparation theorem and 2) from the implicit function theorem. ■

Remark 6.15 The condition that $\partial_\gamma(0)/\partial\mu$ is of full rank is sometimes called a *genericity condition*. If this condition obtains, then the original family is *richly parameterized* in the sense that variation of the original parameters can produce all possible zero patterns that can be generated by small perturbations of . around the singular point. If $\partial_\gamma(0)/\partial\mu$ is degenerated, then even small changes in the model $F(X, \mu)$ are likely to alter the co-dimension of the singularity and change the way in which bifurcations associated with the singularity manifest themselves. We say the singularity is generic or nongeneric, respectively, if the genericity condition is satisfied or not satisfied.

In applications, the transformations are typically approximated using the Taylor expansion of $f(u, \mu)$ [51]. Sequentially compute[5]

$$\alpha_i(\mu) = D_u^i f(0, \mu) = \lim_{X \to 0} D_X^i \left[w_0^T F(X, \mu) \right], \quad i = 0, \ldots, r \tag{6.49}$$

stopping at the index $i = r$ corresponding to the first nonzero $\alpha_i(0)$, i.e.,

$$\alpha_0(0) = 0, \ldots, \alpha_{r-1}(0) = 0, \alpha_r(0) \neq 0 \tag{6.50}$$

The following proposition provides the required construction.

Proposition 6.16 *Suppose* $f(u, \mu)$ *is smooth and (6.50) holds. Then* $f(u, \mu) = 0$ *is equivalent to* $g(z, \mu) = 0$*, where*

$$g(z, \mu) = \gamma_0(\mu) + \gamma_1(\mu) z + \cdots + \gamma_{r-2}(\mu) z^{r-2} + z^r + O\left(|z|^{r+1}\right) \tag{6.51}$$

and

$$z = u - \frac{\alpha_{r-1}(\mu)}{\alpha_r(\mu)} + O\left(\|\mu\|^2\right) \tag{6.52}$$

$$\gamma_i(\mu) = r! \frac{\delta_i(\mu)}{\alpha_r(\mu)}, \quad i = 0, \ldots, r-2 \tag{6.53}$$

[5]In the following computation, with $\tilde{r} = 1$, W has a single column denoted w_0.

$$\delta_i(\mu) = \sum_{j=i}^{r} \frac{\alpha_j(\mu)}{j!} \binom{j}{i} \left(\frac{-\alpha_{r-1}(\mu)}{\alpha_r(\mu)} \right)^{j-i} + O\left(\|\mu\|^{r-i+1}\right) \tag{6.54}$$

Proof Direct computation. ■

Remark 6.17 Notice that the unfolding given above describes a bifurcation of co-dimension $r-1$. The bifurcation surfaces in the unfolding parameter (γ) space are simply $\gamma_i = 0, \quad i = 0, \ldots, r-2$. In the physical parameter (μ) space, the surfaces are defined by $\gamma_i(\mu) = 0, \quad i = 0, \ldots, r-2$. Each of these $r-1$ functions defines a co-dimension one surface. These surfaces intersect to define higher co-dimension manifolds, the highest co-dimension being $r-1$ which is the intersection of all of them.

6.6.2 *Hopf Bifurcation*

As will be discussed below, Hopf bifurcations are frequently encountered in power system dynamics. Hopf bifurcations can be analyzed using the same basic tools as static bifurcations, namely the Newton–Raphson–Seydel method and the Lyapunov–Schmidt reduction. The required modifications will now be discussed.

Consider the DAE (6.1) under the assumption that the equilibrium point being sought is causal. Thus, it is necessary to locate a causal solution of the equilibrium equations (6.20), repeated here

$$F(X, \mu) = 0$$

which has the property that the linearized dynamics of (6.1) has a pair of purely imaginary conjugate roots, which implies

$$\det\left[F_X(X, \mu) - j\omega I_{n,m}\right] = 0, \quad I_{n,m} = \text{diag}\left(I_{n\times n}, 0_{m\times m}\right) \tag{6.55}$$

As in the case of static bifurcation, it is preferable to reformulate (6.55) as

$$\left[F_X(X, \mu) - j\omega I_{n,m}\right]\nu = 0, \quad \|\nu\| = 1 \tag{6.56}$$

In summary, when seeking Hopf bifurcation points the necessary conditions can be expressed $\psi(X, \nu, \mu) = 0$, where

$$\psi(X, \nu, \mu) = \begin{bmatrix} F(X, \mu) \\ \left[F_X(X, \mu) - j\omega I_{n,m}\right]\nu \\ \|\nu\| - 1 \end{bmatrix} \tag{6.57}$$

– a form that is comparable to (6.29) of the static case.

There are a number of computational approaches to locating solutions satisfying (6.57). A straight forward approach, especially convenient in the one-parameter case is to compute the equilibrium surface and then seek those points in which the linearized dynamics have eigenvalues on the imaginary axis [15, 16, 123]. Variants of the Newton–Raphson–Seydel method provide a direct approach to locating Hopf bifurcation points [169, 170].

As in the case of static bifurcation, it is important to identify the key qualitative characteristics associated with Hopf bifurcation, example the number and stability characteristics of period trajectories, as well as to characterize the bifurcation surfaces in the physical parameter space. There are two basic approaches: compute the center manifold and normal form and study the unfolding of the normal form (see Tables 6.1 and 6.2), or use a variant of the Lyapunov–Schmidt reduction in which a reduced bifurcation equation needs to be computed, and from which the desired information can also be obtained. Hopf bifurcations are studied from the former point of view in [12, 83] and from the latter in [51, 81].

Consider the Lyapunov–Schmidt Reduction. In dealing with Hopf bifurcations, a more general formulation of the Lyapunov–Schmidt reduction that applies to mappings on infinite dimensional spaces is required. Let $\mathcal{H}$ and $\mathcal{Y}$ be complete linear (possibly infinite) vector spaces and suppose $\Phi(x, \mu)$ is a map $\Phi : \mathcal{H} \times R^p \to \mathcal{Y}$. Furthermore, assume $(0, 0) \in \mathcal{H} \times R^p$ satisfies the necessary conditions for a bifurcation point, Equation (6.57). The goal is to characterize the solution set of $\Phi(x, \mu) = 0$ (in this slightly more generalized context) around a Hopf bifurcation located at (0, 0). As before, write

$$\Phi(x, \mu) = Lx + N(x, \mu) = 0, \quad L = D_x F(0, 0) \tag{6.58}$$

Let $\mathcal{P} : \mathcal{H} \to \mathcal{H}$ and $\mathcal{Q} : \mathcal{Y} \to \mathcal{Y}$ denote projection operators[6] with $\mathrm{Im}\mathcal{P} = \ker \mathcal{L}$ and $\mathrm{Im}\mathcal{Q} = \mathrm{Im}\mathcal{L}$. Then, both $\mathcal{H}$ and $\mathcal{Y}$ can be divided into direct sums

$$\mathcal{H} = \mathrm{Im}\mathcal{P} \oplus \mathrm{Im}(I - \mathcal{P}) \tag{6.59}$$

$$\mathcal{Y} = \mathrm{Im}\mathcal{Q} \oplus \mathrm{Im}(I - \mathcal{Q}) \tag{6.60}$$

The following is a well-known theorem.

Proposition 6.18 *Let $\mathcal{P}$ and $\mathcal{Q}$ be defined as above. Then*

1. (6.58) is equivalent to

$$Lv + QN(u + v, \mu) = 0 \tag{6.61}$$

$$(I - Q)\,N(u + v, \mu) = 0 \tag{6.62}$$

---

[6] Recall that if $\mathcal{H} = \mathcal{R} \otimes \mathcal{S}$, then there is a map $\mathcal{Q} : \mathcal{H} \to \mathcal{H}$ such that for each $X = R + S$, $\mathcal{Q}X = R$. Clearly, $\mathrm{Im}\mathcal{Q} = \mathcal{R}$ and $\mathrm{Ker}\mathcal{Q} = \mathcal{S}$. $\mathcal{Q}$ is the projection on $\mathrm{Im}\mathcal{R}$ along $\mathrm{Ker}\mathcal{Q}$ if $\mathcal{Q}^2 = \mathcal{Q}$.

with

$$z = u + v, u = Px \in \mathrm{Im}P, v = (I - P)x \in \mathrm{Im}(I - P) \tag{6.63}$$

2. *There exists a linear map $K : Im\mathcal{Q} \rightarrow Im(I - \mathcal{P})$ called the right inverse of L such that $LK = I$ on $Im\mathcal{Q}$ and $KL = I - \mathcal{P}$ on $\mathcal{H}$ so that (6.59) is equivalent to*

$$v + KQN(u + v, \mu) = 0 \tag{6.64}$$

Proof proof See [51, 81]. ■

Once again, the implicit function theorem applied to (6.64) assures the existence of a unique function $v^*(u, \mu)$ defined on a neighborhood of $(0, 0) \in H \times R^p$ that satisfies it. Moreover, $v^*(0, 0) = 0$. Substituting this function in (6.62) yields the reduced bifurcation equation

$$(I - Q)N\left(u + v^*(u, \mu), \mu\right) = 0 \tag{6.65}$$

In applications below the subspaces, $\mathrm{Im}P$ of $\mathcal{H}$ and $\mathrm{Im}(I - Q)$ of $\mathcal{Y}$ will be of finite dimension $\tilde{r}$, in which case (6.66) reduces to $\tilde{r}$ equations in $\tilde{r}$ unknowns.

As shown in [81], the above theorems provide a generalization of the Lyapunov–Schmidt method that can be applied to obtain the reduced bifurcation equation once the Hopf bifurcation point has been located. For convenience, suppose the necessary conditions for a Hopf bifurcation point (6.57) are satisfied at the point $(X, \mu) = (0, 0)$ and that the frequency is $\omega = 1$. Now rescale via the transformation $s = (1 + \tau)t$ is a new parameter introduced to adjust the timescale so as to keep the frequency at unity as parameters vary. In terms of s, the system dynamical equations can be written

$$\Phi(X, \mu, \tau) = -(1 + \tau)I_{n.m}\frac{dX}{dt} + JX + N(X, \mu) = 0 \tag{6.66}$$

We seek a 2π-periodic solution of (6.66). To proceed, consider $\mathcal{H}$ and $\mathcal{Y}$ to be spaces of smooth 2π-periodic functions of s and Φ is a map from $\mathcal{H} \times R^p \times R$ to $\mathcal{Y}$. Its linear part is defined by the operator $L : \mathcal{H} \rightarrow \mathcal{Y}$ defined by

$$L = \Phi_X(0, 0, 0) = -I_{n.m}\frac{d}{dt} + J \tag{6.67}$$

so that (6.66) can be written

$$\Phi(X, \mu, \tau) = -(1 + \tau)LX + N(X, \mu) = 0 \tag{6.68}$$

To compute the reduced bifurcation equation, it is necessary to determine a basis for each of the 2-D subspaces $\mathrm{Im}P$ of $\mathcal{H}$ and $\mathrm{Im}(I - Q)$ of $\mathcal{Y}$. Recall that $\mathrm{Im}P = \ker L$, and note that the linear equation $LX = 0$ has two 2π-periodic solutions,

$$w_1(s) = \mathrm{Re}\left(e^{js}w\right), \quad w_2(s) = \mathrm{Im}\left(e^{js}w\right) \tag{6.69}$$

where w is a complex eigenvector that satisfies

$$\left[-jI_{n,m} + J\right] w = 0 \tag{6.70}$$

Similarly, $\text{Im}\,(I - Q) = \ker L^*$ where the adjoint map L^* is

$$L^* = jI_{n,m} + J^T \tag{6.71}$$

Let w^* denote the eigenvector of L^* associated with eigenvalue j so that the two basis vectors for $\text{Im}\,(I - Q)$ are

$$w_1^*(s) = \text{Re}\left(e^{js} w^*\right), \quad w_2^*(s) = \text{Im}\left(e^{js} w^*\right) \tag{6.72}$$

Now, write $u = w_1(s)\, u_1 + w_2(s)\, u_2$ and consider the inner product

$$\phi_i(u_1, u_2, \mu, \tau) = \left\langle w_i^*(s), \tilde{N}\left(w_1(s)\, u_1 + w_2(s)\, u_2, v^*(u_1, u_2, \mu, \tau), \mu, \tau\right)\right\rangle \quad i = 1, 2 \tag{6.73}$$

where

$$\tilde{N}(u, v, \mu, \tau) = -\tau I_{n,m} \frac{d}{ds}(u + v) + N(u + v, \mu) \tag{6.74}$$

and the inner product is defined as

$$\langle v_1, v_2 \rangle = \frac{1}{2\pi} \int_o^{2\pi} v_1^T(s)\, v_s(s)\, ds$$

The following proposition establishes the form of (6.73).

Proposition 6.19 *The reduced bifurcation equation has the form*

$$\phi(u_1, u_2, \mu, \tau) = p\left(u_1^2 + u_2^2, \mu, \tau\right) \begin{bmatrix} u_1 \\ u_2 \end{bmatrix} + q\left(u_1^2 + u_2^2, \mu, \tau\right) \begin{bmatrix} -u_1 \\ u_2 \end{bmatrix} \tag{6.75}$$

where the scalar functions $p(a, \mu, \tau), q(a, \mu, \tau)$

$$p(0, 0, 0) = 0, \quad \frac{\partial}{\partial a} p(0, 0, \tau) = 0 \tag{6.76}$$

$$q(0, 0, 0) = 0, \quad \frac{\partial}{\partial a} q(0, 0, \tau) = -1 \tag{6.77}$$

Proof [81], see Proposition 2.3 in Chapter 8. ■

It follows from (6.75) that nontrivial solutions of the reduced bifurcation equation ($\phi = 0$) exist only if $p = q = 0$. The form of p and q suggests a transformation to polar coordinates: $u_1 = \beta \cos\theta$, $u_2 = \beta \sin\theta$ so that solutions are defined by

$$p\left(\beta^2, \mu, \tau\right)\beta = 0, \quad q\left(\beta^2, \mu, \tau\right)\beta = 0 \tag{6.78}$$

These equations should be viewed as defining β and τ. In view of (6.62), the implicit function theorem provides that $q\left(\beta^2, \mu, \tau\right) = 0$ can be solved for $\tau = \tau\left(\beta^2, \mu\right)$ which leaves us with the requirement that

$$g(\beta, \mu) = p\left(\beta^2, \mu, \tau\left(\beta^2, \mu\right)\right)\beta = r\left(\beta^2, \mu\right)\beta = 0 \tag{6.79}$$

where

$$r\left(\beta^2, \mu\right) = p\left(\beta^2, \mu, \tau\left(\beta^2, \mu\right)\right) \tag{6.80}$$

Solutions of $r(z, \mu) = 0$ with $z > 0$ are in one-to-one correspondence with the nontrivial periodic solutions of (6.79). The function $r(z, \mu)$ can be approximately computed using a combination of Fourier and power series expansions. This can be accomplished using the above formulas, but there are many variants [37, Section 9.4]. Another alternative is based on the *frequency domain* formulation of the Hopf theory [122, 147, 152].

The existence and number of solutions of (6.78), or (6.79), near the origin can be investigated in terms of the unfolding parameters of Proposition 6.14. However, it is also important to identify the stability of periodic solutions. Stability depends on the sign of the first nonzero derivative $g(\beta, \mu) = r\left(\beta^2, \mu\right)\beta$ with respect to β, or equivalently, its unfolding. The sign can be preserved by introducing the parameter $\varepsilon = \pm 1$, then normal forms associated with $r(z, \mu)$ are as follows:

Normal form: εz^k
Unfolding: $\gamma_0 + \gamma_1 z + \cdots + \gamma_{k-2} z^{k-2} + \varepsilon z^k$
for $k \geq 2$, $\varepsilon = \pm 1$

and hence for $g(\beta, \mu)$

Normal form: $\varepsilon \beta^{2k-1}$
Unfolding: $\gamma_0 \beta + \gamma_1 \beta^3 + \cdots + \gamma_{k-2} \beta^{2k-3} + \varepsilon \beta^{2k-1}$
for $k \geq 2$, $\varepsilon = \pm 1$

Stability is established by the following proposition:

Proposition 6.20 *Suppose that* $(\beta, \mu) = (0, 0)$ *is an equilibrium point with a simple pair if imaginary eigenvalues,* $\lambda_{1,2} = \pm i$, *and with all other eigenvalues having negative real parts. Then, the periodic solution corresponding to* (β, μ) *near* $(0, 0)$ *and satisfying* $g(\beta, \mu) = 0$ *is stable if* $D_\beta g(\beta, \mu) > 0$ *and unstable if* $D_\beta g(\beta, \mu) < 0$.

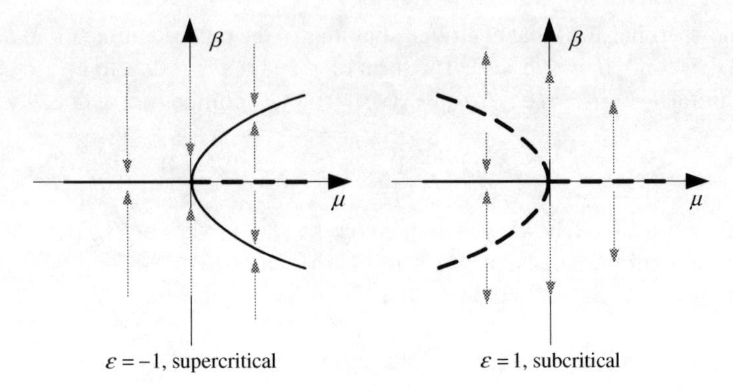

Fig. 6.3 Bifurcation diagrams for the co-dimension 1 supercritical and subcritical Hopf bifurcations. $\beta = 0$ corresponds to the equilibrium point. Stable equilibria and orbits are indicated by solid lines. Unstable by dashed

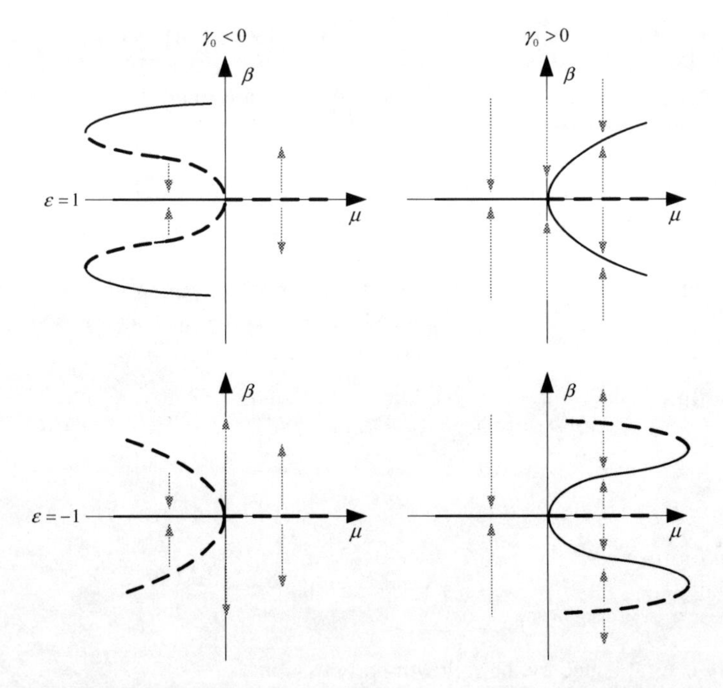

Fig. 6.4 Bifurcation diagrams for the co-dimension 2 ($k = 3$) Hopf bifurcations. Stable equilibria and orbits are indicated by solid lines. Unstable by dashed. The adjustable parameter is $\mu = \gamma_1$. Each diagram plots the oscillation amplitude a versus μ. Note that in some regions there are two periodic trajectories, one stable and one unstable

Proof This proposition follows from the results of [81], Section 8.5, and [51], Section 9.5. ■

Consider the case of the co-dimension one bifurcation, i.e., $k = 2$. There are two cases, $\varepsilon = \pm 1$. $\mu = \gamma_0$ is the adjustable parameter. A pictorial depiction of the Hopf bifurcation is provided by the bifurcation diagrams of Figure 6.3. The oscillation amplitude β is plotted versus the parameter μ. Observe that in the supercritical case, the stable equilibrium becomes unstable as μ changes from negative to positive and a stable periodic trajectory emerges. In the subcritical case, the periodic trajectory is unstable and exists for negative values of μ. For applications, this is a crucial distinction. In the supercritical case, even though the origin is unstable after bifurcation, trajectories are still attracted into its vicinity if the oscillation amplitude is small. The subcritical case can be very dangerous because the unstable limit cycle bounds the domain of attraction of the stable origin and this domain shrinks as the bifurcation point is approached from the left.

Graphical representations of the co-dimension 2 ($k = 3$) Hopf bifurcations are shown in Figure 6.4.

6.7 Applications

Static Bifurcation and Voltage Stability: By the early 1980s, various factors had caused power system operators to seek maximum utilization of the transmission network. Previously unobserved stability-related difficulties emerged and a vigorous effort was made to understand them and find remedies [96]. Typically, these problems involved the inability to maintain load bus voltage magnitudes and became referred to as voltage instabilities. Static bifurcation of the load flow equations is an appealing formalization of voltage stability because it embodies key characteristics associated with real-voltage instability events: the presence of multiple equilibria, a relationship between dynamic stability and voltage collapse, a high degree of sensitivity of certain bus variables to control parameters, and a relatively long period of drift prior to collapse. Of course, static bifurcation intrinsically involves multiple equilibria so the following remarks are confined to the latter three points.

Static Bifurcation and Dynamic Stability: Bifurcating equilibria can be (asymptotically) stable in the sense of Lyapunov. We have seen examples of this. This type of stability is not meaningful in a practical sense because even an arbitrarily small perturbation of the dynamics can render it unstable in the sense of Lyapunov or can annihilate the equilibrium point altogether. A practical concept of stability of an equilibrium point requires that the flow is locally structurally stable at the equilibrium point and that the equilibrium point is stable in the sense of Lyapunov. The term *practical stability*[7] is used for equilibria that satisfy these conditions. Recall that Theorem 6.6 states that local structural stability implies that the equilibrium

[7] Practical stability as used here is related to but not identical with the notion of practical stability in [127].

point is causal and hyperbolic. Thus practical stability of an equilibria point implies that it is causal and hyperbolic. But for a causal, hyperbolic equilibria (x^*, y^*, μ^*) Lyapunov stability can be ascertained from the ordinary differential (6.11) which has linearization

$$\delta \dot{x} = A\, \delta x \tag{6.81}$$

$$A = \left[D_x f - D_y f\left(D_y g\right)^{-1} D_x g\right]^* \tag{6.82}$$

where the right-hand side is evaluated at the equilibrium point as denoted by "*". The equilibrium point is stable in the sense of Lyapunov only if all of the eigenvalues of A have nonpositive real parts, and hyperbolicity further restricts the eigenvalues to the open left-hand plane. Thus, practical stability implies exponential stability.

It can easily be shown that static bifurcation points are not stable in the sense of practical stability. Practical stability implies causality which allows us to apply Schur's formula to write the determinant of the load flow equations at the bifurcation point

$$\det\left[J^*\right] = \det\left[A\right] \det\left[\left(D_y g\right)^*\right] \tag{6.83}$$

From this, it is seen that a (causal) static bifurcation point corresponds to $\det[A] = 0 = \det[-A]$. Hence, static bifurcation points are not stable in the sense of practical stability, since the latter requires $\det[-A] > 0$.

Venikov et al. [192] recognized the significance of a degeneracy in J with respect to the steady-state stability of a power system. They observed that under certain conditions a change the sign of $\det[J]$ during a continuous variation of system states and parameters coincides with the movement of a real characteristic root of the linearized swing equations across the imaginary axis into the right half of the complex plane. Thus, they recommended tracking det[J] during load flow calculations and proposed a modification of Newton's method which allows precise determination of the parameter value where such a sign change occurs. Tamura et al. [181] discuss some computational experience using this method.

Bus Variable Sensitivities: A discussion of the significance of a degeneracy in J in terms of bus variable sensitivities was given by Abe et al. [1]. It can be seen from the Lyapunov–Schmidt analysis, that in the (u, v) coordinates, the sensitivities of the v variables are well behaved. Simply differentiate the Equation (6.43) to obtain (at (0,0))

$$D_\mu v\,(0,0) = -L^* D_\mu N\,(0,0)$$

However, derivatives $D_\mu u$ are indeterminant at the bifurcation point - (0, 0). These are the sensitive variables. It follows that the sensitive physical variables can be identified from the basis vectors for ker J. Specifically, components of X that correspond to nonvanishing elements in the basis vectors are highly sensitive to parameter variations-indeed their sensitivities to parameter change tend to infinity as the parameter value approaches its bifurcation value. The magnitudes of the elements in the basis vectors are useful for identifying the relative participation of the system-dependent variables.

Slow Timescale Behavior: Dobson and Chiang [65] discuss the dynamical behavior associated with a saddle-node bifurcation. At a generic saddle-node bifurcation point, the equilibrium point has a one-dimensional center manifold. Motion on the center manifold is such that trajectories beginning on it from one side of the equilibrium point approach the equilibrium point, and trajectories on the other side diverge from it. The convergent trajectory, of course takes infinite time to reach the equilibrium, and the rate of divergence of the divergent trajectory approach zero near the equilibrium. Trajectories that begin off the center manifold but near it exhibit similar behavior-a slow approach followed by a slow divergence. Post-bifurcation behavior is similar as well even though there is no longer an equilibrium point, its "fingerprint" is clearly evident.

Example 6.21 Example 6.3 Revisited. Consider, once again, the three-bus network of Figure 6.1 along with the governing equations (6.12), rewritten here as

$$\dot{\theta} = \omega$$

$$\dot{\omega} = -V\left(\sin\left(\theta - \phi\right) + \sin\left(\phi\right)\right) + \Delta P$$

$$0 = V\left(\sin\left(\phi\right) + \sin\left(\phi - \theta\right)\right) - P_3$$

$$0 = -V\left(\cos\left(\phi\right) + \cos\left(\phi - \theta\right)\right) + (2 - B)\,V^2 - Q_3$$

Equilibrium requires that θ is constant and the load flow equations are satisfied, i.e.,

$$\begin{aligned} 0 &= -V\left(\sin\left(\theta - \phi\right) + \sin\left(\phi\right)\right) + \Delta P \\ 0 &= V\left(\sin\left(\phi\right) + \sin\left(\phi - \theta\right)\right) - P_3 \\ 0 &= -V\left(\cos\left(\phi\right) + \cos\left(\phi - \theta\right)\right) + (2 - B)\,V^2 - Q_3 \end{aligned} \tag{6.84}$$

With the parameters $(\Delta P, B, P_3, Q_3)$ specified, it is possible to solve for the variables (θ, ϕ, V). Figure 6.5 shows three curves generated using a combination of NR and NRS methods. The variable V is plotted versus the parameter P_3 for three different values of Q_3 with ΔP and B set to zero. Voltage collapse occurs at the nose of each curve which is a bifurcation point.

If the parameter vector is $\mu = (\Delta P, B, P_3, Q_3) = (0, 0, -1, 0)$, it is easily verified that $(\theta^*, \phi^*, V^*) = \left(0, -\pi/4, 1/\sqrt{2}\right)$ satisfies (6.84), and it is therefore an equilibrium point. This is indeed the nose of the solid curve in Figure 6.5. The Jacobian of the load flow equations is

$$J = \begin{bmatrix} -V\cos\left(\phi - \theta\right) & V\left(\cos\left(\phi - \theta\right) - \cos\left(\phi\right)\right) & -\sin\left(\theta - \phi\right) - \sin\left(\phi\right) \\ -V\cos\left(\phi - \theta\right) & V\left(\cos\left(\phi - \theta\right) + \cos\left(\phi\right)\right) & -\sin\left(\theta - \phi\right) + \sin\left(\phi\right) \\ \sin\left(\theta - \phi\right) & -V\left(\sin\left(\theta - \phi\right) - \sin\left(\phi\right)\right) & 2\,(2 - B)\,V - \cos\left(\phi - \theta\right) - \cos\left(\phi\right) \end{bmatrix}$$

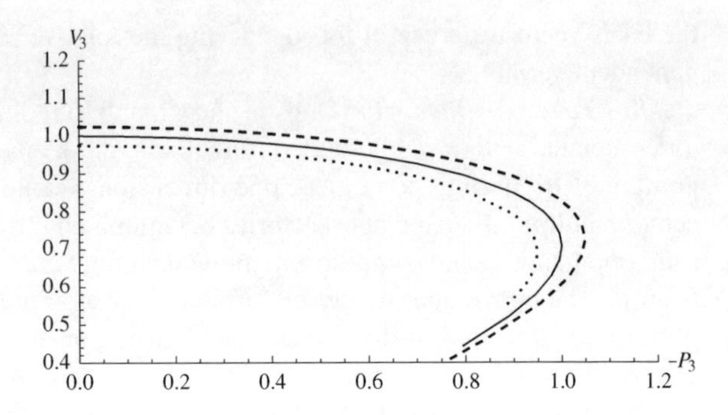

Fig. 6.5 Load bus voltage. V as a function of load real power, P_3 with $\Delta P = 0$, $B = 0$, and three different values of Q_3: 1) dotted, $Q_3 = -0.05$; 2) solid, $Q_3 = 0$; and 3) dashed, $Q_3 = 0.05$

which evaluates at the equilibrium point to

$$J = \begin{bmatrix} -1/2 & 0 & 0 \\ -1/2 & 1 & -\sqrt{2} \\ 1/2 & -1 & \sqrt{2} \end{bmatrix}$$

Inspection of the last two rows shows the matrix to be degenerate. Thus, the equilibrium point is a static bifurcation point. Moreover, the null space is of dimension one, and the null space spanning vector is $v = \begin{bmatrix} 0 & \sqrt{2} & -1 \end{bmatrix}^T$. From this, it can be concluded that the behavior around the bifurcation point under variation of the parameter vector μ can be observed in the variables ϕ or V, but not in θ.

In terms of Proposition 6.16, direct construction of the reduced bifurcation equation and Taylor expansion shows that $r = 2$ (the bifurcation is of co-dimension 1) and the unfolding (see Equation (6.51)) is

$$g(z, \mu) = \gamma_0(\mu) + z^2$$

with

$$\gamma_0(\mu) = -\frac{\Delta P}{4} - \frac{B}{4}\left(1 + \frac{B}{4}\right) + \frac{P_3 - 1}{2} + \frac{Q_3}{2}$$

Notice that the equations are *richly parameterized* with respect to this bifurcation point, as mentioned above, because the single parameter, γ_0 can be changed by manipulating any one of the four physical parameters.

Example 6.22 Four-Bus Example:Higher-Order Singularities and Load Models. Consider the example from [123] that consists of three generators feeding a single-load bus with combined constant admittance and PQ as shown in Figure 6.6.

With only the constant admittance load, the load bus can be treated as an internal bus and eliminated in the usual way by defining a reduced network admittance matrix. Uniform damping is assumed. The translational symmetry in the three resulting second-order differential equations allows reduction to two equations by specifying bus one as a swing bus and defining the relative angles

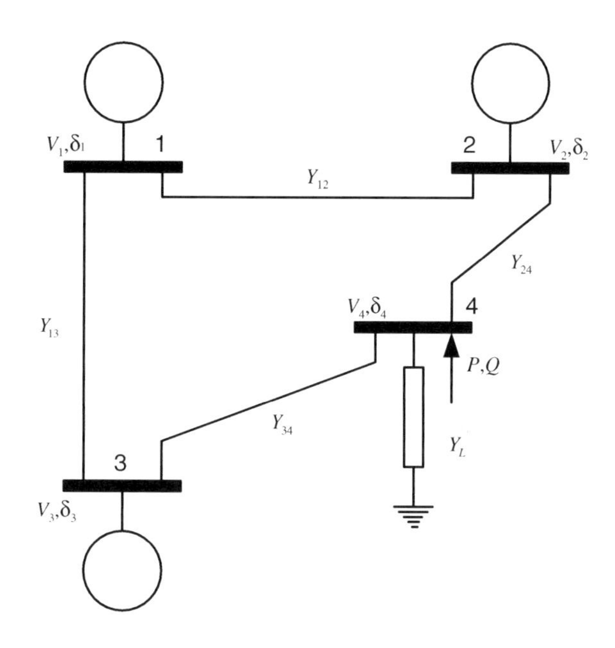

Fig. 6.6 A four-bus network with three generators feeding a single complex load

$$\theta_1 = \delta_2 - \delta_1, \quad \theta_2 = \delta_3 - \delta_1$$

The reduced bus admittance matrix is symmetric and composed of elements

$$y_{ik} = g_{ik} + jb_{ik}, \quad i, k = 1, 2, 3$$

The resulting equations are

$$\begin{aligned}\ddot{\theta}_1 + \gamma\dot{\theta}_1 + 2b_{12}\sin\theta_1 + b_{23}\sin(\theta_1 - \theta_2) + b_{12}\sin\theta_2 \\ +g_{23}\cos(\theta_1 - \theta_2) - g_{13}\cos\theta_2 = \Delta P_1\end{aligned} \tag{6.85}$$

$$\begin{aligned}\ddot{\theta}_2 + \gamma\dot{\theta}_2 + 2b_{13}\sin\theta_2 + b_{23}\sin(\theta_2 - \theta_1) + b_{12}\sin\theta_1 \\ +g_{23}\cos(\theta_1 - \theta_2) - g_{12}\cos\theta_1 = \Delta P_2\end{aligned} \tag{6.86}$$

where $\Delta P_1 = P_2 - P_1$ and $\Delta P_2 = P_3 - P_1$, P_1, P_2, P_3 represent, respectively, the real power supplied by generators 1,2, and 3.

Tavora and Smith [183, 184] analyze the equilibrium solution structure of these equations without transfer conductances as a function of the two parameters ΔP_1 and ΔP_1. They show that the parameter space partitions into regions and within each region the number of equilibrium solutions is 0, 2, 4, or 6.[8] The system studied by

[8] Another approach to the study of equilibrium point structure of lossless power systems is given by Baillieul and Byrnes in [20].

Tavora and Smith is obtained from (6.85), (6.86) by ignoring transfer conductances $g_{ik} = 0$ and setting $b_{ik} = 1$. The equilibrium equations are

$$\begin{aligned} 2\sin\theta_1 + \sin(\theta_1 - \theta_2) + \sin\theta_2 - \Delta P_1 = 0 \\ 2\sin\theta_2 + \sin(\theta_2 - \theta_1) + \sin\theta_1 - \Delta P_2 = 0 \end{aligned} \quad (6.87)$$

There are several bifurcation points that appear as cusps in the $\Delta P_1 - \Delta P_2$ plane. One of these corresponds to $\theta_1 = \pi/2$, $\theta_2 = \pi$, $\Delta P_1 = 1$, $\Delta P_2 = 2$. The static bifurcation search algorithm locates this point precisely from appropriate initial conditions. Next, the Jacobian, J in (6.39) is computed

$$J = \begin{bmatrix} 0 & -1 \\ 0 & -2 \end{bmatrix} = \begin{bmatrix} -1 \\ -2 \end{bmatrix} \begin{bmatrix} 0 & 1 \end{bmatrix}$$

From this, it is clear that the kernel of J has dimension one ($\tilde{r} = 1$). Now compute

$$L = \begin{bmatrix} -1 \\ -2 \end{bmatrix}, R = \begin{bmatrix} 0 & 1 \end{bmatrix} \Rightarrow U = \begin{bmatrix} 1 \\ 0 \end{bmatrix}, W = \begin{bmatrix} -2 & 1 \end{bmatrix}$$

The objective now is to compute the reduced bifurcation equation by applying the construction of Proposition 6.16. The first step is to compute the α_i's of Equations (6.49) and (6.50). In this case, it is found that it is not possible to find a k such that $\alpha_k(0) \neq 0$. This is a co-dimension ∞ bifurcation, indicating the high degree of symmetry and degeneracy associated with this idealized system.

Now, we consider a case with transfer conductances, retaining $B_{ij} = 1$, set $C_{ij} = 0.15$. A bifurcation point is found with $\theta_1 = 1.64962$, $\theta_2 = 3.13274$, $\Delta P_1 = 1.16961$, $\Delta P2 = 2.03570$. The sequence of α coefficients are determined to be $\alpha_0(0) = 0.000$, $\alpha_1(0) = 0.000$, $\alpha_2(0) = 0.000$, $\alpha_3 = 0.026$, indicating a singularity of co-dimension $r - 1 = 2$. Thus, the reduced bifurcation equation is of the form (see Proposition 6.16)

$$g(z, u) = \gamma_0(\mu) + \gamma_1(\mu) z + z^3 + O\left(|z|^4\right) \quad (6.88)$$

The parameter sensitivity matrix can also be obtained

$$\frac{\partial \gamma(\mu^*)}{\partial \mu} = \begin{bmatrix} 167.22 & -100.22 \\ 0 & 0 \end{bmatrix}, \quad \gamma = \begin{bmatrix} \gamma_0 & \gamma_1 \end{bmatrix}, \quad \mu = \begin{bmatrix} \Delta P_1 & \Delta P_2 \end{bmatrix} \quad (6.89)$$

Notice that although the singularity is of co-dimension 2, and hence, it is generic in two parameter families, the particular two parameter family defined by (6.85), (6.86) does not represent a versal unfolding of the singularity. This means that not all changes in the equilibrium point structure that might be induced by small perturbations of the equations will be observed by varying the parameters ΔP_1, ΔP_2.

Consider a combination of a (lossless) constant admittance load and a constant power load on bus 4. In this case, the load bus must be retained, but the translational symmetry allows specification of a swing bus. Assume the transmission lines themselves are lossless and the generator terminal bus voltages are regulated so that $V_1 = V_2 = V_3 = 1$. The resultant set of differential-algebraic equations take the form

$$\begin{aligned}
&\ddot{\theta}_1 + \gamma\dot{\theta}_1 + 2b_{12}\sin\theta_1 + b_{24}V_4\sin(\theta_1 - \theta_3) + b_{13}\sin\theta_2 = \Delta P_1 \\
&\ddot{\theta}_2 + \gamma\dot{\theta}_2 + 2b_{13}\sin\theta_2 + b_{34}V_4\sin(\theta_2 - \theta_3) + b_{12}\sin\theta_1 = \Delta P_2 \\
&b_{42}V_4\sin(\theta_3 - \theta_1) + b_{43}V_4\sin(\theta_3 - \theta_2) - P_4 = 0 \\
&b_{42}V_4\cos(\theta_3 - \theta_1) + b_{43}V_4\cos(\theta_3 - \theta_2) - b_{44}V_4^2 + Q_4 = 0
\end{aligned} \tag{6.90}$$

where again, $\Delta P_1 = P_2 - P_1$, $\Delta P_2 = P_3 - P_1$.

Consider the case in which $b_{12} = b_{13} = 1$, $b_{24} = b_{34} = 2$, $b_{44} = -4$, and $b_{ij} = b_{ji}$. Computation identifies a singular point at $\theta_1 = 1.5709005$ $(\approx \pi/2)$, $\theta_2 = 3.1415926$ $(\approx \pi)$, $\theta_3 = 2.3562465$, $V_4 = 0, 70714361$, $\Delta P_1 = 1$, $\Delta P_2 = 2$, $P_4 = 0$, $Q_4 = 0$. The null space spanning vector of J is

$$v = \begin{bmatrix} 0.8528 & 0.0000 & 0.4264 & 0.3015 \end{bmatrix}^T$$

and the sequence of α-coefficients is $\alpha_0(0) = 0.000$, $\alpha_1(0) = 0.000$, $\alpha_2(0) = 0.000$, $\alpha_3(0) = 0.372$, indicating a singularity co-dimension 2. The parameter sensitivity matrix is

$$\frac{\partial\gamma(\mu^*)}{\partial\mu} = \begin{bmatrix} -8.596 & 4.298 & -2.149 & -6.446 \\ 0 & 0 & 0 & 0 \end{bmatrix}$$

$$\mu = \begin{bmatrix} \Delta P_1 & \Delta P_2 & P_4 & Q_4 \end{bmatrix}$$

Once again, the family defined by (6.90) is not versal since variation of the parameters μ does not show all possible variations of the singular point. It is easy to vary that will always be the case when the parameters enter the equations as an affine translation.

Finally, consider voltage-dependent reactive loading. It has long been recognized that the constant admittance and constant power load models represent only crude approximations to actual load behavior and that more precise load models may be necessary for accurate characterization of voltage stability issues. Proposed refinements include static representation of frequency and voltage dependence and, in some cases, the inclusion of load dynamics. Such modifications are readily accommodated within the bifurcation framework. In the present instance, we wish to show that even a simple expansion of the load model can have a significant qualitative effect. To see this, replace the constant reactive load with the voltage-dependent model

$$Q_4(V_4) = kV_4$$

with constant k. By so doing, it is determined that the singular point identified above remains a singular point, the null space spanning vector and the co-dimension are unchanged. However, the parameter sensitivity matrix is

$$\frac{\partial \gamma(\mu^*)}{\partial \mu} = \begin{bmatrix} -8.596 & 4.298 & -2.149 & -11.005 \\ 0 & 0 & 0 & -1.944 \end{bmatrix}$$

$$\mu = \begin{bmatrix} \Delta P_1 & \Delta P_2 & P_4 & k \end{bmatrix}$$

The significance of this result is that the system with the voltage-dependent load represents a versal unfolding of the singularity.

Hopf and Generalized Hopf Bifurcation: Oscillations associated with instability in the power systems are well known and frequently described in an extensive literature which spans many decades. The connection with bifurcation analysis was made in the 1980s. Van Ness et al. [190] suggest that an observed oscillation is associated with a Hopf bifurcation. Abed and Varaiya [2] illustrate subcritical Hopf bifurcations in several electric power system models. Alexander [4] provides a thorough local stability analysis of Hopf bi furcations for a model of two machines connected with a lossy transmission line and demonstrates the occurrence of both subcritical and supercritical Hopf bifurcations. An example of Hopf bifurcation in a three-machine classical network with lossy lines is given by Kwatny et al. [123, 122]. Another example is given by Rajagopalan et al. [166] in which a three-machine system is modeled with a two-axis representation and excitation is included. Iravani and Semlyen [99] show Hopf bifurcation in a single machine system with a flexible turbine generator shaft.

Chen and Varaiya [42] illustrate a degenerate Hopf bifurcation in a two-generator network with excitation and an infinite bus. Venkatasubramanian et al. [194] illustrate a different type of degenerate Hopf (double zero eigenvalues) in a single-line network including a generator with voltage control (either excitation or a thyristor-controlled reactance) and a constant power load. Subcritical and supercritical Hopf bifurcations are naturally encountered in studies of chaos in power systems [49, 197].

One could justifiably conclude that Hopf bifurcations are pervasive in power system dynamics.

Example 6.23 Example 6.22 Revisited. Return now to the system of Example 6.22 with constant admittance loads as characterized by Equations (6.85) and (6.86). Hopf bifurcations occur in this system as described in [122, 123]. First, focus on the static bifurcation point at $(\theta_1, \theta_2, \Delta P_1, \Delta P_2) = (1.549, 0.759, 4.261, 2.794)$. The Lyapunov–Schmidt reduction indicates that this bifurcation has co-dimension 1 and can be efficiently observed in the variable θ_1 as the parameter ΔP_1 is varied, as shown in Figure 6.7. The bifurcation point noted above is a classical saddle-node bifurcation. Thus, stability of the equilibrium solution changes from stable to unstable at the fold. It should be emphasized that this example has constant admittance loads and the system therefore reduces to a system of ordinary differential equations, so the saddle-node is typical for a fold in the equilibrium manifold. This is not the case in more power systems with general load models.

The curve in Figure 6.7 is extended by decreasing ΔP_1. By computing the eigenvalues at equilibria along this curve, a dynamic bifurcation point is encountered at

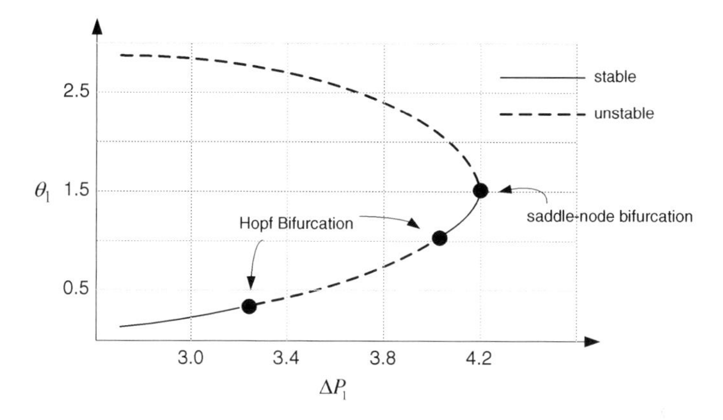

Fig. 6.7 A portion of the equilibrium surface shows θ_1 as a function of ΔP_1

$\Delta P_1 = 4.042$ with damping parameter $\gamma = 0.214$. Another bifurcation is encountered by further decreasing the parameter ΔP_1. Both are supercritical Hopf bifurcations, so that a stable limit cycle emerges as ΔP_1 increases from 3.244, and as ΔP_1 decreases from 4.042. The above is information summarized in the bifurcation diagram of Figure 6.7.

Chapter 7
Elements of Power Systems Control

"When you have a pair of interesting matrices study the pencil that they generate, or even the algebra."
—Olga Taussky, "How I Became a Torchbearer for Matrix Theory"

7.1 Introduction

Control of voltage, frequency and load is central to power network operation. The following sections review the basics of these systems. Each generator has two basic controllers, a frequency (or load frequency) controller and a voltage controller. The two are generally analyzed independently. The primary voltage control system involves using the generator excitation system to change the field voltage in order to regulate the generator terminal bus voltage. This problem is treated first in Section 7.2. The load frequency control problem has two goals: 1) regulate the electrical frequency (equivalently, rotor speed) supplied by each generator to synchronous frequency, and 2) insure that the total real power supplied by the generators is distributed among them in accordance with a specified set of *distribution factors*. This problem is treated in Section 7.3. Section 7.4 addresses the multi-area control problem of regulating frequency along with the power flow exchanges between control areas, known as automatic generation control (AGC). The integration of AGC with economic dispatch of generation within each area is considered in Section 7.4.3.

7.2 Primary Voltage Control

The central issue is control of voltage throughout the power network. This comes down to the problem of controlling the voltage on a selection of network buses – usually a large number.

H.G. Kwatny and K. Miu-Miller, *Power System Dynamics and Control*,
Control Engineering, DOI 10.1007/978-0-8176-4674-5_7

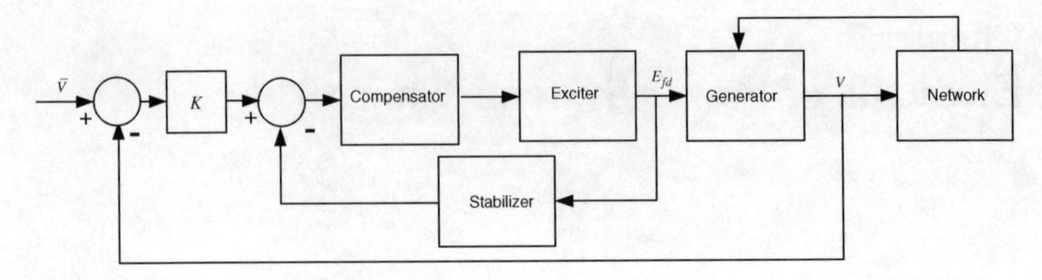

Fig. 7.1 A generic block diagram for an excitation voltage control system shows the key components and their interconnections

7.2.1 *Excitation Systems*

A first step is the control of voltage on the terminal bus of each generator. This is typically accomplished using an *excitation* system associated with each generator. The exciter itself supplies a (variable) DC voltage to the generator field winding. The exciter itself can be thought of as some form DC generator. Associated with the excitor are various control components such as the compensator and stabilizer as shown in Figure 7.1.

For the purposes of voltage control, it is assumed that frequency is constant. The generator model of Section 4.4.2 will be employed attached to a constant admittance load. A single-line diagram illustrating the configuration is shown in Figure 7.2. Note that the two impedance variables z_{12}, z_{22} convert to admittance variables

$$y_{12} = -j/X_d, \quad y_{22} = \frac{R_L}{R_L^2 + X_L^2} - j\frac{X_L}{R_L^2 + X_L^2}$$

and the bus admittance matrix can be written

$$Y = \begin{bmatrix} Y_{11} & Y_{12} \\ Y_{21} & Y_{22} \end{bmatrix} = \begin{bmatrix} y_{12} & -y_{12} \\ -y_{12} & y_{12} + y_{22} \end{bmatrix}$$

Fig. 7.2 Single-line depiction of a generator feeding a constant admittance load

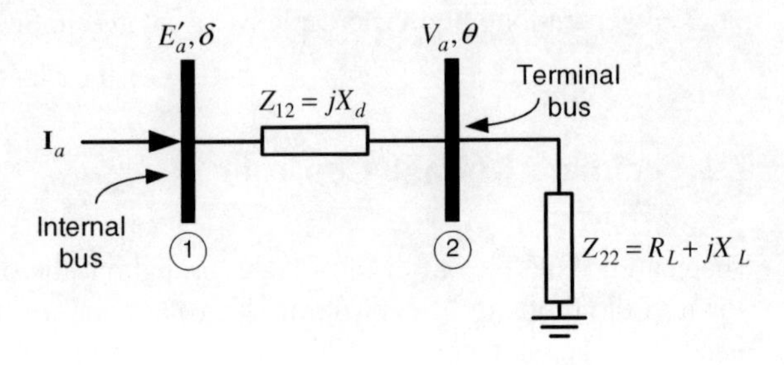

The bus current injection relation is

$$\begin{bmatrix} \mathbf{I}_a \\ 0 \end{bmatrix} = Y \begin{bmatrix} \mathbf{E}'_a \\ \mathbf{V}_a \end{bmatrix}$$

and the second equation yields

$$\mathbf{V}_a = \frac{y_{12}}{y_{12} + y_{22}} \mathbf{E}'_a$$

Substituting the admittance values produces

$$\mathbf{V}_a = \left(\frac{R_L^2 + X_L (X_L - X_d)}{R_L^2 + (X_L - X_d)^2} + j \frac{R_L X_d}{R_L^2 + (X_L - X_d)^2} \right) \mathbf{E}'_a \tag{7.1}$$

from which

$$V_a = k E'_a, \quad k = \sqrt{\frac{R_L^2 + X_L^2}{R_L^2 + (X_L - X_d)^2}} \tag{7.2}$$

$$\theta = \delta - \arctan \left(\frac{R_L X_d}{R_L^2 + (X_L - X_d)^2} \right) \tag{7.3}$$

Now, recall Equation (4.96), rewritten here

$$\tau_f \frac{dE'_a}{dt} = -E'_a - \sigma V_a \cos\theta + (1 + \sigma) E_{fd}$$

In the feedback analysis of the excitation system, the term $\sigma V_a \cos\theta$ can be treated as a disturbance, or decoupled completely by redefining the field input voltage. In either case, the voltage controller feedback system is reduced to the block diagram shown in Figure 7.3. Conventional linear models are used for the various components.

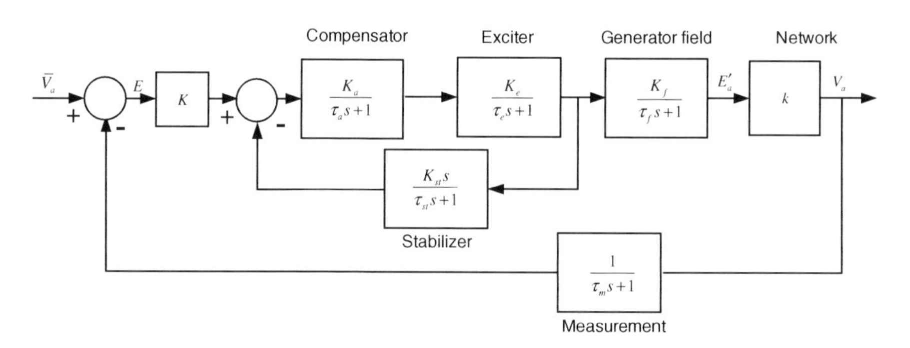

Fig. 7.3 Block diagram of the excitation system

A little block diagram algebra yields the voltage command to error ($\bar{V}_a \rightarrow E$) transfer function:

$$G_e(s) = \frac{1}{1 + \frac{K K_a K_e K_f k}{((\tau_a s+1)(\tau_e s+1)(\tau_f s+1)+K_a K_e K_{st} s)(\tau_m s+1)}} \tag{7.4}$$

Now, the *Final Value Theorem* asserts [156], providing the closed loop is stable, that the ultimate state error in response to a step command is

$$e(\infty) = \lim_{s \to 0} s G_e(s) \frac{1}{s} \tag{7.5}$$

Thus,

$$e(\infty) = \frac{1}{K K_a K_e K_f k} = \frac{1}{K_{dc}} \tag{7.6}$$

The first observation is that the error is inversely proportional to the loop zero frequency (or DC) gain – as anticipated for a type zero system. Consequently, it is desired to achieve as high a gain as possible without losing stability. The purpose of the stabilizer is to increase the achievable gain. Note, by the absence of k_{st} in (7.6), that the stabilizer does not directly affect the loop DC gain because of the presence of "s" in the stabilizer numerator. Nevertheless, it has a very profound effect. To see this, take the numerical example with system time constants $\tau_a = 0.05$, $\tau_m = 0.075$, $\tau_e = 1.0$, $\tau_f = 5.0$. The controller design parameters are K, k_{st} and τ_{st}. The parameter K will establish the DC gain, which is ultimately limited by stability. Now, we will fix $\tau_{st} = 5$ and consider various values of k_{st} along with its effect on the stability bound on the DC gain.

The value $k_{st} = 0$ corresponds to the absence of the stabilizer. The root locus for this case is shown in Figure 7.4. The maximum achievable gain is $K_{dc} = 51.534$. Table 7.1 shows a range of k_{st} values up to $k_{st} = 100$. The root locus for this case is shown in Figure 7.5. Obviously, the stabilizer can be very effective for increasing the bound on the DC gain and thereby improving voltage regulator accuracy.

Example 7.1 Hybrid Electric Ship Power System. Consider the system described in Appendix A. The no-load admittance matrix is given in Equations (A.1) and (A.2). Identical resistive loads on buses 7 through 12 can accommodated with Equation (A.3). Assume the system is operating with only generators 1 and 2 connected, the resistive loading totals 0.60 p.u., and there is a 0.15 p.u. constant real power propulsion load on bus 3. Each of the two generators is equipped with identical voltage controllers as depicted above in Figure 7.3, with the following parameters. The time constants are $\tau_a = 0.2$, $\tau_e = 0.314$, $\tau_f = 1$, $\tau_{st} = 1$. The measurement delay is neglected, $\tau_m = 0$. The gains are $K = 1000$, $K_a = 1$, $K_e = 1$, $K_f = 5$, $K_{st} = 500$.

Selected responses from an arbitrary initial state near the equilibrium point are shown in Figures 7.6, 7.7, and 7.8. Notice that Figure 7.6 shows the field voltage at the internal buses of generators 1 and 2. These, of course, are the control variables

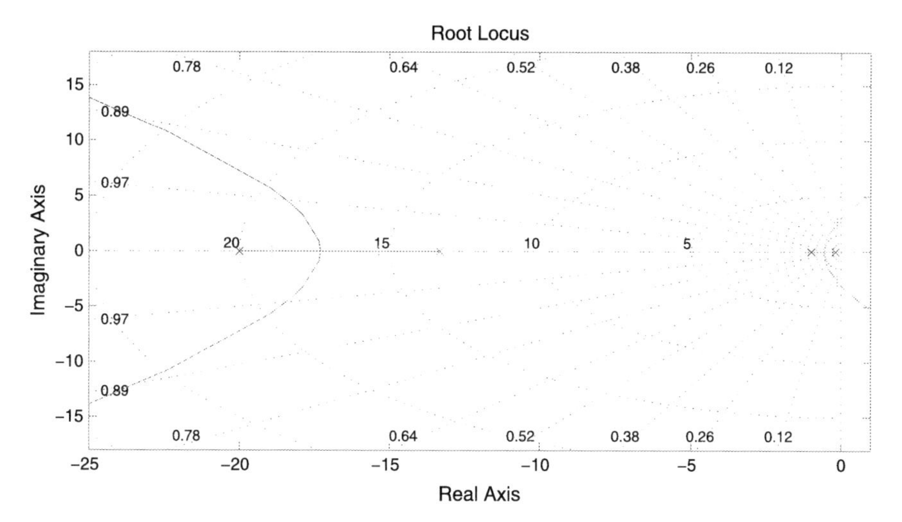

Fig. 7.4 Root locus for the unstabilized system ($k_{st} = 0$). The upper bound for DC gain is $K_{dc} = 51.534$

Table 7.1 Upper stability bound of K_{dc} for various values of stabilizer gain, k_{st}

k_{st}	0	1	10	50	100
K_{dc}	51.534	65.244	144.356	679.67	1621.7

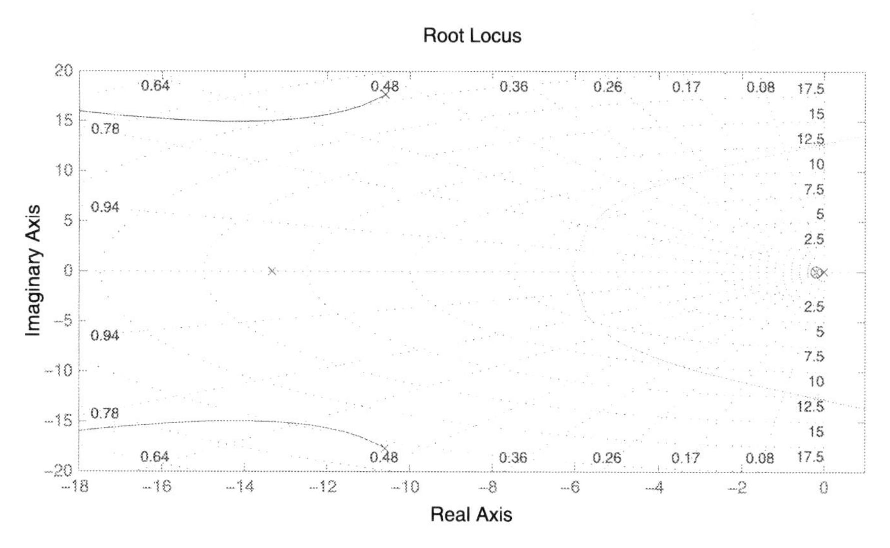

Fig. 7.5 Root locus for the stabilized system with $k_{st} = 100$. The upper bound for DC gain is $K_{dc} = 1621.7$

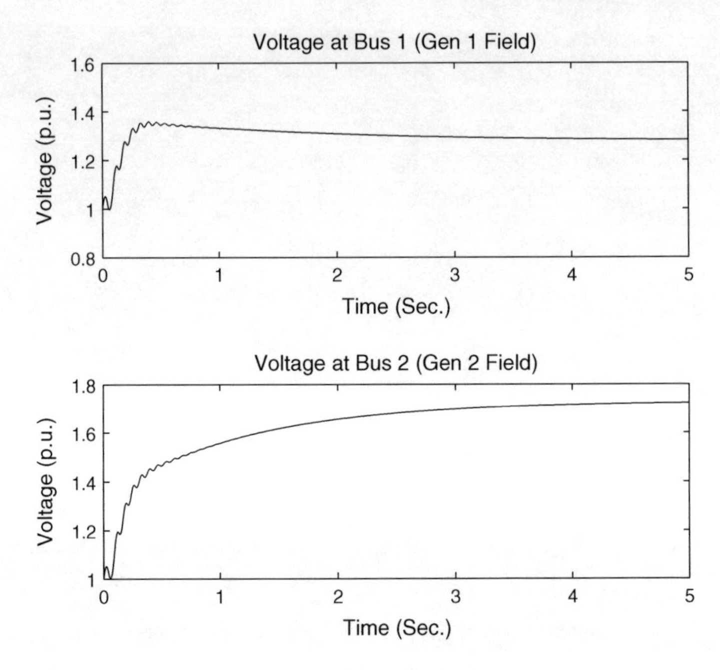

Fig. 7.6 The field voltages of generators 1 and 2 are shown

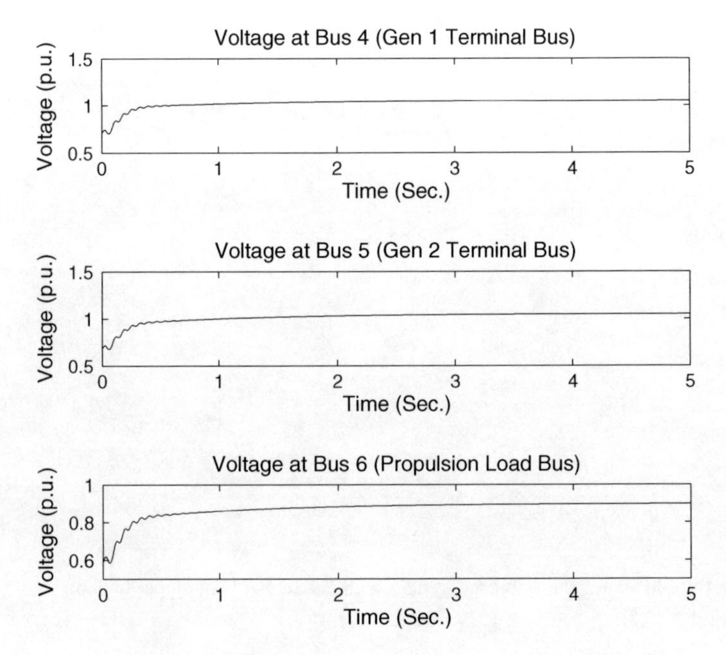

Fig. 7.7 The top two figures show the generator 1 and generator 2 terminal voltages while the bottom figure illustrates the voltage on bus 3, a load bus

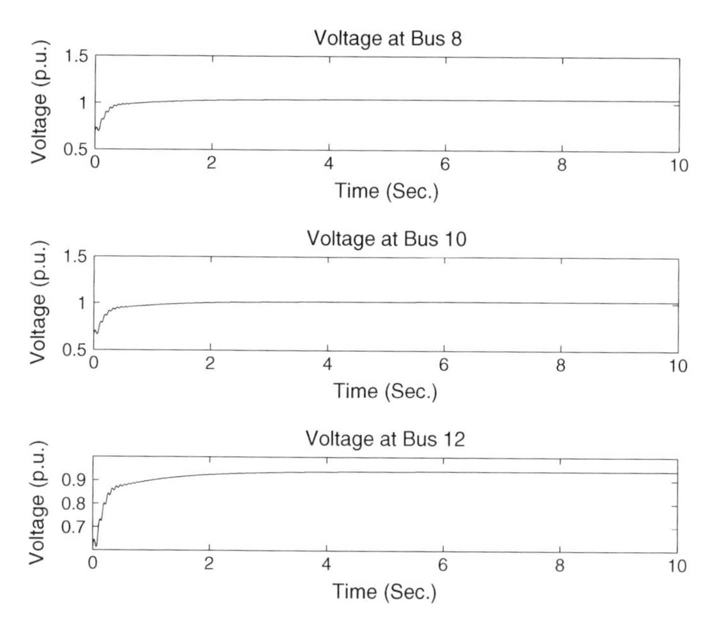

Fig. 7.8 The voltage transients on load buses 8, 10, and 12

that are being manipulated to control the terminal bus voltages for the two generators. These controlled bus voltages are shown in the first two figures of Figure 7.7. The generator field voltage is variation is intended to regulate the associated generator terminal bus voltage (in this case, 1.05 p.u. is the target voltage for both terminal buses. The ultimate intent is to keep all load bus voltages close to 1 p.u.

The third graph of Figure 7.7 shows the voltage on bus 3 which is the load bus that supplies constant power to the electric propulsion drive. In other configurations, it is the terminal bus of generator 3. In this example, generator 3 is disconnected. The bus voltage is low, ultimately reaching a value of just over 0.9 p.u. Figure 7.8 shows load buses 8, 10, and 12. Bus 8 reaches a steady-state voltage of about 1.05 p.u.—a bit high, bus 10 achieves about 1.01 p.u. and bus 12 is somewhat low at 0.95 p.u. In this example, the system is loaded symmetrically so it is no surprise that buses 7, 9, and 11 mirror 8,10, and 12.

7.3 Load Frequency Control

The central goal of load frequency control (LFC) is to balance generation and load. In this context, "load" should be understood as the total power consumed by the totality of electrical devices and the transmission/distribution network together at the synchronous frequency, ω_s. "Generation" means the total power delivered to the network by all power supply components. The control of frequency and power supply is inextricably connected as the typical load element's power consumption is frequency dependant.

Consider the case where power is supplied by steam turbine-driven generators. Then in order to change the generator power output, it is necessary to change the mechanical power into the turbine generator rotor which is accomplished by opening or closing the throttle valve, thereby changing the flow rate of steam into the turbine. The device that regulates the flow of steam into the turbine is referred to as a *speed governor* or simply a governor. The earliest speed governors were purely mechanical and contained both a speed-sensing mechanism and a valve actuator. The centrifugal governor, patented by James Watt in 1787, is such a device. Such a governor, which moves the valve in proportion to the speed error, gives rise to a *proportional* control compensator. It was the now well-known deficiencies of proportional control that inspired Maxwell to invent the proportional plus integral compensator described in his pioneering paper "On Governors" [143].

Consider a system composed of two generators that feed a network of constant admittance loads. If the voltage behind reactance generator model is employed, then regardless of the size or complexity of the network it can be reduced to a two-bus representation where the two buses correspond to the generator internal buses. The network is then characterized by the 2×2 reduced bus admittance matrix:

$$Y_{red} = \begin{bmatrix} g_{11} & g_{12} \\ g_{21} & g_{22} \end{bmatrix} + i \begin{bmatrix} b_{11} & b_{12} \\ b_{21} & b_{22} \end{bmatrix} \tag{7.7}$$

The admittance matrix, Y_{red}, depends, of course, on the load. An example of this dependency is shown in Table 7.2.

The dynamics of the system are described by the differential equations

$$\begin{aligned} \dot{\theta}_1 &= \omega_1 \\ \dot{\theta}_2 &= \omega_2 \\ M\dot{\omega}_1 + D\omega_1 &= P_{m1} - g_{11}V_1^2 - g_{12}V_1V_2\cos(\theta_1 - \theta_2) - b_{12}V_1V_2\sin(\theta_1 - \theta_2) \\ M\dot{\omega}_2 + D\omega_2 &= P_{m2} - g_{22}V_2^2 - g_{21}V_1V_2\cos(\theta_1 - \theta_2) - b_{21}V_1V_2\sin(\theta_2 - \theta_1) \end{aligned} \tag{7.8}$$

Table 7.2 Reduced bus admittance matrix as a function of load (pu)

Load	g_{11}	g_{11}	g_{11}	g_{11}	b_{11}	b_{11}	b_{11}	b_{11}
1.10	0.006605	0.008415	0.008415	0.019477	0.410675	0.234430	0.234430	0.644844
0.75	0.004920	0.005631	0.005631	0.014342	0.410769	0.234612	0.234612	0.645206
0.60	0.004197	0.004437	0.004437	0.012137	0.410798	0.234669	0.234669	0.645323
0.45	0.003474	0.003241	0.003241	0.009931	0.410820	0.234715	0.234715	0.645418
0.30	0.002750	0.002045	0.002045	0.007723	0.410836	0.234748	0.234748	0.645491

Because these equations have a translation symmetry with respect to θ_1, θ_2, it is useful to define $\delta_1 = \theta_2 - \theta_1$ so that these equations become

$$\begin{aligned} \dot{\delta}_1 &= \omega_2 - \omega_1 \\ M\dot{\omega}_1 + D\omega_1 &= P_{m1} - g_{11}V_1^2 - g_{12}V_1V_2\cos(\delta_1) + b_{12}V_1V_2\sin(\delta_1) \\ M\dot{\omega}_2 + D\omega_2 &= P_{m2} - g_{22}V_2^2 - g_{21}V_1V_2\cos(\delta_1) - b_{21}V_1V_2\sin(\delta_1) \end{aligned} \tag{7.9}$$

Now, assume that the system is in steady state, by which it is meant that $\dot{\delta}_1 = 0, \dot{\omega}_1 = 0, \dot{\omega}_2 = 0$. This implies that the rotational speeds of both machines must be constant and they must be equal since δ is also constant. If it is assumed that the equilibrium speed is synchronous speed, then

$$\omega_1 = \omega_2 = \omega_s$$

It is also necessary that

$$\begin{aligned} P_{m1} &= D\omega_s + g_{11}V_1^2 + g_{12}V_1V_2\cos(\delta_1) - b_{12}V_1V_2\sin(\delta_1) \\ P_{m2} &= D\omega_s + g_{22}V_2^2 + g_{21}V_1V_2\cos(\delta_1) + b_{21}V_1V_2\sin(\delta_1) \end{aligned}$$

Notice that if the last two equations are summed, then the result is

$$P_{m1} + P_{m2} = 2D\omega_s + (g_{11} + g_{22})V_1^2 + (g_{12} + g_{21})V_1V_2\cos(\delta_1) \tag{7.10}$$

This result is as expected. The total mechanical power delivered to the turbine generators is equal to the total mechanical friction power (the first term on the right) and the total electrical power consumption (the sum of the second and third terms on the right). Table 7.2 clearly shows how the dissipative admittance terms, g_{ij}, and, thus, how electrical consumption vary with load.

The primary goals of load frequency control are to steer the system to a stable equilibrium in which the machine speed is synchronous speed, ω_s and the total power $P_T = P_{m1} + P_{m2}$ as given in (7.10) is distributed such that

$$P_{m1} = k_1P_T,\ P_{m2} = k_2P_T, \quad k_1 + k_2 = 1 \tag{7.11}$$

The specified constants k_1, k_2 are called distribution factors. A conventional control configuration that achieves these goals is shown in Figure 7.9. Here, linear models (the transfer functions G_1, G_2) are used to characterize the steam turbines and the governors. Such models can be of different orders of complexity. These include a first-order lag characterizing the dominant turbine response time constant, or the dominant time constant plus a first-order lag for the valve actuator. Occasionally, a second-order turbine model is employed along with the actuator lag. The regulator is shown as a proportional plus integral compensator.

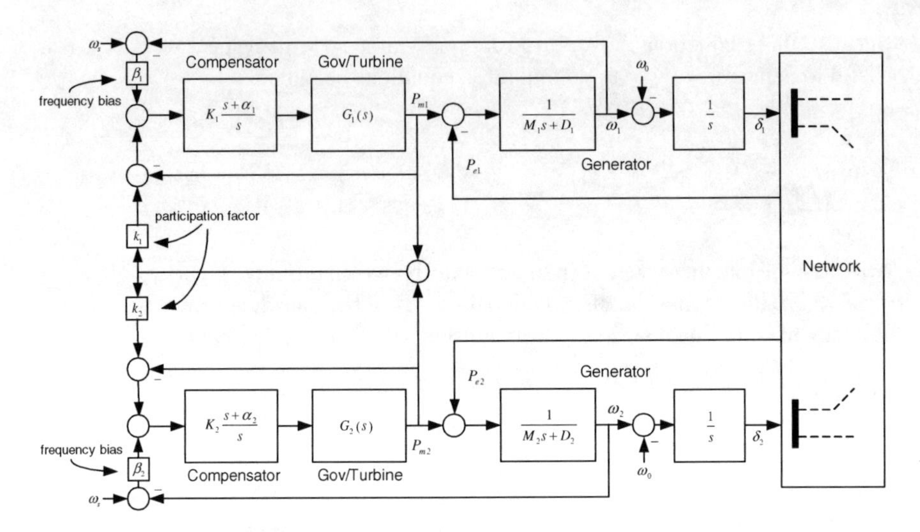

Fig. 7.9 A load frequency control configuration for two generators feeding a power network

Recall that the steady-state conditions insure $\omega_1 = \omega_2$. The integral compensator requires the steady-state conditions:

$$\begin{aligned} \beta_1 (\omega_1 - \omega_0) + (P_{m1} - k_1 (P_{m1} + P_{m2})) = 0 \\ \beta_2 (\omega_2 - \omega_0) + (P_{m2} - k_2 (P_{m1} + P_{m2})) = 0 \end{aligned} \tag{7.12}$$

Adding these equations, using the fact that the equilibrium frequencies are the same and that $k_1 + k_2 = 1$, it is easy to see that in steady state $\omega_1 = \omega_2 \equiv \omega_s$. From the swing equations, in steady state $P_{mi} = P_{ei}$, $i = 1, 2$, so that

$$P_{mi} = k_i P_T, \quad i = 1, 2$$

where P_T is the total real power supplied and is the sum of turbine mechanical friction losses plus the real power load including line losses.

Example 7.2 Example 7.1 Revisited. Once again, the system of Appendix A is used. In the simulation of Example 7.1, a load frequency controller as shown in Figure 7.9 was employed. The generator p.u. mechanical parameters are $M = 5$, $D = 0.5$, the compensator is proportional plus integral, and the turbine is modeled as first-order lag, specifically

$$G_c = 2.2\frac{s + 0.5}{s}, \; G_t (s) = \frac{1}{0.5s + 1}$$

The frequency bias parameters are $\beta_1 = \beta_2 = 5$ and the participation factors are $k_1 = k_2 = 0.5$. The same simulation as Example 7.1 produces the results for generator frequency and power flow as shown below in Figures 7.10 and 7.11. As expected,

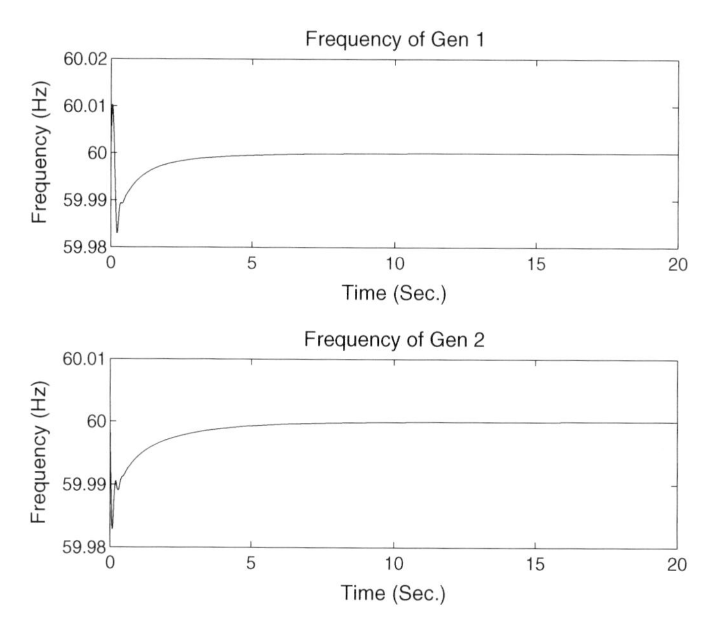

Fig. 7.10 Generators 1 and 2 frequencies are shown for the same initial conditions as in Example 7.1

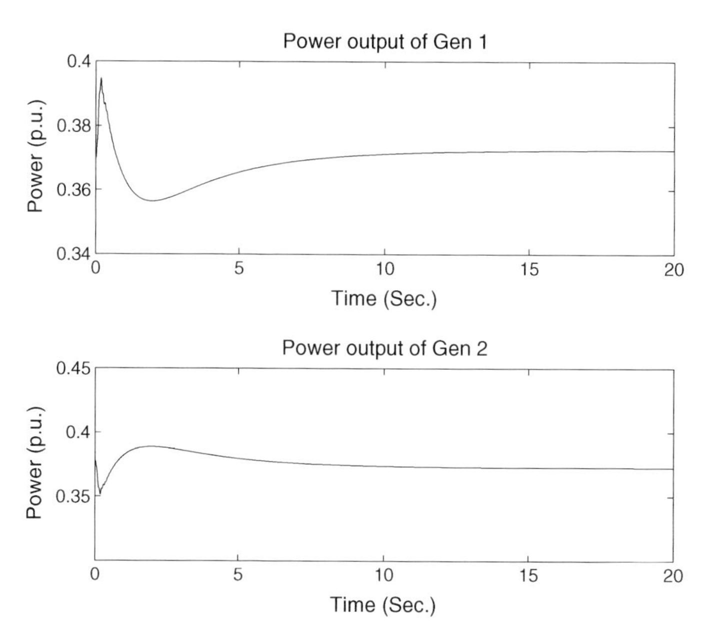

Fig. 7.11 The power outputs of Generators 1 and 2 are shown. Notice the somewhat longer timescale than voltage or frequency responses

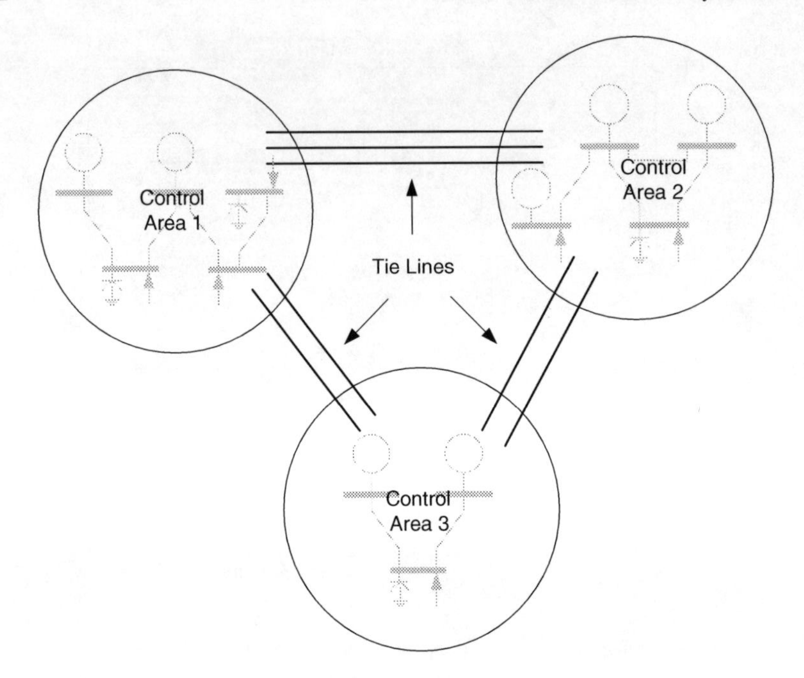

Fig. 7.12 Illustrates a system with three independent control areas interconnected by tie-lines

Figure 7.10 shows both generator frequencies stabilize at 60 Hz. Similarly, in Figure 7.11 it is seen the total load power of 0.75 p.u. is ultimately equally divided between the two generators at 0.375 p.u. each.

7.4 Automatic Generation Control (AGC)

The AGC problem involves two or more independently operated power systems that are interconnected via transmission lines with scheduled real power transfers [55, 56][1]. Figure 7.12 illustrates a three-area system. The generation in each area is independently regulated to provide inter-area tie flows to match a predefined schedule negotiated by the participating areas.

There are four principle regulation goals of classical AGC address the steady-state operating condition. They are:

1. match total generation to load
2. regulate system frequency error to zero

[1] An interesting discussion of AGC operations and issues some three decades after its first inception can be found in [101]

3. distribute generation to areas so that inter-area tie flows match a prescribed schedule
4. distribute generation within areas to minimize area operating costs

The first of these objectives is classically associated with system governing or "primary speed control" and is usually achieved in a time span of several seconds. The latter three objectives are accomplished by "supplementary" controls with objectives 2) and 3) associated with the "regulation" function and 4) with the "economic dispatch" function of these controls. The regulation and dispatch functions are executed in a time span on the order of minutes.

On most modem interconnections, the basis for accomplishing the regulation function is the tie-line bias control concept introduced in [55]. When operating in accordance with the principles of tie-line bias control, each area attempts to regulate its "area control error" (ACE), defined for area i by

$$ACE_i = \Delta P_i + \beta_i \Delta\omega_i \tag{7.13}$$

where for the ith area, ACE_i is the area control error, ΔP_i is the net tie flow error, $\Delta\omega_i$ is the frequency error and β_i is the frequency bias constant. The success of the tie-line bias control strategy is based on the twofold premise that: 1) independent regulation of its own ACE by each area implies achievement of the overall objectives of system regulation, and 2) each area can, in fact, achieve its assigned subgoal (regulation of its ACE) without some sort of inter-area coordination.

Originally, AGC was viewed as a static problem. The objectives of AGC regulation were defined in terms of steady-state conditions and the first part of the above premise was validated on the basis of this definition. In fact, from this point of view it is merely necessary to show that the requirement of zero area control error for each area is sufficient to establish the desired steady-state frequency and tie flows. The second part of the premise is then verified by demonstrating that integral control of ACE within each area results in a stable structure with the desired equilibrium state.

But interconnection does have its attendant problems and they are dynamical in nature. The interconnection of neighboring systems establishes the possibility for disturbances in one system to produce effects in others which may be quite remote from the source of the difficulty. In view of this, the dynamic aspects of power system AGC became the focus of considerable interest, for example [72, 73, 38, 61]. The underlying motivation was the growing recognition of the importance of the control of bulk power flows across potential separation boundaries within the network, particularly following large disturbances or during generation shifts to accommodate security constraints or tie-line schedule changes. In addition, in order that economic dispatch be effective it is necessary that each area generation requirement be established quickly and this is essentially the function of AGC. Since load is persistently changing, with rapid and sustained variations expected to occur during certain periods of normal operation, AGC must provide a "load tracking" capability [71].

7.4.1 Elements of the Classical AGC Problem

The classical model used in the analysis of AGC is based on the central assumption that electrical interconnections within a control area are very strong relative to the tie-line interconnections between control areas. Thus, all machines within an area operate at a common frequency and within each area all bus frequencies vary in unison. As a consequence, all turbine generators in the area can be aggregated into a single machine.

A second assumption is that voltage variations can be neglected. The justification for this is that voltage regulation takes place on a much shorter timescale that AGC regulation. This assumption can also be challenged in some applications.

Finally, the tie-lines themselves are considered external to the independent control areas. As a consequence, the power that exits one area over the interconnection enters the connecting area with losses. With these assumptions, a classical two-area model is shown in Figure 7.13. Note the term a_{12} in the figure accounts for tie-line losses. In the lossless case, $a_{12} = -1$.

Now, consider the steady state of a system with N independent control areas. In equilibrium, the frequencies of all areas are identical, $\omega_1 = \omega_2 = \cdots = \omega_N$, as would be the case of any multi-machine interconnection. Furthermore, the integral action in the secondary control compensation insures that $ACE_i = 0,\ i = 1, \ldots, N$. It will be assumed that the tie-line power flow schedule accounts for losses in the tie-lines. That is, $P_{ij}^0 + P_{ji}^0 + P_{ij}^{loss} = 0$, where P_{ij}^0 is the scheduled tie power flow from area i to area j and P_{ij}^{loss} is the power loss associated with this power transfer. As a result,

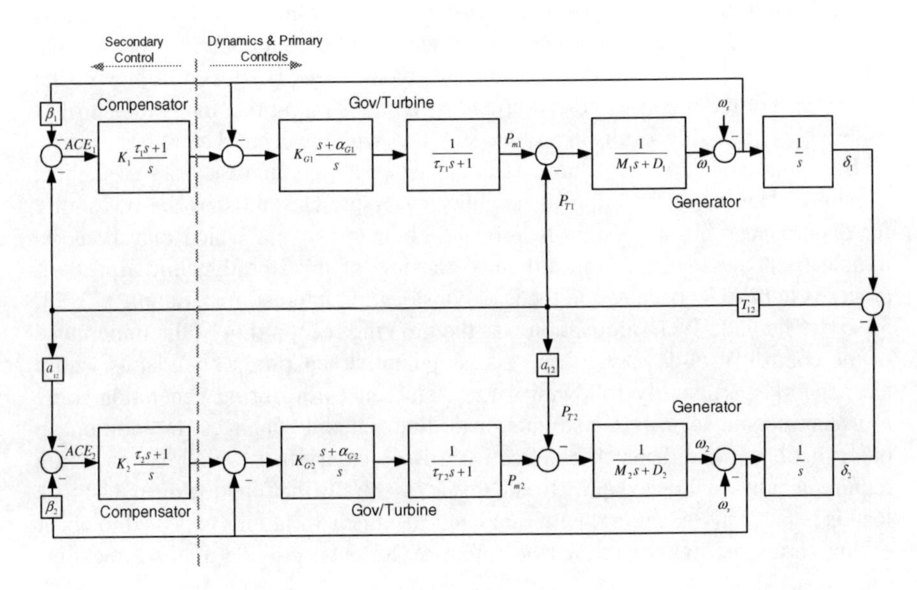

Fig. 7.13 The classical linear model of the dynamics of a two-area system

the sum of the area net tie transfer deviations from schedule is zero, $\sum_{i=1}^{N} \Delta P_i = 0$. These equations can be assembled into a set of $N+1$ linear equations

$$\begin{array}{l} \sum_{i=1}^{N} \Delta P_i = 0 \\ \Delta P_i + B_i \Delta\omega_i = 0, \; i = 1, \ldots, N \end{array} \tag{7.14}$$

in terms of the $N+1$ independent variables $\Delta\omega_1$ and $\Delta P_i, \; i = 1, \ldots, N$. In matrix form

$$\begin{bmatrix} 1 & 1 & \cdots & 1 & 0 \\ 1 & 0 & \cdots & 0 & B_1 \\ 0 & 1 & \ddots & \vdots & \vdots \\ 0 & \ddots & \ddots & 0 & B_{N-1} \\ 0 & \cdots & 0 & 1 & B_N \end{bmatrix} \begin{bmatrix} \Delta P_1 \\ \Delta P_2 \\ \vdots \\ \Delta P_N \\ \Delta\omega_1 \end{bmatrix} = 0 \tag{7.15}$$

It is easily confirmed that the matrix is nonsingular so the unique solution is $\Delta\omega_1 = 0, \; \Delta P = 0_i, \; i = 1, \ldots, N$. Consequently, objectives 1), 2), and 3) will be achieved in steady state provided the system is stable.

Objective 4) is relegated to a separate function of AGC called *economic dispatch* [27]. Economic dispatch is the process of determining the distribution of power among plants within a control area in order to minimize fuel cost. The earliest version was a static optimization problem which distributed the total generation among m plants to minimize the total area fuel cost,

$$C(P_1, \ldots, P_m) = \sum_{i=1}^{m} C_i(P_i) \tag{7.16}$$

under the constraint that the sum of the plant power outputs is equal to the required area total demand real power

$$\sum_{i=1}^{m} P_i = P_{Tot} \tag{7.17}$$

and, in addition, each plant power is constrained according to a specified minimum and maximum power output,

$$P_{i,\min} \le P_i \le P_{i,\max}, \quad i = 1, \ldots, m \tag{7.18}$$

Over the years, the dispatch function has evolved to accommodate changing realities of the system [52, 202]. Thus, load dynamics, emission constraints, load management, and other factors have become relevant considerations.

7.4.2 AGC Control Strategies

While the goals of AGC as originally formulated focused mainly on achieving certain steady-state behaviors, dynamics and stability are central to any feedback control structure. However, load varies substantially in the course of a day and depends on many factors including weather. Furthermore, for many systems dominated by steam-powered generation systems, the generation response capability can be a significant limiting factor in matching generation to load. So dynamical behavior as well as load prediction and tracking is important as is the coordination of the economic dispatch function with the ACE regulation function [171].

The dynamical behavior of the classical model, Figure 7.13, was studied in [72]. In another paper, [73], these authors proposed a new approach to controller design based on the linear quadratic regulator (LQR) design method [13]. Although better dynamic performance is achieved, a key deficiency of this approach is that it does not provide the precise steady-state ACE zeroing objective of AGC because the LQR design does not include the equivalent of integral control. The method was expanded in [38] and [116] to remedy this. Moreover, in [116] generation plant response limitations were considered and a form of load estimation was incorporated to improve load tracking capabilities. The approach of [116] was further extended in [110] to include coordination of economic dispatch and AGC.

Another distinction in the discussion below with respect to the early applications of LQR design is that each control area has its own regulator that uses only local information other than the shared tie flow schedule. This preserves an important decentralized structure of the original AGC controllers.

The approach used in [116, 110] is based on the theory of the linear multivariable regulator as described in [103, 74, 124, 200]. The linear system dynamics are augmented by a set of disturbance variables which are themselves described by a systems of linear differential equations so that the augmented system is described by the equations

$$\begin{aligned} \dot{x} &= Ax + Ew + Bu \\ \dot{w} &= Zw \\ e &= Cx + Fw + Du \\ y &= \tilde{C}x + \tilde{F}w + \tilde{D}u \end{aligned} \tag{7.19}$$

In this model, $x \in R^n$ is the state of the original dynamical system, $u \in R^m$ is the control input, $e \in R^m$ is the error output to be regulated, $y \in R^p$ is the measurement, and $w \in R^q$ is the disturbance vector. The goal was to design a feedback controller that provides

1. **internal stability**: the feedback system is stable
2. **regulation**: $\lim_{t\to\infty} e(t) \to 0$ for any $w(t)$ generated by the disturbance model.

Before considering the regulator design details, consider the following example that sets up an AGC regulator design problem in the framework of Equation (7.19).

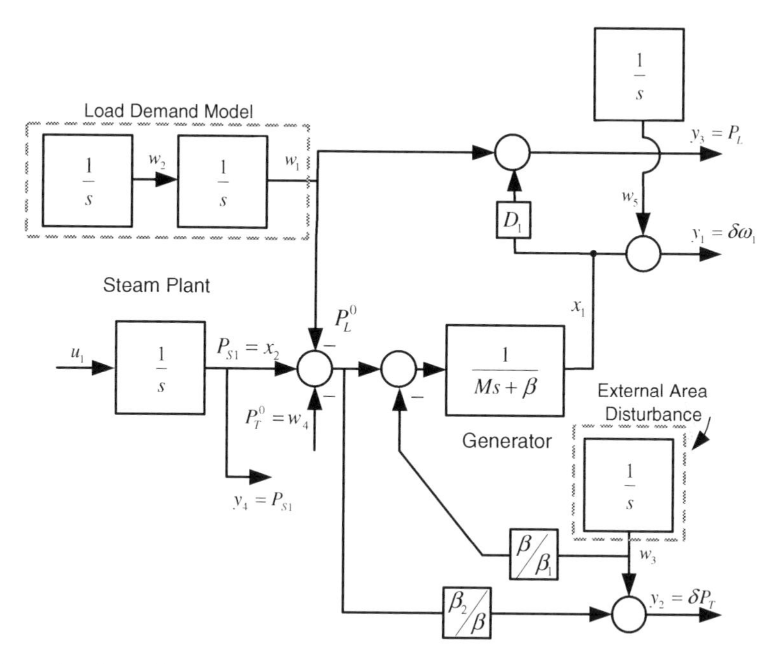

Fig. 7.14 The figure illustrates the system dynamics to be used in design of the area regulator. Note the model of the aggregated steam plant involves an integrator whose input, u_1, is the commanded rate of change of power input. Disturbance inputs w_1, w_2, w_3, w_4, w_5 are also shown

Example 7.3 AGC Regulator Design. Consider the two-area system of Figure 7.13. An AGC controller will be designed for control area 1 based on the extracted model shown in Figure 7.14. This system includes a single aggregated power plant, an integrator, with input u_1 represents the rate of change of power output. Thus, u_1 can be maintained below the allowable rate of change of power output. It is assumed that this rate limit is sufficiently low that the power output responds accordingly. The goal of the AGC regulator was to track the relatively slow variations in load demand by regulating ACE_1 to zero. It is not intended to manage the considerably faster inter-area frequency oscillations. Consequently, the area 2 is not included in the model. The effects of area 2 are included with the constant disturbance w_3 that affects the tie flow. Also, in view of the slower timescale of AGC relative to inter-area swings, the areas are assumed to operate at the same frequency. The impact of internal and external area load changes with frequency is included. The area 1 electrical load is represented by a constant plus ramp disturbance with states w_1, w_2. The regulator employs three measured outputs, $\Delta\omega_1$, P_L, δP_T, and the commanded scheduled tie flow P_T^0. Note that a disturbance variable w_4 is used to characterize P_T^0 as a constant, measurable command.

The system differential equations are

$$\delta\dot{\omega}_1 = \left(-\beta\delta\omega_1 + P_{S1} - w_1 - P_T^0 - (\beta/\beta)w_3\right)/M$$
$$\dot{P}_{S1} = u_1$$

with error

$$e = ACE_1 = \delta P_T + B\delta\omega_1$$

and measurements

$$y_1 = \delta\omega_1$$
$$y_2 = \delta P_T = (\beta_2/\beta)\left(P_{S1} - P_T^0 - w_1\right) + w_3$$
$$y_3 = P_L = w_1 + D_1\delta\omega_1$$
$$y_4 = P_{S1}$$

These equations can be written in the form of (7.19)

$$\frac{d}{dt}\begin{bmatrix} x_1 \\ x_2 \end{bmatrix} = \begin{bmatrix} -\beta/M & 1/M \\ 0 & 0 \end{bmatrix}\begin{bmatrix} x_1 \\ x_2 s \end{bmatrix} + \begin{bmatrix} -1/M & 0 & -\beta/M\beta_1 & -1/M & 0 \\ 0 & 0 & 0 & 0 & 0 \end{bmatrix}\begin{bmatrix} w_1 \\ w_2 \\ w_3 \\ w_4 \\ w_5 \end{bmatrix} + \begin{bmatrix} 0 \\ 1 \end{bmatrix} u_1 \quad (7.20)$$

$$\frac{d}{dt}\begin{bmatrix} w_1 \\ w_2 \\ w_3 \\ w_4 \\ w_5 \end{bmatrix} = \begin{bmatrix} 0 & 1 & 0 & 0 & 0 \\ 0 & 0 & 0 & 0 & 0 \\ 0 & 0 & 0 & 0 & 0 \\ 0 & 0 & 0 & 0 & 0 \\ 0 & 0 & 0 & 0 & 0 \end{bmatrix}\begin{bmatrix} w_1 \\ w_2 \\ w_3 \\ w_4 \\ w_5 \end{bmatrix} \quad (7.21)$$

$$e = \begin{bmatrix} B & \beta_2/\beta \end{bmatrix}\begin{bmatrix} x_1 \\ x_2 \end{bmatrix} + \begin{bmatrix} -\beta_2/\beta & 0 & 1 & -\beta_2/\beta & B \end{bmatrix}\begin{bmatrix} w_1 \\ w_2 \\ w_3 \\ w_4 \\ w_5 \end{bmatrix} \quad (7.22)$$

$$\begin{bmatrix} y_1 \\ y_2 \\ y_3 \\ y_4 \end{bmatrix} = \begin{bmatrix} 1 & 0 \\ 0 & \beta_2/\beta \\ D_1 & 0 \\ 0 & 1 \end{bmatrix}\begin{bmatrix} x_1 \\ x_2 \end{bmatrix} + \begin{bmatrix} 0 & 0 & 0 & 0 & 1 \\ -\beta_2/\beta & 0 & 1 & -\beta_2/\beta & 0 \\ 1 & 0 & 0 & 0 & 0 \\ 0 & 0 & 0 & 0 & 0 \end{bmatrix}\begin{bmatrix} w_1 \\ w_2 \\ w_3 \\ w_4 \\ w_5 \end{bmatrix} \quad (7.23)$$

The design of a regulator for the system (7.19) proceeds in three steps

1. **Regulation**: determination of an *ultimate state trajectory*,
2. **Stabilization**: design of a stabilizing state feedback controller,
3. **Observation**: design of a state and disturbance variable observer.

For step 1, regulation, a solution pair $\bar{x}(t), \bar{u}(t)$ is sought that satisfies the differential equations in (7.19) for some initial state $x(0)$, and insures that $e(t) \equiv 0$ for any initial disturbance $x(0)$. Hence, it is required that

$$\begin{aligned} \dot{\bar{x}} &= A\bar{x} + Ew + B\bar{u} \\ \dot{w} &= Zw \\ 0 &= C\bar{x} + Fw + D\bar{u} \end{aligned} \tag{7.24}$$

Now, assume a solution of the form $\bar{x} = Xw$, $\bar{u} = Uw$, where X, U are matrices to be determined. Substitution into (7.24) leads to the requirement that X and U must satisfy the matrix equations

$$\begin{aligned} XZ - AX - BU &= E \\ CX + DU &= F \end{aligned} \tag{7.25}$$

These equations constitute a large set of linear equations in the unknowns, i.e., the elements of the matrices X and U. Furthermore, they are typically under determined so there are multiple solutions. Define the variables $\delta x = x - \bar{x}$ and $\delta u = u - \bar{u}$. Upon substitution into (7.19), the state equations become

$$\delta\dot{x} = A\delta x + B\delta u \tag{7.26}$$

Thus, so long as (A, B) is a stabilizable pair it is a simple matter to determine a matrix K, so that the control $\delta u = K\delta x$ solves step 2 stabilization. Since $u = \bar{u} + \delta u$, we have

$$u = Uw + K(x - Xw) = \begin{bmatrix} K & (U - KX) \end{bmatrix} \begin{bmatrix} x \\ w \end{bmatrix} \tag{7.27}$$

Finally, for step 3, observation, consider the composite system

$$\frac{d}{dt}\begin{bmatrix} x \\ w \end{bmatrix} = \begin{bmatrix} A & E \\ 0 & Z \end{bmatrix} \begin{bmatrix} x \\ w \end{bmatrix} + \begin{bmatrix} B \\ 0 \end{bmatrix} u \tag{7.28}$$

$$y = \begin{bmatrix} \tilde{C} & \tilde{F} \end{bmatrix} \begin{bmatrix} x \\ w \end{bmatrix} + \tilde{D}u \tag{7.29}$$

If (7.28) and (7.29) comprise an observable system, then an observer is easily determined in the form

$$\frac{d}{dt}\begin{bmatrix}\hat{x}\\ \hat{w}\end{bmatrix}=\begin{bmatrix}A & E\\ 0 & Z\end{bmatrix}\begin{bmatrix}\hat{x}\\ \hat{w}\end{bmatrix}+\begin{bmatrix}B\\ 0\end{bmatrix}u+L\left(\tilde{C}\hat{x}+\tilde{F}\hat{w}+\tilde{D}u-y\right) \tag{7.30}$$

The matrices K and L can be determined by pole placement or by solving the relevant Riccati equations. Equations (7.30) and (7.27) constitute the controller as illustrated in Figure 7.15.

Example 7.4 AGC Regulator Design, Continued. Consider the system of Example 7.3. The data provided in Table 7.3, for the system shown in Figure 7.14, is taken from [72].

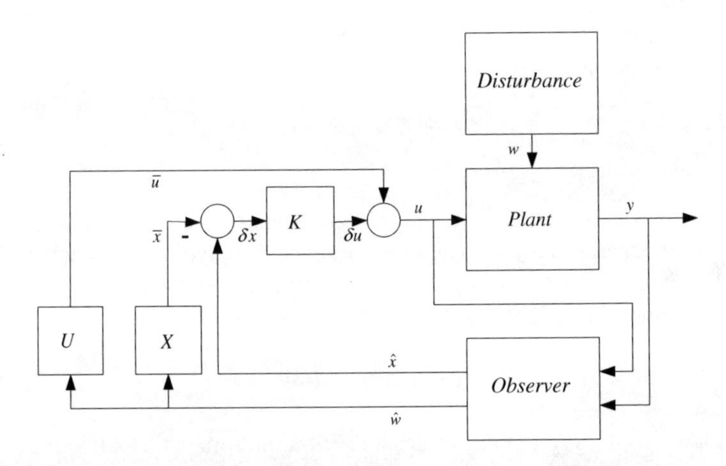

Fig. 7.15 This diagram illustrates the implementation of Equations (7.30) and (7.27)

Table 7.3 AGC Nomenclature

Symbol	**Quantity**	**Value in Example** 7.4
$\delta\omega_1, y_1$	frequency deviation from synchronous	
$\delta P_T, y_2$	area net tie flow deviation from schedule	
P_T^0	area net tie flow schedule	
P_L, y_3	area electrical load at prevailing frequency	
P_{S1}, y_4	area mechanical Power	
M	total system inertia	10 sec
β_1, β_2	local area regulation, $\beta_i = D_i + R_i^{-1}$	
β	total system regulation $\beta = \beta_1 + \beta_2$	
D_1, D_2	local area load characteristic	0.00833 pu MW/Hz
R_1, R_2	local area load characteristic	2.4 Hz/pu MW
B	frequency bias constant	

Now, with this data, let us consider design of a regulator for the system defined by equations (7.20) through (7.23).

7.4.2.1 Regulation

First, consider Equation (7.25). These equations are solved for $X \in R^{2\times 5}$ and $U \in R^{1\times 5}$:

$$X = \begin{bmatrix} 0 & 0 & \frac{-0.85}{0.425B+0.181} & 0 & \frac{B}{B+0.425} \\ 1 & 0 & \frac{2(B-0.425)}{B+0.425} & 1 & \frac{0.85B}{B+0.425} \end{bmatrix}, \quad U = \begin{bmatrix} 0 & 1 & 0 & 0 & 0 \end{bmatrix}$$

All parameter values are as indicated in Table 7.3. Notice that the frequency bias constant, B, has not been specified. A commonly used value is $B = 1$. In this case,

$$X = \begin{bmatrix} 0 & 0 & -1.404 & 0 & -0.702 \\ 1 & 0 & 0.807 & 1 & -0.596 \end{bmatrix}, \quad U = \begin{bmatrix} 0 & 1 & 0 & 0 & 0 \end{bmatrix}$$

7.4.2.2 Stabilization

Stabilization requires choosing a control matrix K, such that the matrix $A + BK$ has poles in the left half plane. In this example,

$$A = \begin{bmatrix} -\beta/M & 1/M \\ 0 & 0 \end{bmatrix}, \quad B = \begin{bmatrix} 0 \\ 1 \end{bmatrix}$$

with numerical values for the parameters given above. First, consider the eigenvalues of A which are

$$\{-0.17, 0.\}$$

The primary goal was to move the eigenvalue at the origin into the left half plane. The remaining real eigenvalue represents relatively *fast*, stable dynamics of the system. Modification of the fast dynamics is generally considered the function of the primary controls (governors), not the AGC system. Although the AGC system could be used for that purpose, to be effective would require high-speed interchange of measurement data between areas. One of the key features of AGC is that its primary goals can be achieved using only local area data. In other words, classical AGC is a *decentralized* control structure. To preserve this desirable property, K should be chosen, so that only the dynamics associated with the zero eigenvalue are affected by the controller.

One simple approach to achieving the desired goal is based on determination of the linear state feedback control, $u = Kx$, for the linear system

$$\dot{x} = Ax + Bu$$

that minimizes the quadratic cost function

$$J(x) = \lim_{T \to \infty} \frac{1}{T} \int_0^T \left[x^T(\tau) Q x(\tau) + u^T(\tau) R u(\tau) \right] d\tau$$

with $Q = Q^T \geq 0$, $R = R^T > 0$. If the pair (A, B) is stabilizable, then the optimal K exists and is given by

$$K = -R^{-1} B^T S \tag{7.31}$$

where $S = S^T \geq 0$ satisfies the algebraic Riccati equation

$$A^T S + SA - SBR^{-1} B^T S + Q = 0 \tag{7.32}$$

In the application of this result to the current design problem, we set $Q = 0$, and replace A in (7.32) by $A + \gamma I$, with the scalar $\gamma > 0$ to be specified. Note that the addition of the term γI shifts all the eigenvalues of A to the right by and amount γ. Equation (7.32) now becomes

$$(A + \gamma I)^T S + S(A + \gamma I) - SBR^{-1} B^T S = 0 \tag{7.33}$$

With S determined from (7.33) and K obtained from (7.31) the closed loop system matrix $(A + BK)$ has some useful properties. First, only eigenvalues of A with real parts to the right of $-\gamma$ are affected by feedback. Note that all of these eigenvalues would be destabilized by the γI shift. The feedback control stabilizes these eigenvalues by reflection with respect to the line $s = -\gamma$ in the complex s-plane.

As an example, by choosing $\gamma = 0.0125$, Equations (7.33) and (7.31) yield

$$K = \left[0 \ -0.025 \right]$$

and the eigenvalues of $(A + BK)$ are

$$\{-0.17, -0.025\}$$

Thus, only the dynamics associated with the zero eigenvalue have been affected and the eigenvalue has been shifted from 0 to -0.025 corresponding to a time constant of 40 seconds.

Note that

$$\mathbf{K} = \left[K \ \ (U - KX) \right] = \left[0 \ -0.025 \ 0.025 \ 1 \ \frac{0.05(B-0.425)}{B+0.425} \ 0.025 \ \frac{-0.02125B}{B+0.425} \right] \tag{7.34}$$

7.4.2.3 Observation

Implementation of the controller (7.34) requires only three states x_5, w_2, w_3. Two of these x_5 and w_3 are directly measured. The third, w_2, is easily estimated. In general, however, with different design parameters in (7.34) more states will be required to implement the controller and an observer needs to be designed.

To design the observer, it is first necessary to assemble the composite system (7.28) and (7.29). Label the composite matrices:

$$A_c = \begin{bmatrix} A & E \\ 0 & Z \end{bmatrix}, \quad C_c = \begin{bmatrix} \tilde{C} & \tilde{F} \end{bmatrix}$$

It is readily confirmed that the composite system is observable. There are several alternatives to designing the observer L matrix in Equation (7.30). In this example, the design process is essentially the same as that used above for stabilization. L is given by

$$L = SC_c^T V^{-1} \tag{7.35}$$

where S is the unique positive semi-definite solution of the Riccati equation

$$S(A_c + \gamma I)^T + (A_c + \gamma I)\, S - SC_c^T V^{-1} C_c S = -W, \quad S \geq 0 \tag{7.36}$$

Here, the scalar parameters γ, $V \in R^{5\times 5}$, $V > 0$, and $W \in R^{7\times 7}$, $W \geq 0$, are design parameters. Note that γ plays the same role as in the stabilization problem. The primary goal was to place the observer eigenvalues (the eigenvalues of $(A_c + LC_c)$) sufficiently into the left half plane to insure that the observer dynamics are fast relative to the stabilized plant dynamics.

For example, take $\gamma = 2.0$, $V = I_{5\times 5}$, and $W = 0.1 I_{7\times 7}$. Then, obtaining S from (7.38) and computing L from (7.37) yields

$$L = \begin{bmatrix} -9.911 & 0 & 0.151 & 0 & 5.304 \\ 0 & -0.571 & -0.159 & -5.694 & 0 \\ 0.226 & 1.928 & -9.319 & -0.159 & 0 \\ 0.773 & 7.891 & -21.674 & -0.643 & 0 \\ 30.555 & -3.323 & -1.275 & 0.399 & -7.378 \\ 60.874 & 4.924 & 2.754 & -3.595 & -14.709 \\ -0.415 & 0 & 0 & 0 & -5.719 \end{bmatrix}$$

The observer eigenvalues are

$$\{-5.47, -5.05, -3.67, -3.04, -2.74, -2.26, -2.04\}$$

The smallest observer eigenvalue, -2.04, is over 10 times larger that the fastest stabilized plant eigenvalue, -0.17.

7.4.3 *Coordination of Economic Dispatch and AGC*

The distribution of power generation within an area in order to achieve minimal fuel consumption, i.e., economic dispatch, was originally formulated as a static optimization problem as described above in Section 7.4.1. The realities of the evolving power system, notably the significant variations in system load along with the plant power output rate change limitations, prompted consideration of a dynamic dispatch formulation [22, 113, 172, 202].

Optimal generation allocation typically evolves on a timescale significantly slower than ACE regulation. Moreover, if load estimates used in the generation allocation process are in error or if the generating plant outputs fail to follow their prescribed trajectories, a mismatch between total generation and load will result. The regulation function of AGC must then restore the generation-load balance by modification of the plant outputs. This natural coupling between the dispatch and the regulation functions can significantly degrade performance of the AGC system. One approach to mitigating this problem is the application of a *permissive* control strategy [171]. This strategy assigns a priority to regulation with the central objective of maintaining a balance of load and generation. Thus, any change in unit output commanded by the economic dispatch computation is inhibited unless it is consistent with restoration of the generation-load balance. The problem with this approach is that it can negate any economic realignment of generation.

An alternative approach to coordinating dispatch and regulation was proposed in [110]. It is an extension of the method described in Section 7.4.2 and will be discussed in the following sections.

7.4.3.1 Modeling

In the discussion of Section 7.4.2, the steam-generating plant is represented by a single unit, i.e., a single integrator, as shown in Figure 7.14. As a result, of course, there is no discussion of economic dispatch. For the present discussion, the model will be expanded to include multiple steam plants. The turbine generator will remain as a single aggregated system. Thus, introduce the k plants with states $P_{S1} = x_2, \ldots, P_{Sk} = x_{k+1}$ and inputs u_k with governing equations

$$\dot{x}_2 = u_1, \dot{x}_2 = u_2, \ldots, \dot{x}_{k+1} = u_k$$

Define the vectors,

$$\mathbf{x}_2 = \begin{bmatrix} x_2 \\ \vdots \\ x_{k+1} \end{bmatrix}, \quad \mathbf{u} = \begin{bmatrix} u_1 \\ \vdots \\ u_k \end{bmatrix}$$

Now, Equations 7.20, 7.22 and 7.23 are replaced by

$$\frac{d}{dt}\begin{bmatrix} x_1 \\ \mathbf{x}_2 \end{bmatrix} = \begin{bmatrix} -\beta/M & (1/M)_{1\times k} \\ 0 & 0_{1\times k} \end{bmatrix}\begin{bmatrix} x_1 \\ \mathbf{x}_2 \end{bmatrix} + \begin{bmatrix} -1/M & 0 & -\beta/M\beta_1 & -1/M & 0 \\ 0_{k\times 1} & 0_{k\times 1} & 0_{k\times 1} & 0_{k\times 1} & 0_{k\times 1} \end{bmatrix}\begin{bmatrix} w_1 \\ w_2 \\ w_3 \\ w_4 \\ w_5 \end{bmatrix} + \begin{bmatrix} 0_{1\times k} \\ I_{k\times k} \end{bmatrix}\mathbf{u} \tag{7.37}$$

$$e = \begin{bmatrix} B & (\beta_2/\beta)_{1\times k} \end{bmatrix}\begin{bmatrix} x_1 \\ \mathbf{x}_2 \end{bmatrix} + \begin{bmatrix} -\beta_2/\beta & 0 & 1 & -\beta_2/\beta & B \end{bmatrix}\begin{bmatrix} w_1 \\ w_2 \\ w_3 \\ w_4 \\ w_5 \end{bmatrix} \tag{7.38}$$

and

$$\begin{bmatrix} y_1 \\ y_2 \\ y_3 \\ \mathbf{y}_4 \end{bmatrix} = \begin{bmatrix} 1 & 0_{1\times k} \\ 0 & (\beta_2/\beta)_{1\times k} \\ D_1 & 0_{1\times k} \\ 0_{k\times 1} & I_{k\times k} \end{bmatrix}\begin{bmatrix} x_1 \\ \mathbf{x}_2 \end{bmatrix} + \begin{bmatrix} 0 & 0 & 0 & 0 & 1 \\ -\beta_2/\beta & 0 & 1 & -\beta_2/\beta & 0 \\ 1 & 0 & 0 & 0 & 0 \\ 0_{k\times 1} & 0_{k\times 1} & 0_{k\times 1} & 0_{k\times 1} & 0_{k\times 1} \end{bmatrix}\begin{bmatrix} w_1 \\ w_2 \\ w_3 \\ w_4 \\ w_5 \end{bmatrix}, \tag{7.39}$$

respectively. Equation 7.21 remains the same.

7.4.3.2 Control

Design of the controller proceeds via the three-step procedure as described above in Section 7.4.2. The key distinction, other than the expansion of the model to include multiple generating plants, is in the solution of the *regulation* step. The following paragraphs discuss this step in detail.

The ultimate state trajectory is computed using the modified model of Equations (7.37) and (7.38). The essential difference in the models of this Section 7.4.3 and Section 7.4.2 is the expansion from 1 to k generating plants, so that $x_2 \rightarrow \mathbf{x}_2$ and $u_1 \rightarrow \mathbf{u}$. Consequently, it is easy to verify that the only change to the ultimate state trajectory alters $\bar{\mathbf{x}}_2$ and $\bar{\mathbf{u}}$, so that the following relations are obtained.

$$\bar{x}_1 = \begin{bmatrix} 0 & 0 & \frac{-0.85}{0.425B+0.181} & 0 & \frac{B}{B+0.425} \end{bmatrix}\begin{bmatrix} w_1 \\ w_2 \\ w_3 \\ w_4 \\ w_5 \end{bmatrix}$$

$$\bar{\mathbf{x}}_2 = \eta \left[1 \; 0 \; \tfrac{2(B-0.425)}{B+0.425} \; 1 \; \tfrac{0.85B}{B+0.425} \right] \begin{bmatrix} w_1 \\ w_2 \\ w_3 \\ w_4 \\ w_5 \end{bmatrix}$$

$$\mathbf{u} = \eta \left[0 \; 1 \;\; 0 \; 0 \; 0 \right] \begin{bmatrix} w_1 \\ w_2 \\ w_3 \\ w_4 \\ w_5 \end{bmatrix}$$

where

$$= \left[\eta_1 \cdots \eta_k \right]^T, \quad \eta_i \geq 0, \quad \eta_1 + \cdots + \eta_k = 1$$

Note that the generation is distributed among the k plants according to the distribution vector η.

Lemma 7.1 [Superposition in Regulation] *A key observation for the coordination of economic dispatch within AGC is the following. Suppose $\bar{x}^*$, $\bar{u}^*$ represent a particular ultimate state trajectory corresponding to any particular disturbance trajectory w^*. If X, U are known, then*

$$\bar{x}^* = Xw^*, \quad \bar{u}^* = Uw^*$$

The ultimate state trajectory corresponding to an arbitrary disturbance trajectory, w, is, similarly,

$$\bar{x} = Xw, \quad \bar{u} = Uw$$

Thus, $\bar{x}$ and $\bar{u}$ can be written

$$\bar{x} = \bar{x}^* + X\left(w - w^*\right), \quad \bar{u} = \bar{u}^* + U\left(w - w^*\right)$$

Lemma 7.1 provides a natural method to coordinate economic dispatch within AGC. Suppose the economic distribution of generation is based on the disturbance trajectory w^*, where w^* is consistent with the disturbance model being used. For example, suppose the optimization is based on an estimated constant load and with all other disturbances assumed zero: $w_1^* =$ constant, $w_2^* \equiv 0$, $w_3^* \equiv 0$, $w_4^* =$ constant, $w_5^* \equiv 0$. Also assume the following:

1. The plant generation commands generated by the economic dispatch function are provided directly to the plant controllers and not through the rate restricted AGC regulator, i.e., $\bar{\mathbf{u}}^* = 0$.
2. It is assumed that both the AGC regulator and the dispatch computation are provided the same tie-line scheduling data, $P_T^0 = w_4$. Since w_4 is the scheduled tie flow, it follows that $w_4 = w_4^*$.
3. At any time t, the electrical load at synchronous frequency is $P_L = w_1$. Economic dispatch distributes generation based on an estimate of load, $P_L^* = w_1^*$.

With w^* specified as above, from Lemma 7.1 we obtain:

$$\bar{x}_1 = \frac{-0.85}{0.425B + 0.181} w_3 + \frac{B}{B + 0.425} w_5 \tag{7.40}$$

$$\bar{\mathbf{x}}_2 = \left(\bar{\mathbf{x}}_2^* - \eta w_1^*\right) + \eta \left(w_1 + \frac{2\,(B - 0.425)}{B + 0.425} w_3 + \frac{0.85B}{B + 0.425} w_5 \right) \tag{7.41}$$

$$\bar{\mathbf{u}} = \eta w_2 \tag{7.42}$$

The stabilizer and observer are designed as in Section 7.4.2, accounting for the additional plant states. Again use the equation (7.33). When applied to the current situation, k poles are shifted from zero to -0.025 with

$$K = \left[\, 0_{k\times 1} \;\; -0.025 I_{k\times k} \,\right]$$

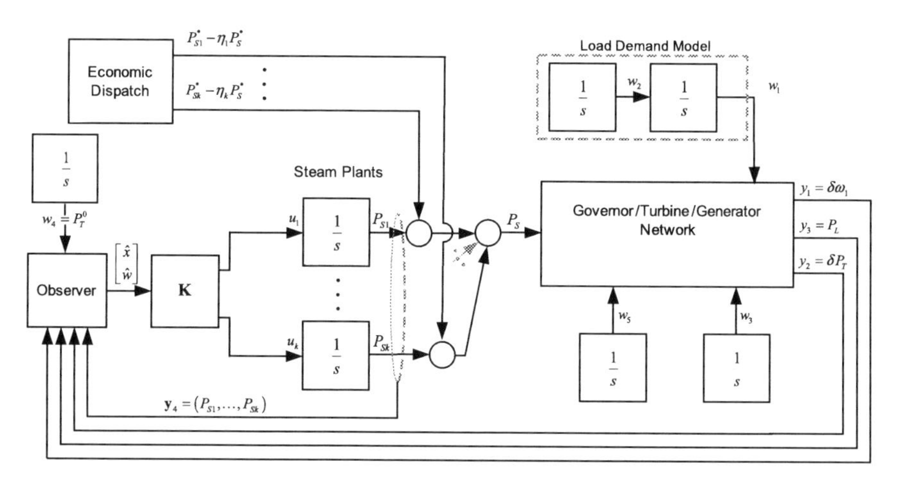

Fig. 7.16 Coordinated AGC and economic dispatch controller

The result is that

$$\mathbf{u} = \mathbf{K}\begin{bmatrix} \hat{x} \\ \hat{w} \end{bmatrix}, \quad \mathbf{K} = \begin{bmatrix} \mathbf{K}_x \ \mathbf{K}_w \end{bmatrix}$$

$$\mathbf{K}_x = \begin{bmatrix} 0_{k\times 1} & -0.025 I_{k\times k} \end{bmatrix}$$

$$\mathbf{K}_w = 0.025\eta \begin{bmatrix} 1 & 1 & \frac{2(B-0.425)}{B+0.425} & 0 & \frac{0.85B}{B+0.425} \end{bmatrix}$$

The structure of the coordinating controller is illustrated in Figure 7.16.

Chapter 8
Power System Management

> *"It must be considered that there is nothing more difficult to carry out, nor more doubtful of success, nor more dangerous to handle, than to initiate a new order of things."*
> — Niccolò Machiavelli, "The Prince and The Discourses"

8.1 Introduction

Many systems undergo reconfiguration or switching during normal and abnormal operations. Such systems can function in different *modes* or *discrete states* in each of which the system may exhibit distinct dynamical behavior. Transitions between modes are defined by logical conditions that can depend on continuous dynamical states or external signals. Such systems are called *hybrid systems* [174] or *Mixed Logical-Dynamical systems* (MLD) [24, 77]. The relevance of such problems to power systems was clearly noted by Dy Liacco in [131, 132, 133]. This chapter is concerned with power systems that operate in this way.

The class of control problems described herein derives from specific applications in power systems, specifically systems that involve operation in highly nonlinear regimes where failure events cause abrupt changes in the controlled system behavior, which, in turn, require a change in control strategy.

All of the applications of interest herein involve both continuous and discrete dynamics and are conveniently conceived as a *hybrid automaton*. Such a model is composed of a description of the discrete transition behavior from one mode to another along with models of continuous dynamic behavior within each mode. The hybrid automaton model has proved to be an important theoretical tool and is a key conceptual device for model building. However, other forms of models, like the MLD, are far more convenient for control system design. The ability to convert from one form of model to another is important.

H.G. Kwatny and K. Miu-Miller, *Power System Dynamics and Control*,
Control Engineering, DOI 10.1007/978-0-8176-4674-5_8

In the following approach, the transition behavior of a hybrid automaton is modeled by a logical statement (or *specification*). The logical specification can be converted into a set of mixed-integer formulas (IP formulas)[1]. Thus, the transition specification for the automaton is converted into a set of inequalities involving Boolean variables. Logical constraints other than the transition dynamics can also be added to the specification, making this a powerful approach to formulating an optimal control problem. The authors in [187] describe a tool for building MLD models that allows the inclusion of Boolean equivalents to logical specifications. So one could use the tools noted above tool to create those expressions from an arbitrary logical specification.

The IP formulas are used in computing the optimal control strategy. Our approach derives a feedback policy based on finite horizon *dynamic programming* [23]. Dynamic programming has been used extensively in control system design and has recently been explored as a tool for designing hybrid system feedback controls. Its popularity derives from the generality and broad applicability of the *principle of optimality* on which it is based. A drawback of dynamic programming is the *curse of dimensionality* - a term coined by Bellman about 50 years ago, well before the advent of powerful desktop computers.

Branicky et al [30] laid the groundwork for the use of dynamic programming in hybrid systems. They focused on the existence of optimal and near-optimal controls and the establishment of a taxonomy for hybrid systems. In [86], the authors introduced a discrete version of Bellman's inequality to compute a lower bound on the optimal cost function using linear programming. In this way, an approximation of the optimal feedback control law is derived. Another innovative work, [134], considers the problem of approximating the value function. They called their procedure value iteration from which a suboptimal solution is found within a user-specified distance from the optimal solution. They have applied this *relaxed dynamic programming* approach to design a switched power controller for a DC-DC converter.

The hybrid systems study most closely related to our approach is the one described by Bemporad et al in their recent paper [29]. They consider the optimal control of constrained discrete time linear hybrid systems with quadratic or linear performance criteria. The associated Hamilton-Jacobi-Bellman equations are solved backward in time using a multi-parametric quadratic (or linear) programming solver. Two cases are considered, one without binary inputs and the other one with binary inputs. In the latter case, all possible combinations of binary inputs are enumerated.

In our case, we consider nonlinear discrete time hybrid dynamics with a general convex cost function with primarily binary controls. A central feature of our formulation is that it applies to systems with complex logical constraints, defined either by the transition system or by the auxiliary considerations. We exploit the fact that the system is highly constrained and most of the constraints are linear in Boolean variables. Thus, we use the *Mathematica* function `Reduce` to determine feasible points from which we identify those of minimum cost by enumeration. `Reduce` is a

---

[1] A computational tool for this purpose has been constructed in *Mathematica*. This work, described in [118, 119, 120], extends earlier work in this area reported in [130, 199].

powerful function that finds feasible solutions by solving equations and inequalities and eliminating quantifiers. The method used depends on the specific structure of the expressions involved.

In Section 8.2, we provide a specific definition of the problems considered herein. Sections 8.3 and 8.4 describe the main concerns of this paper, namely the reduction of a logical specification for the discrete subsystem to a set of inequalities and the use of this model of a hybrid system to design optimal feedback controllers via dynamic programming. An example is given in Section 8.7.3. The example shows how additional logical constraints - other than the transition behavior - can be incorporated into the control problem.

8.2 Problem Definition

8.2.1 Modeling

The class of hybrid systems to be considered is defined as follows. The system operates in one of the m modes denoted $q_1, \ldots, q_m$. We refer to the set of modes $Q = \{q_1, \ldots, q_m\}$ as the discrete state space. The discrete time difference-algebraic equation (DAE) describing operation in mode q_i is

$$\begin{aligned} x_{k+1} &= f_i(x_k, y_k, u_k) \\ 0 &= g_i(x_k, y_k, u_k) \end{aligned} \qquad i = 1, \ldots, m \tag{8.1}$$

where $x \in X \subseteq R^n$ is the system continuous state, $y \in Y \subseteq R^p$ is the vector of algebraic variables, and $u \in U \subseteq R^m$ is the continuous control. Transitions can occur only between certain modes. The set of admissible transitions is $E \subseteq Q \times Q$. It is convenient to view the mode transition system as a graph with elements of the set Q being the nodes and the elements of E being the edges. We assume that transitions are instantaneous and take place at the beginning of a time interval. So, if a system transitions from mode q_1 to q_2 at time k, we would write $q(k) = q_1, q(k^+) = q_2$. We do allow resets. State trajectories are assumed continuous through events, i.e., $x(k) = x(k^+)$, unless a reset is specified.

Transitions are triggered by external *events* and *guards*. We denote the finite set of events Σ. It is convenient to partition the events into two types: those that are controllable (they can be assigned a value by the controller) and those that are not. The latter are exogenous and occur spontaneously. Such an event might correspond to a component failure, or a high-level change of operational mode. We will use the symbols s to represent controllable events and p to represent uncontrollable events. Thus, $\Sigma = S \times P$ where $s \in S$ and $p \in P$. A guard is a subset of the continuous state space X that enables a transition. A transition enabled by a guard might represent a protection device. Not all transitions have guards, and some transitions might require simultaneous satisfaction of a guard and the occurrence of an event. The guard assignment function is $G : E \to 2^X$.

We consider each discrete state label, $q \in Q$, and each event, $\sigma \in \Sigma$, to be logical variables that take the values true or false. Guards also are specified as logical conditions. In this way, the transition system, including guards, can be defined by a logical specification (formula) $\mathcal{L}$.

In summary, a hybrid control system is composed of the following:

1. Q, discrete space,
2. X, continuous state space,
3. E, set of transitions,
4. Σ, event set,
5. G, guard assignment function,
6. $\mathcal{L}$, logical specification,
7. F, family of controlled vector fields.

Example 8.1 (*Three-mode system*) Consider the simple three-mode hybrid system shown in Figure 8.1. Each mode, q_1, q_2, q_3, is characterized by continuous dynamics $x_{k+1} = f_{q_i}(x_k, u_k)$, $i = 1, 2, 3$.

Discrete transitions are associated with the events represented by logical variables p, s_1, s_2, s_3, i.e., $\Sigma = \{p, s_1, s_2, s_3\}$. For example, if the system is in mode q_1 and s_1 evaluates to true, then a mode transition occurs in which the mode changes from q_1 to q_2. In this example, we use two different symbols s and p to denote transition variables to underscore the fact that some transitions are controllable and others not so.

In our formulation, the transition system behavior is defined by the logical specification:

$$\begin{aligned}\mathcal{L} = & \textit{exactly}\,(1, \{q_1(t), q_2(t), q_3(t)\}) \wedge \textit{exactly}\left(1, \left\{q_1\left(t^+\right), q_2\left(t^+\right), q_3\left(t^+\right)\right\}\right) \wedge \\ & \left(q_1(t) \wedge s_1 \Rightarrow q_2\left(t^+\right)\right) \wedge \left(q_1(t) \wedge p \Rightarrow q_3\left(t^+\right)\right) \wedge \left(q_1(t) \wedge \neg(s_1 \vee p) \Rightarrow q_1\left(t^+\right)\right) \wedge \\ & \left(q_2(t) \wedge s_2 \Rightarrow q_1\left(t^+\right)\right) \wedge \left(q_2(t) \wedge \neg s_2 \Rightarrow q_2\left(t^+\right)\right) \wedge \\ & \left(q_3(t) \wedge s_3 \Rightarrow q_2\left(t^+\right)\right) \wedge \left(q_3(t) \wedge \neg s_3 \Rightarrow q_3\left(t^+\right)\right)\end{aligned} \tag{8.2}$$

Let us dissect this specification. The first line expresses the fact that the system can only be in one discrete state before the transition (at time t) and after the transition

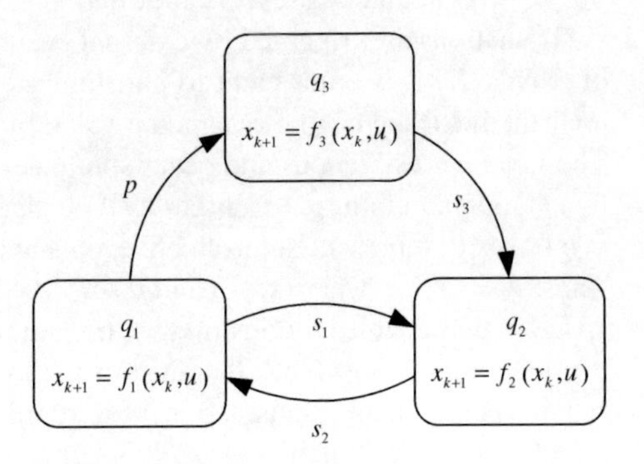

Fig. 8.1 Three-mode hybrid system with controllable and uncontrollable events

(at time t^+). The next line describes all possible transitions from state q_1. Similarly, the last line characterizes all possible transitions from states q_2 and q_3, respectively.

For computational purposes, it is useful to associate with each logical variable, say α, Boolean variable or indicator function, δ_α, such that δ_α assumes the values 1 or 0 corresponding, respectively, to α being true or false. It is convenient to define the discrete state vector $\delta_q = [\delta_{q_1}, \ldots, \delta_{q_m}]$, the control event vector $\delta_s = [\delta_{s_1}, \ldots, \delta_{s_{m_S}}]$, and the exogenous event vector $\delta_p = [\delta_{p_1}, \ldots, \delta_{p_{m_P}}]$. Precisely, one of the elements of δ_q will be unity and all others will be zero.

Notice that with the introduction of the Boolean variables, we can replace the set of dynamical equations (8.24) with the single relation

$$\begin{aligned} x(k+1) &= f\left(x(k), \delta_q(k), u(k)\right) \\ &= \delta_{q_1} f_{q_1}(x(k), u(k)) + \cdots \\ &\quad \cdots + \delta_{q_m} f_{q_m}(x(k), u(k)) \\ 0 &= g\left(x(k), \delta_q(k), u(k)\right) \\ &= \delta_{q_1} g_{q_1}(x(k), u(k)) + \cdots \\ &\quad \cdots + \delta_{q_m} g_{q_m}(x(k), u(k)) \end{aligned} \tag{8.3}$$

8.2.2 *The Control Problem*

Assume that the system is observed in operation over some finite time horizon T that is divided into N discrete time intervals of equal length. A control policy is a sequence of functions

$$\pi = \left\{\mu_0\left(x_0, \delta_{q,0}\right), \ldots, \mu_{N-1}\left(x_{N-1}, \delta_{q,(N-1)}\right)\right\}$$

such that

$$\left[u_k, \delta_{s,k}\right] = \mu_k\left(x_k, \delta_{q,k}\right)$$

Thus, μ_k generates the continuous control u_k and the discrete control $\delta_{s,k}$ that are to be applied at time k, based on the state $\left(x_k, \delta_{q,k}\right)$ observed at time k.

Consider the set of m-tuples $\{0, 1\}^m$. Let Δ_m denote the subset of elements $\delta \in \{0, 1\}^m$ that satisfy $\delta_1 + \cdots + \delta_m = 1$. Denote by Π the set of sequences of functions $\mu_k : X \times \Delta_m \to U \times \{0, 1\}^{m_S}$ that are piecewise continuous on X.

Now, given the initial state $(x_0, \delta_{q,0})$, the problem is to find a policy, $\pi^* \in \Pi$, that minimizes the cost functional

$$J_\pi\left(x_0, \delta_{q,0}\right) = g_N\left(x_N, \delta_{q,N}\right) + \sum_{k=0}^{N-1} g_k\left(x_k, \delta_{q,k}, \mu_k\left(x_k, \delta_{q,k}\right)\right) \tag{8.4}$$

Specifically, the *optimal feedback control problem* is defined as follows. For each $x_0 \in X$, $\delta_{q,0} \in \Delta_m$ determine the control policy $\pi^* \in \Pi$ that minimizes the cost (8.26) subject to the constraints (8.24) and the logical specification, $\mathcal{L}$, i.e.,

$$J_{\pi}^{*}\left(x_0, \delta_{q,0}\right) \leq J_{\pi}\left(x_0, \delta_{q,0}\right) \quad \forall \pi \in \bar{\Pi} \tag{8.5}$$

where $\bar{\Pi} \subset \Pi$ is the subset of policies that steer (8.24) along trajectories that satisfy $\mathcal{L}$.

Notice that if a receding horizon optimal control is desired, once the optimal policy is determined, we need only to implement the state feedback control

$$[u, \delta_s] = \mu_0\left(x, \delta_q\right) \tag{8.6}$$

8.3 Logical Specification to IP Formulas

The first step in solving the optimal control problem is to transform the logical specification $\mathcal{L}$ into a set of inequalities involving integer (in fact, Boolean) variables and possibly real variables, the so-called *IP formulas*. The idea of formulating optimization problems using logical constraints and then converting them to IP formulas has a long history. This concept was recently used as a means to incorporate qualitative information in process control and monitoring [189] and generally introduced into the study of hybrid systems in [24].

McKinnon, [145], proposed the inclusion of logical constraints in optimization methods. They suggested a sequence of transformations that bring a logical specification into a set of IP formulas. Li, [130], presents a systematic algorithm for transforming logic formulas into IP formulas. Those methods have been modified and extended in order to obtain simpler and more compact IP formulas with other modifications to enhance their applicability to hybrid systems.

8.3.1 Logical Modeling Language

We use a logical specification to describe the transition behavior of a hybrid automaton. The specification is simply a logical formula. Here, we describe the set of formulas, i.e., the language, to be employed. A *propositional variable* is a variable that can assume the values true or false. A *propositional formula* is composed of propositional variables, *logical connectives* (specifically $\wedge$, $\vee$, $\Rightarrow$, $\Leftrightarrow$, $\neg$), predicates (Boolean-valued functions of propositional variables), and constraints. Specifically, we will use the predicates: atleast(m, S), atmost(m, S), exactly(m, S), and none(S), where $m \geq 1$ is an integer and S is a list of propositional variables or formulas. A constraint is an arithmetic equality or inequality involving integer or real numbers and variables. Constraints evaluate to true (satisfied) or false (not satisfied).

Formulas are defined by the following statements:

1. A propositional variable or a constraint is an atomic formula,
2. An atomic formula is a formula,

3. $F_1 \sim F_2$ is a formula if F_1 and F_2 are formulas and $\sim$ is one of the logical connectives,
4. $\neg F$ is a formula if F is a formula,
5. atleast(m, S), atmost(m, S), exactly(m, S), and none(S) are formulas if S is a list of formulas and $m \geq 1$ is an integer.

8.3.2 *Transformation to IP Formulas*

Logical formulas are convenient for problem formulation. However, in order to compute efficiently, it is often convenient to convert a logical formula into a set of the so-called IP formulas[2], that is, a set of linear equalities or inequalities involving Boolean variables. To do this, we use the transformation procedure defined in [129]. Following [145], the process involves first transforming the original formula into an intermediate form called a Γ-form, and then, a series of transformations are applied that reduce the Γ-form to a set of IP formulas.

The Γ-form is a logically equivalent normal form that leads to a more compact set of IP formulas than better known normal forms like the CNF (conjunctive normal form) or DNF (disjunctive normal form).

8.3.3 *Implementation*

The basic function in our *Mathematica* implementation is `GenIP` which takes as two arguments, the specification and a list of variables, either propositional variables or bounded real or integer variables. The latter are specified in the form $a \leq x \leq b$. `GenIP` performs a series of transformations and simplifications and returns the IP formulas. A typical usage would look like:

$$\texttt{GenIP}[(q1 \oplus q2) \wedge (qq1 \oplus qq2) \wedge ((q1 \wedge (x > 0)) \Rightarrow qq2) \wedge \\ ((q2 \wedge s) \Rightarrow qq1), \{q1, q2, qq1, qq2, s, -2 \leq x \leq 2\}]$$

$$\{1 - \delta_{q1} - \delta_{q2} \geq 0, -1 + \delta_{q1} + \delta_{q2} \geq 0, 1 - \delta_{qq1} - \delta_{qq2} \geq 0, \\ d7 - \delta_{q1} + \delta_{qq2} \geq 0, -1 + \delta_{qq1} + \delta_{qq2} \geq 0, \\ 1 - \delta_{q2} + \delta_{qq1} - \delta_s \geq 0, -2 + 2d7 + x \leq 0, -2 \leq x \leq 2, \\ 0 \leq d7 \leq 1, 0 \leq \delta_{q1} \leq 1, 0 \leq \delta_{q2} \leq 1, 0 \leq \delta_{qq1} \leq 1, \\ 0 \leq \delta_{qq2} \leq 1, 0 \leq \delta_s \leq 1\}$$

[2] While, generally, computing with IP formulas is preferred, [92] shows that there are instances when it is an advantage to compute using the original logical constraint.

Notice that propositional variables are replaced by Boolean indicator functions, example q_1 is replaced by δ_{q_1} and new auxiliary variables may be introduced, in this case $d7$.

If all of the guards are linear (set boundaries are composed of linear segments), then the IP formulas are system of linear constraints involving the Boolean variables $\delta_q, \delta_{q^+}, \delta_s, \delta_p$, respectively, the discrete state before transition, the discrete state after transition, the controllable events, and the exogenous events. They also involve a set of auxiliary Boolean variables, d, introduced during the transformation process and the real state variables, x. The general form is[3]

$$E_5\delta_{q^+} + E_6 d \leq E_0 + E_1 x + E_2\delta_q + E_3\delta_s + E_4\delta_p \tag{8.7}$$

where the matrices have appropriate dimensions. As we will see in examples below, with $x, \delta_q, \delta_s, \delta_p$ given, these inequalities typically provide a unique solution for the unknowns δ_{q^+} and d. The system evolution is described by the closed system of equations (8.7) and (8.25).

8.4 Constructing the Optimal Solution

An optimal policy π^* is one that satisfies (8.27). Now, we are in a position to apply Bellman's principle of optimality: Suppose $\pi^* = \left\{\mu_1^*, \ldots, \mu_{N-1}^*\right\}$ is an optimal control policy. Then, the subpolicy $\pi_i^* = \left\{\mu_i^*, \ldots, \mu_{N-1}^*\right\}$, $1 \leq i \leq N-1$ is optimal with respect to the cost function (8.26).

Let us denote the optimal cost of the trajectory beginning at $x_i, \delta_{q,i}$ as $J_i^*\left(x_i, \delta_{q,i}\right)$. It follows from the principle of optimality that

$$J_{i-1}^*\left(x_{i-1}, \delta_{q,(i-1)}\right) = \min_{\mu_{i-1}}\left\{g_{i-1}\left(x_{i-1}, \delta_{q(,i-1)}, \mu_{i-1}\right) + J_i^*\left(x_i, \delta_{q,i}\right)\right\} \tag{8.8}$$

Equation (8.8) provides a mechanism for backward recursive solution of the optimization problem. To begin the backward recursion, we need to solve the single-stage problem with $i = N$. The end point $x_N, \delta_{q,N}$ is free, so we begin at a general terminal point

$$J_{N-1}^*\left(x_{N-1}, \delta_{q,(N-1)}\right) = \min_{\mu_{N-1}}\left\{\begin{array}{r} g_{N-1}\left(x_{N-1}, \delta_{q,(N-1)}, \mu_{N-1}\right) \\ +g_N\left(f_{N-1}, \delta_{q^+,(N-1)}\right) \end{array}\right\} \tag{8.9}$$

Once the pair μ_{N-1}^*, J_{N-1}^* is obtained, we compute μ_{N-2}^*, J_{N-2}^*. Continuing in this way, we obtain

[3]Linearity only obtains if the conditions in the specification involving real variables are themselves linear.

$$J^*_{N-i}\left(x_{N-i}, \delta_{q,(N-i)}\right) = \min_{\mu_{N-i}} \left\{ \begin{array}{c} g_{N-i}\left(x_{N-i}, \delta_{q,(N-i)}, \mu_{N-i}\right) \\ +J^*_{N-i+1}\left(f_{N-i}, \delta_{q^+,(N-i)}\right) \end{array} \right\} \tag{8.10}$$

for $2 \leq i \leq N$.

We need to solve (8.10) recursively backward, for $i = 2, \ldots, N$ after initializing with (8.9). We begin by constructing a discrete grid on the continuous state space. The discrete space is denoted $\bar{X}$. At each iteration, the optimal control and the optimal cost are evaluated at discrete points in $Q \times \bar{X}$. To continue with the next stage, we need to set up an interpolation function to cover all points in $Q \times X$.

We exploit the fact that the system is highly constrained, and almost all of the constraints are linear in Boolean variables. The basic approach is as follows:

1. Before beginning the time iteration:
   a. Separate the inequalities into binary and real sets, binary formulas contain only binary variables, and real formulas can contain both binary and real variables.
   b. For each $q \in Q$, obtain all feasible solutions of the binary inequalities, a list of possible solutions of pairs $\left(\delta_{q^+}, d\right)$. Our implementation employs the *Mathematica* function `Reduce`.
   c. Define projection $\bar{X} \rightarrow \bar{X}_{\mathcal{P}}$ where $\bar{X}_{\mathcal{P}}$ is the subspace of real states actually appearing in the real equations.
   d. For each $x_{\mathcal{P}} \in \bar{X}_{\mathcal{P}}$
      i. prescreen the binary solutions to eliminate those that do not produce solutions to the real inequalities - typically a very large fraction is dropped.
      ii. for every feasible combination of binary variables obtained above, solve the real inequalities for the real variables.
   e. Lift real solutions to entire $\bar{X}$.
2. For each i,
   a. For each pair $(q, x) \in Q \times \bar{X}$
      i. enumerate the values of the cost to go using the feasible sets of binary and real variables.
      ii. select the minimum.

In step 1b above, the number of solutions corresponding to each q can be very large because there are numerous redundant solutions associated with nonactive transitions. Thus, we add additional logical constraints that specify the inactive transitions. Step 1c exploits the fact that some real states do not appear in the real formulas. Because a large fraction of the binary solutions do not lead to real solutions, the prescreening in step 1(d)i is very effective in reducing computing time. Finally, we note that the inequalities are independent of the stage of the dynamic programming recursion. Thus, step 1d, which is by far the most intensive computational element of the optimization, is done only once before the recursion step 2a begins.

8.5 Example: Load Shedding

This section provides a simple illustration of the formulation and solution of a power management optimal control problem. For simplicity of exposition, load shedding is used as a means for accommodating transmission line faults.

8.5.1 Network and Load Dynamics

A relatively simple system that is known to exhibit interesting voltage stability characteristics is a single generator feeding an aggregated load composed of constant impedance loads and induction motors. The system has been used to study the effect of tap-changing transformers and capacitor banks in voltage control, example [21, 161, 164].

Consider the system shown in Figure 8.2. The system consists of a generator, a transmission line, an on-load tap-changing transformer (OLTC), and an aggregated load. The generator is characterized by a "constant voltage behind reactance" model. The generator internal bus voltage E is used to maintain the voltage at bus 2, as long as E remains within the limits imposed by the excitation current limits. The OLTC ordinarily moves in small discrete steps over a narrow range. The load is an aggregate composed of parallel induction motors and constant impedance loads. An induction motor can be characterized as an impedance with slowly varying resistance; consequently, the aggregate load is represented by constant impedance - actually, a slowly varying impedance, where the impedance depends on the aggregate induction motor slip.

The network equations are easily obtained. Suppose δ_1, δ_2 denote the voltage angles at buses 1 and 2. Define the relative angle $\theta_2 = \delta_2 - \delta_1$. The network equations are

$$\begin{aligned} I_1 \omega_0 \dot{\omega} &= P_g - cV_2^2 \\ 0 &= (a/n)\, E V_2 \sin\theta_2 + cV_2^2 \\ 0 &= (a/n)\, E V_2 \cos\theta_2 + dV_2^2 \end{aligned}$$

Fig. 8.2 System configuration

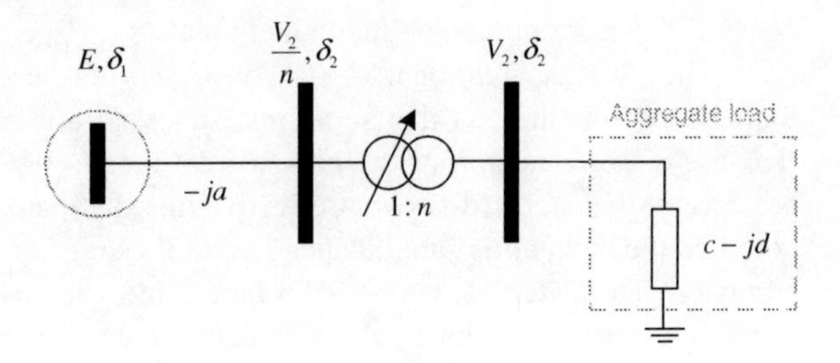

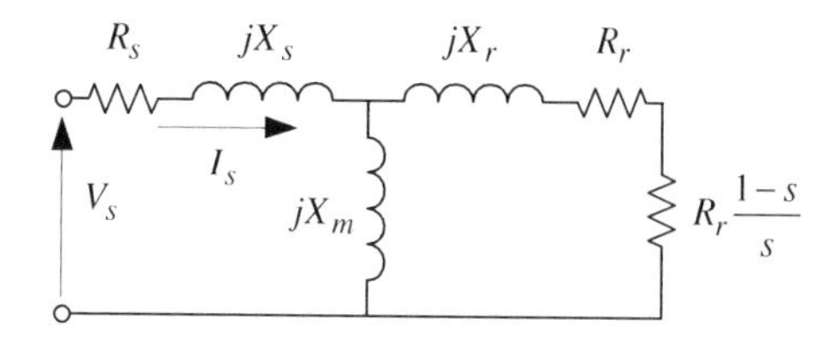

Fig. 8.3 Induction motor equivalent circuit

From the last two equations, we obtain

$$V_2 = \frac{a/n}{\sqrt{c^2 + d^2}} E, \ \theta_2 = \tan^{-1} \frac{c}{d}$$

The power absorbed by the load is

$$P_L = -V_2^2 c, \quad Q_L = V_2^2 d$$

Now, let us turn to the induction motors. An equivalent circuit for an induction motor is shown in Figure 8.3. Here, the parameters R_s, X_s denote the resistance and inductance of the stator, X_m denotes the magnetizing inductance, and R_r, X_r the rotor resistance and inductance, respectively. The resistance $R_r(1-s)/s$ represents the motor electrical output power. We will neglect the small stator resistance and inductance. We also assume the approximation of large magnetizing inductance is acceptable.

Under these conditions, obtain the following. The real power delivered to the rotor, P_d, and the power delivered to the shaft, P_e, are

$$P_d = V_s^2 \frac{R_r s}{R_r^2 + s^2 X_r^2} \quad P_e = P_d(1-s)$$

The dynamical equation for the motor (Newton's law) is

$$\dot{\omega}_m = \frac{1}{I_m \omega_0}(P_e - P_m)$$

Introducing the slip, s, $s = (\omega_0 - \omega_m)/\omega_0$, the motor dynamics take the form:

$$\dot{s} = \frac{1}{I_m \omega_0^2}(P_m - P_e) = \frac{1}{I_m \omega_0^2}\left(P_m - V_s^2 \frac{R_r s(1-s)}{R_r^2 + s^2 X_r^2}\right)$$

8.5.2 System Operation

In the following, we allow for shedding a fraction, η, of the load. In the present example, we allow three different values of η including zero, so $\eta \in \{0, \eta_1, \eta_2\}$.

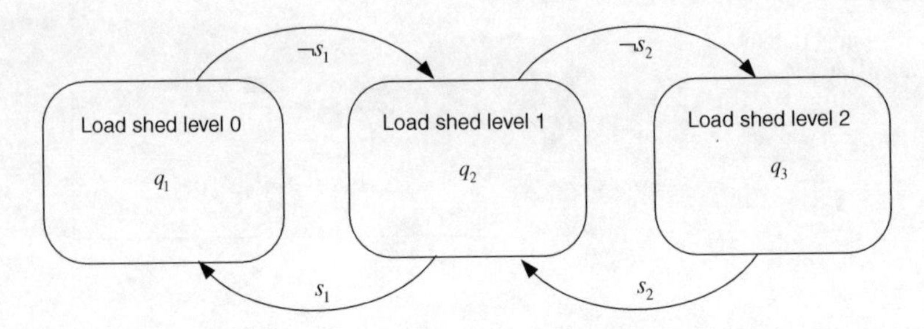

Fig. 8.4 Transition diagram for load shedding optimization

Consequently, there is normal operation and two prioritized blocks of load that can be dropped in accordance with the transition behavior defined in Figure 8.4. The corresponding logical specification is

$$\begin{aligned}\mathcal{L} = &\, exactly\,(1, \{q_1(t), q_2(t), q_3(t)\}) \wedge exactly\left(1, \left\{q_1\left(t^+\right), q_2\left(t^+\right), q_3\left(t^+\right)\right\}\right) \wedge \\ &\left(q_1(t) \wedge \neg s_1 \Rightarrow q_2\left(t^+\right)\right) \wedge \left(q_1(t) \wedge s_1 \Rightarrow q_1\left(t^+\right)\right) \wedge \\ &\left(q_2(t) \wedge \neg s_2 \Rightarrow q_3\left(t^+\right)\right) \wedge \left(q_2(t) \wedge s_1 \Rightarrow q_1\left(t^+\right)\right) \wedge \left(q_2(t) \wedge \neg(s_1 \vee \neg s_2) \Rightarrow q_2\left(t^+\right)\right) \wedge \\ &\left(q_3(t) \wedge s_2 \Rightarrow q_2\left(t^+\right)\right) \wedge \left(q_3(t) \wedge \neg s_2 \Rightarrow q_3\left(t^+\right)\right)\end{aligned}$$

In the present case, assume the blocks are sized such that

$$q_1 \Rightarrow \eta = 0, \quad q_2 \Rightarrow \eta = 0.4, \quad q_3 \Rightarrow \eta = 0.8$$

Assume also that the OLTC ratio is fixed, i.e., the OLTC is not being used for control, so $n = \text{const}$. If the OLTC is to be employed, the dynamics of tap change must be added.

$$I_1 \omega_0 \dot{\omega} = P_g - cV_2^2 \tag{8.11}$$

$$E = (1 - \eta) \frac{\sqrt{c_0^2 + d_0^2}}{a/n} V_2 \tag{8.12}$$

$$\dot{s} = \frac{(1-\eta)}{I_m \omega_0^2} \left(P_m - V_2^2 \frac{R_r s (1-s)}{R_r^2 + s^2 X_r^2}\right) \tag{8.13}$$

$$c = (1-\eta)\, c_0, \;\; c_0 = \left(\frac{1}{R_L} + \frac{R_r s}{R_r^2 + s^2 X_r^2}\right) \tag{8.14}$$

$$d = (1-\eta)\, d_0, \;\; d_0 = \left(\frac{X_r s^2}{R_r^2 + s^2 X_r^2}\right) \tag{8.15}$$

Equation (8.11) represents turbine generator dynamics. Ordinarily, the power input P_g is adjusted to regulate the speed ω which is to be maintained at the value ω_0.

Assume that regulation is fast and accurate. It is possible to investigate the impact of frequency variation on system behavior. If it were assumed that frequency variations were small, then the effect on all impedances could be approximated, and this is often done. That has not been included here, so there is no apparent coupling between (8.11) and the remaining equations; consequently, it can be dropped. Equation (8.12) represents the network voltage characteristic. The field voltage E is used to control the load bus voltage V_2. It will be assumed that it is desired to maintain $V_2 = 1$. If the exciter dynamics are ignored, then (8.12) allows the determination of the field voltage that yields the desired load bus voltage. However, the field voltage is strictly limited, $0 \leq E \leq 2$. Assume that only the upper limit is a binding constraint. There are two possibilities for satisfying (8.12):

$$V_2 = 1, \ E = \frac{\sqrt{c^2+d^2}}{a/n} \text{ or } E = 2, \ V_2 = 2\frac{a/n}{\sqrt{c^2+d^2}}$$

Equation (8.21) represents the aggregated motor dynamics, and the load admittance is given by the last two equations. The system data are $R_L = 2, \ R_r = 0.25, \ X_r = 0.125, \ a = 1 \ \text{(nominal)}, \ I_m\omega_0^2 = 4$.

8.5.3 The Optimal Control Problem Without OLTC, $n = 1$

The problem is formulated as an N step moving horizon optimal control problem, in which the slip dynamics are written in discrete time form. The control variables are $E(k), \eta(k)$. The goal is to keep the load voltage V_2 close to 1; specifically, it is required that $0.95 \leq V_2 \leq 1.05$. Our intent is to use the field voltage, E, to regulate the terminal voltage, V_2, to 1 p.u. Because $0 < E \leq 2$ is constrained, specify that solutions must satisfy

$$(V_2 = 1 \wedge 0 < E < 2) \vee (E = 2)$$

If the field voltage saturates, the only remaining option is to shed some load. We seek an optimal control policy, i.e., a sequence of controls $u(0), \ldots, u(N-1)$, $u(k) = \eta(k)$ that minimize the cost function

$$J = \sum_{k=0}^{N-1} \left(\| V_2(k) - 1 \|^2 + r_1 \| \eta(k) \|^2 \right)$$

subject to the system constraints. Some rough assessments of appropriate weighting constants r_1 can be made. Load shedding should be avoided with respect to regulating V_2 unless the V_2 tolerance is violated. Hence, it is desired that $r_1 > 0.25^2/0.05^2 = 1/25$.

In summary, the following equations are obtained:

1. The slip dynamics in discrete time form (with $s_k = s(t_k)$, $t_k = t_{k-1} + h$)

$$s_{k+1} = f(s_k, V_2, \eta)$$

2. The transition specification in IP form

$$\begin{gathered}
1-\delta_{q_1}-\delta_{q_2}-\delta_{q_3}\geq 0,\quad -1+\delta_{q_1}+\delta_{q_2}+\delta_{q_3}\geq 0\\
1-\delta_{q_1^+}-\delta_{q_2^+}-\delta_{q_3^+}\geq 0,\quad -1+\delta_{q_1^+}+\delta_{q_2^+}+\delta_{q_3^+}\geq 0\\
1-\delta_{q_1}+\delta_{q_1^+}-\delta_{s_1}\geq 0,\quad 1-\delta_{q_2}+\delta_{q_1^+}-\delta_{s_1}\geq 0\\
1-\delta_{q_2}+\delta_{q_2^+}-\delta_{s_2}\geq 0,\quad 1-\delta_{q_3}+\delta_{q_2^+}-\delta_{s_2}\geq 0\\
-\delta_{q_1}+\delta_{q_2^+}+\delta_{s_1}\geq 0\\
-\delta_{q_2}+\delta_{q_3^+}+\delta_{s_2}\geq 0,\quad -\delta_{q_3}+\delta_{q_3^+}+\delta_{s_2}\geq 0\\
0\leq\delta_{q_1}\leq 1, 0\leq\delta_{q_2}\leq 1, 0\leq\delta_{q_3}\leq 1\\
0\leq\delta_{q_1^+}\leq 1, 0\leq\delta_{q_2^+}\leq 1, 0\leq\delta_{q_3^+}\leq 1\\
0\leq\delta_{s_1}\leq 1, 0\leq\delta_{s_2}\leq 1
\end{gathered}$$

3. The IP formulas for the logical constraint

$$\begin{gathered}
3-d_1-E>0,\quad 1-d_1+E>0,\quad -2d_2+E\geq 0\\
-2d_1+V_2\geq 0,\quad -2+d_1+V_2\leq 0\\
0\leq d_1, d_2\leq 1,\quad 0\leq E, V_2\leq 2
\end{gathered}$$

4. And the IP formulas for the load shed parameter η

$$\begin{gathered}
-0.4d_4+\eta\geq 0,\quad -0.8d_5+\eta\geq 0,\\
d_3-\delta_{q_1^+}\geq 0,\quad d_4-\delta_{q_2^+}\geq 0,\quad d_5-\delta_{q_3^+}\geq 0\\
-1+d_3+\eta\leq 0,\quad -1+0.6d_4+\eta\leq 0,\\
-1+0.2d_5+\eta\leq 0\\
0\leq d_3\leq 1,\quad 0\leq d_4\leq 1,\quad 0\leq d_5\leq 1,\quad 0\leq\eta\leq 1
\end{gathered}$$

One result is shown in Figure 8.5. It illustrates the optimal load shedding strategy following a line failure represented as a reduction of a. The feedback control is given as a function of the state - the latter composed of the continuous slip and the three discrete states. At each state, the values of the control actions δ_{s_1}, δ_{s_2} are given. The controlled transitions are also indicated.

Suppose immediately post-failure, the system is in mode q_1, with a reduced slip of 0.1, then the system will respond as follows. Given a mechanical power level of 0.7, the equilibrium slip is about 0.47. As slip increases toward its equilibrium value, the first block of load is dropped at about $s = 0.3$ and the second at about $s = 0.4$.

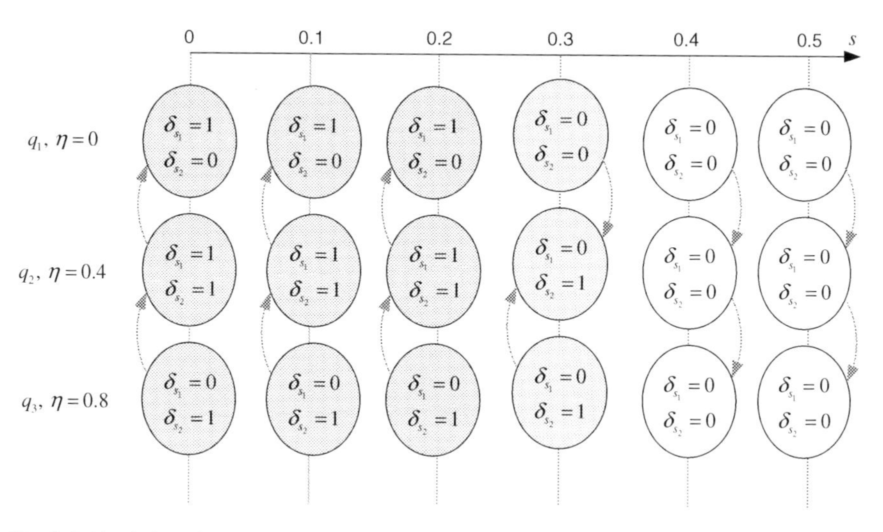

Fig. 8.5 Depiction of the feedback law obtained with $a = 0.25$, $h = 0.5$, and $N = 20$

8.5.4 *Incorporating Time Delays*

Sometimes, it is desirable to insure that there is a finite time duration between two successive controlled transitions. It is easy to do this by incorporating a time "residence" requirement within a discrete state. For example, suppose we wish to insure that a load shedding action will not be followed by another until at least a time Δ has passed. This can be accomplished by requiring that after entry into state q_2, the system must remain in q_2 for at least time Δ.

To accomplish this, we introduce a resetting "clock"

$$\tau(k+1) = \tau\left(k^+\right) + h$$

where $\tau(k^+) = 0$ upon entry into q_2 from q_1 or q_3 or $\tau(k^+) = \tau(k)$ and replace the specification $\mathcal{L}$ by

$$\begin{aligned}
\mathcal{L} = {} & \mathit{exactly}\left(1, \{q_1(t), q_3(t), q_3(t)\}\right) \wedge \\
& \mathit{exactly}\left(1, \left\{q_1\left(t^+\right), q_3\left(t^+\right), q_3\left(t^+\right)\right\}\right) \wedge \\
& \left(q_1(t) \wedge \neg s_1 \Rightarrow q_2\left(t^+\right) \wedge \tau\left(t^+\right) = 0\right) \wedge \\
& \left(q_1(t) \wedge s_1 \Rightarrow q_1\left(t^+\right) \wedge \tau\left(t^+\right) = \tau(t)\right) \wedge \\
& \left(q_2(t) \wedge s_1 \wedge \tau(t) > \Delta \Rightarrow q_1\left(t^+\right) \wedge \tau\left(t^+\right) = \tau(t)\right) \wedge \\
& \left(q_2(t) \wedge \neg s_2 \wedge \tau(t) > \Delta \Rightarrow q_3\left(t^+\right) \wedge \tau\left(t^+\right) = \tau(t)\right) \wedge \\
& \left(\begin{array}{l} q_2(t) \wedge \neg\left(\left(s_1 \wedge \tau(t) > \Delta\right) \vee \left(\neg s_2 \wedge \tau(t) > \Delta\right)\right) \\ \quad \Rightarrow q_2\left(t^+\right) \wedge \tau\left(t^+\right) = \tau(t) \end{array}\right) \wedge \\
& \left(q_3(t) \wedge s_2 \Rightarrow q_2\left(t^+\right) \wedge \tau\left(t^+\right) = 0\right) \wedge \\
& \left(q_3(t) \wedge \neg s_2 \Rightarrow q_3\left(t^+\right) \tau\left(t^+\right) = \tau(t)\right)
\end{aligned}$$

The control law now becomes a function of the discrete state, the two components of the continuous state: the slip s and the clock variable τ. With time delay $\Delta = 1$, the control law is virtually identical to that shown in Figure 8.5 expect that the clock dependence inhibits transitions from q_2 as required.

We will not display the resulting IP formulas, but it is interesting to note that the binary equations involve 24 binary variables, 3 of which are the current state. Consequently, there are $2^{21} = 2,097,152$ possible solutions, but actually only 1000 - 2000 prove to be feasible (depending on the current discrete state). From these emerge about 40-80 feasible real solutions. Finally, the associated cost for these few solutions is enumerated and a minimum cost control is chosen.

8.6 Induction Motor Load with UPS

A relatively simple system that is known to exhibit interesting voltage stability characteristics is a single generator feeding an aggregated load composed of constant impedance loads and induction motors [164]. By expanding this system to include a vital load with a UPS, as shown in Figure 8.6, we obtain one of interest to us.

The primary means for voltage control is the field voltage. However, in the event of a transmission line fault, it may be necessary to shed load in order to avoid a system collapse. This can be accomplished by dropping nonvital load in discrete blocks and, if necessary, switching the vital load to battery supply.

Assume that two blocks of nonvital load can be dropped independently by opening circuit breakers. Correspondingly, a load shed parameter is introduced $\eta \in \{0, \eta_1, \eta_2\}$ that denotes the fraction of load dropped.

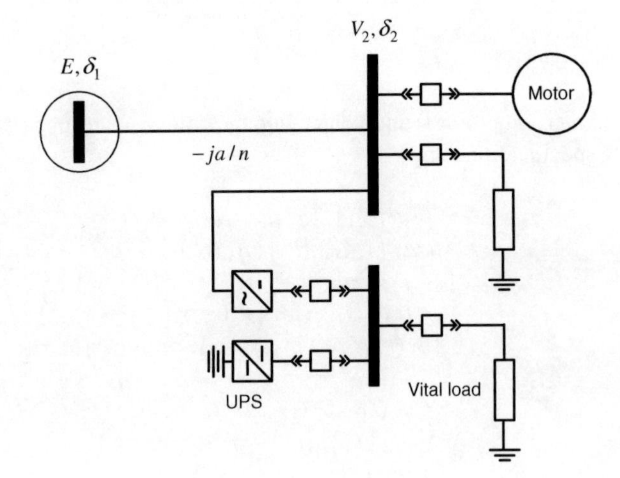

Fig. 8.6 System with vital load and UPS

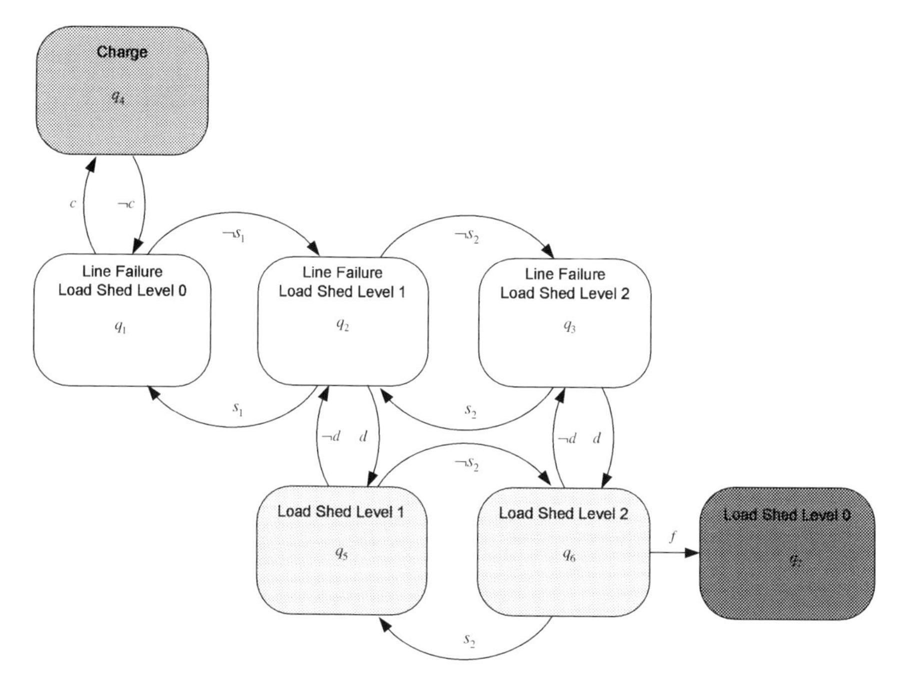

Fig. 8.7 Transition behavior for system with UPS

The battery is connected to the DC load bus through a DC-DC converter. There are three possible UPS operating modes:

1. Battery unconnected.
2. Battery discharging: The battery and vital load are detached from the rest of the network. The battery supplies the load through a voltage-controlled DC-DC converter set up to keep the load voltage constant.
3. Battery charging: In this mode, the battery is charged through a DC-DC converter operated in current-controlled mode – the current is controlled to a specified value.

The overall system transition system is shown in Figure 8.7. It represents operational constraints that are imposed on the system.

8.6.1 Dynamics

8.6.1.1 Battery disconnected, modes q_1, q_2, q_3

The voltage-regulated rectifier controls the voltage on vital load bus. We assume that the rectifier is power factor-corrected so that from the AC side of the rectifier, the vital load looks like a constant power load with unity power fact, $P = P_v$, $Q = 0$.

Let δ_1, δ_2 denote the voltage angles at buses 1 and 2. Define the relative angle $\theta_2 = \delta_2 - \delta_1$. The network equations are

$$\begin{aligned} P_\upsilon &= a E\, V_2 \sin\theta_2 - c\, V_2^2 \\ 0 &= a E\, V_2 \cos\theta_2 + d\, V_2^2 \end{aligned} \tag{8.16}$$

where P_υ is the power consumed by the vital load and $c - jd$ is the admittance of the nonvital aggregate load.

The field voltage E is used to control the load bus voltage V_2 to its desired nominal value of 1. If we ignore the exciter dynamics, then (8.16) allows the determination of the field voltage that yields the desired load bus voltage, provided the resultant E is within its strict limits, $0 \le E \le 2$. It is always the upper limit that is the binding constraint. This implies two possibilities for satisfying (8.16): either $V_2 = 1$ or $E = 2$. These are as follows:

$$V_2 = 1,\ E = \frac{\sqrt{(c+P_\upsilon)^2 + d^2}}{a}, \quad 0 < P_\upsilon \tag{8.17}$$

$$\begin{aligned} &E = 2, \\ &V_2 = \sqrt{\frac{2a^2 - cP_\upsilon - \sqrt{4a^4 - 4a^2 cP_\upsilon - d^2 P_\upsilon^2}}{c^2 + d^2}}, \\ &0 < P_\upsilon < 2a^2\left(\sqrt{c^2 + d^2} - c\right) \end{aligned} \tag{8.18}$$

Once the excitation system saturates, there is an upper limit to P_υ, as seen in (8.18). This is the voltage collapse bifurcation point. Also, these relations are only good for $P_\upsilon > 0$. When $P_\upsilon = 0$, we have

$$V_2 = \frac{a}{\sqrt{c^2 + d^2}} E \tag{8.19}$$

Equation (8.17) (nonsaturated field) does approach the proper limit as $P_\upsilon \to 0$, but the Equation (8.18) (saturated field) does not. This is as it should be.

Remark 8.2 (*Network Solution*) As discussed in Remark 8.5, we can express the network constraints in terms of the logical constraint

$$\mathcal{L}_0 = (V_2 = 1 \Rightarrow E = z_1) \wedge (E = 2 \Rightarrow V_2 = z_2) \tag{8.20}$$

where z_1, z_2 are defined via (8.17), (8.18), and (8.19).

8.6.1.2 Battery Charging, mode q_4

The battery model is composed of a differential equation describing the battery "state of charge" σ and an output map that gives the battery terminal voltage υ_b as a function of the state of charge.

$$\frac{d}{dt}\sigma = \frac{1}{C}i,\ v_b = f(\sigma),\ 0 \le \sigma \le 1$$

where i is the battery charging current and C is the battery effective capacitance. The DC-DC converter operates in current control mode, so the battery is charged with constant current, $i = i_c$. While charging, we have

$$\frac{d\sigma}{dt} = \frac{i_c}{C}$$

Because the AC-DC rectifier maintains constant V_3, from the AC side of the rectifier, charging looks like an additional constant power load, $P_c = V_3 i_c$. The network supplies both the vital load and the power to charge the battery. Thus, the network relation is given by Equations (8.17) and (8.18) with P_v replaced by $P_v + P_c$.

8.6.1.3 Battery Discharging, modes q_5, q_6

The vital loads and battery are separated from the rest of the system and draw no power from the network. Consequently, the relationship between E and V_2 is given by Equation (8.19). The DC-DC converter now maintains constant voltage on bus 3, so that the battery current is $i = -P_v/V_3$ and

$$\frac{d\sigma}{dt} = -\frac{P_v}{C V_3}$$

In the following study, we take $C = 0.5$ and $P_v = 10$.

8.6.1.4 Induction Motors

If we neglect the small stator resistance and inductance and assume a large magnetizing inductance, the equivalent circuit for an induction motor consists of a series rotor resistance and inductance R_r, X_r. Define the slip $s = (\omega_0 - \omega_m)/\omega_0$, and let P_m denote the mechanical load power. Then, the motor dynamics take the form:

$$\dot{s} = \frac{1}{I_m \omega_0^2}\left(P_m - V_s^2 \frac{R_r s(1-s)}{R_r^2 + s^2 X_r^2}\right) \tag{8.21}$$

8.6.1.5 Load Shedding

We assume discrete load shedding blocks and define η to represent the fraction of load shed. Thus, η can assume a finite number of values $0 \le \eta < 1$. The nonvital load admittances, taking into account the load shedding parameter, are as follows:

$$c = (1 - \eta)\, c_0,\ c_0 = \left(\frac{1}{R_L} + \frac{R_r s}{R_r^2 + s^2 X_r^2}\right) \tag{8.22}$$

$$d = (1 - \eta)\, d_0,\ d_0 = \left(\frac{X_r s^2}{R_r^2 + s^2 X_r^2}\right) \tag{8.23}$$

Equation (8.21) represents the aggregated motor dynamics, and the load admittance is given by the last two equations, (8.22), (8.23). The system data are $R_L = 2,\ R_r = 0.25,\ X_r = 0.125,\ a = 1$ (nominal) $,\ I_m \omega_0^2 = 4$.

8.6.2 IP Formulas for UPS System

Four logical constraints need to be converted to IP formulas:

1. the network specification, $\mathcal{L}_0$, Equation (8.20)
2. the transition specification, $\mathcal{L}_1$, of Figure 8.7

(3) the excitation shedding specification

$$\mathcal{L}_2 = (V_2 = 1 \wedge 0 < E < 2) \vee (E = 2)$$

(4) the load shedding specification

$$\mathcal{L}_3 = (q_1^+ \Rightarrow \eta = 0) \wedge (q_2^+ \Rightarrow \eta = 0.4) \wedge \\ (q_3^+ \Rightarrow \eta = 0.8)$$

The corresponding IP formulas are generated automatically. We do not display them here because of space limitations. All of the inequalities derived from $\mathcal{L}_1$ involve only binary variables, while some of those derived from $\mathcal{L}_0$, $\mathcal{L}_2$, and $\mathcal{L}_3$ involve both binary and real variables. The latter also contain auxiliary binary variables d_i introduced during the conversion process. All of the inequalities are linear in all variables.

8.6.3 Optimal Control

An optimal control policy is sought that minimizes the cost function

$$J = \sum_{k=0}^{N-1} \left(\begin{array}{c} \| V_2(k) - 1 \|^2 + r_0 \| \sigma - 1 \|^2 \\ + r_1 \| \eta_L(k) \|^2 \end{array} \right)$$

subject to the system constraints. In the following, we take $r_0 = 1$ *and* $r_1 = 1/25$.

Consider the optimal controller for a line fault that results in a line admittance of $a = 0.375$. This is a severe fault, but one that is manageable. The state space includes the 7 discrete states (modes) and two continuous states induction motor slip, s, indicative of power, and battery state, σ, that represents the fractional battery charge. For computational purposes, the continuous state is discretized $s \in \{.1, .2, .3, .4, .5\}$ and $\sigma \in \{.25, .5, .75.1.0\}$, and the feedback control is computed in terms of these 140 states. In implementation, an interpolation function is used for the continuous states.

Figures 8.8, 8.9, and 8.10 illustrate a particular feedback trajectory in which the initial battery state of charge is 0.1 and the initial slip is 0.

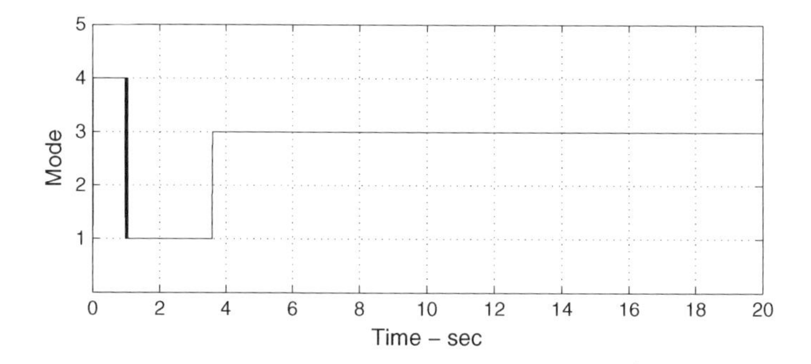

Fig. 8.8 Because of the low battery charge, an initial switch into charging mode 4 occurs before load is dropped, modes 2 and 3

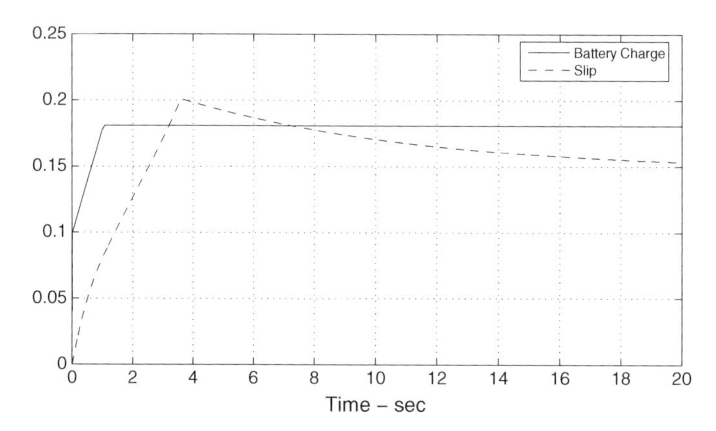

Fig. 8.9 The battery initially charges, but increasing slip, and hence, electrical power, eventually, requires load shedding

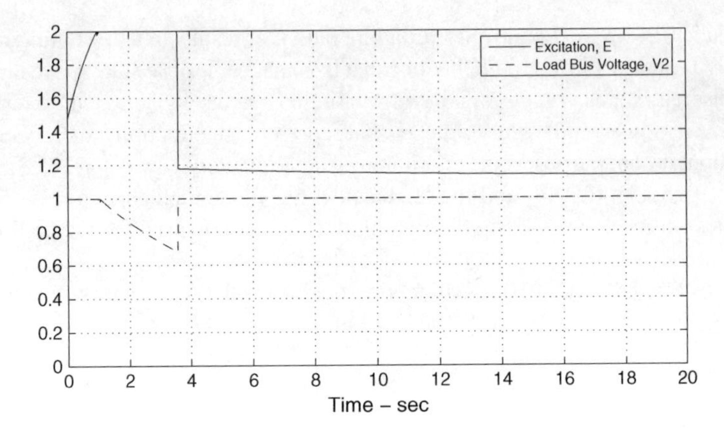

Fig. 8.10 After about 1 second, the excitation saturates and load bus voltage drops. Load voltage regulation is re-established following load shedding

8.7 Ship Integrated Electric Power System

The number of power supply sources available on a ship power system is determined by the need to supply the maximum anticipated electrical and propulsion load. On a naval ship, operational modes that require high level of online resources persist only for a small fraction of the total time a ship is in service. Consequently, a plan for fuel reduction should focus on the low load, normal operations that dominate the ship's lifetime. A significant reduction of fuel consumption can result from running a small number of turbine generators during these periods. However, there is a real risk of contingencies that could lead to the need to curtail load. To insure an acceptable level of reliability of power supply, it is necessary to maintain sufficient online generation and to distribute it appropriately around the network.

In [69], the authors draw an important distinction between *survivability* and *quality of service* (QOS). Survivability addresses the prevention of fault propagation and restoration of service under severe damage conditions, whereas QOS concerns insuring a reliable supply of power to loads during normal operations (see also [70, 97]). QOS is an important consideration during normal operations because equipment malfunction is a relatively common occurrence. Not all loads have the same requirements for continuity of power supply. As used in [69], QOS is quantified as the *mean time between service interruptions* where a service interruption is defined as a degraded network condition that lasts longer than a load can tolerate before losing functionality. In [97], loads are divided into four categories that depend on two time parameters associated with the power network. T_1 is the reconfiguration time: the maximum time to reconfigure the network without bringing on additional generators. T_2 is the generator start time: the time to bring online the slowest generator. Accordingly, four categories of loads are defined:

1. Uninterruptible loads: cannot tolerate a power loss of duration T_1.
2. Short-term interruptible loads: can tolerate a power loss of duration T_1, but not T_2.
3. Long-term interruptible loads: can tolerate a power loss of duration T_2.
4. Exempt loads: loads not considered in evaluating QOS.

Because this QOS metric is intended primarily for DC distribution systems, it does not consider power quality measures such as harmonic content, or voltage fluctuations. In fact, it does not consider dynamics at all. In AC systems, however, dynamics are important.

In [125], the authors formulate the fuel optimization problem with QOS constraints, where QOS has a meaning appropriate for AC system power quality. The problem is formulated as follows. Given a time interval, $[0, T]$, over which the ship is to perform a specified mission with corresponding maximum load, ℓ, having a corresponding distribution over the network, determines a commitment, c_ℓ^*, of generation resources that minimizes fuel costs, supplies the load, and also satisfies QOS constraints. In this case, the QOS constraints are defined as follows.

Definition 8.3 Given

1. a set of contingency events, $\mathcal{R} = \{r_i, i = 1, \cdots, m\}$,
2. a set of performance variables (example bus voltages, line currents, frequency), $\mathcal{Y} = \{y_i, i = 1, \cdots, p\}$, each variable with a corresponding admissible range so that $Y_{i,\min} \leq y_i(t) \leq Y_{i,\max}$ and a time duration, T_i, for which an out-of-range value can be tolerated.

The QOS constraints are satisfied if for every $r \in \mathcal{R}$, occurring at any time $t_r \in [0, T]$, at which time the network is in equilibrium, none of the performance variables $y_i(t)$ experience a constraint violation for a duration longer than its corresponding T_i.

The fuel optimization problem as formulated above is naturally a static optimization problem as meaningful fuel cost savings are obtained when measured over a long period of operation. QOS constraints, on the other hand, involve short-term dynamics. They are incorporated by eliminating from consideration any otherwise feasible commitment configuration. This is accomplished by evaluating the response of the given configuration to the specified contingencies. No attempt is made to optimize that response. In [108], that analysis is expanded to allow the temporary use of load shedding and energy storage to avoid violating contingency constraints. The proposed framework also allows inclusion of load scheduling as a means of fuel conservation.

In the following discussion, an example based on the ship propulsion system described in Appendix A will be employed. The electrical load is assumed constant over the duration of the analysis. Its value varies with the mission and the season and may range from about 2000 KW to 4500 KW.

8.7.1 The Fuel Consumption Model

It is instructive to first consider the operation of the ship in its various configurations in terms of fuel consumption without regard to QOS constraints. The only constraints considered here are the generation capacity of each of the generators and the electric power flow constraints of the network.

Fuel consumption data were obtained from the Navy's Energy Conservation Program Web site http://www.i-encon.com. Based on the DDG 51 CLASS SHIPS data, the associated fuel data and fuel curves for both Allison GTGs and GE LM2500 GTMs can be obtained. Curve fits were used to parameterize the data in terms of ship speed, v, in knots. There are three propulsion alignments with distinct fuel curves.

Trail Shaft One GTM engine online and one shaft windmilling.

$$f_{TS} = 117.17 \exp(0.1087\, v)$$

Split Plant One GTM engine online on each shaft.

$$f_{SP} = 181.74 \exp(0.098\, v)$$

Full Power Two GTM engines online on each shaft.

$$f_{FP} = 334.48 \exp(0.082\, v)$$

For the **Allison 501-K34** GTG fuel consumption, the curve was parameterized in KW electric load, L, and the number of GTGs, N_{GTG}.

$$f_{GTG} = 0.068\, L + 97.4\, N_{GTG}$$

Figures 8.11 and 8.12 show the fuel consumption at low speed (up to 8 knots) and high speed (above 8 knots), respectively, assuming a constant electric load of 3000 KW. Split plant operation has two GTMs operational, one on each shaft with all electric power supplied by two GTGs, as one would not be sufficient. This is the most fuel-efficient configuration. Trail shaft operation is somewhat better as only one GTM is operational. Note that one GTM can comfortably produce 22 knots. The HED motoring configuration with 2 GTGs supplies 3000 KW and 1500 KW (or 2011 HP) for propulsion – so that about 8 knots is achievable – with 500 KW remining. This is the most fuel-efficient configuration for low-speed operation (see Figure 8.11). The HED generation configuration allows all of the GTGs to be shut down, but this configuration is not as efficient as motoring.

The HED motoring configuration can only be used above 8 knots with 3 GTGs operational, thereby increasing fuel consumption and raising it to about the same as trail shaft HED generation. With three GTGs and 3000 KW of electrical load, it can only produce a maximum speed of about 12 knots. Consequently, it is omitted

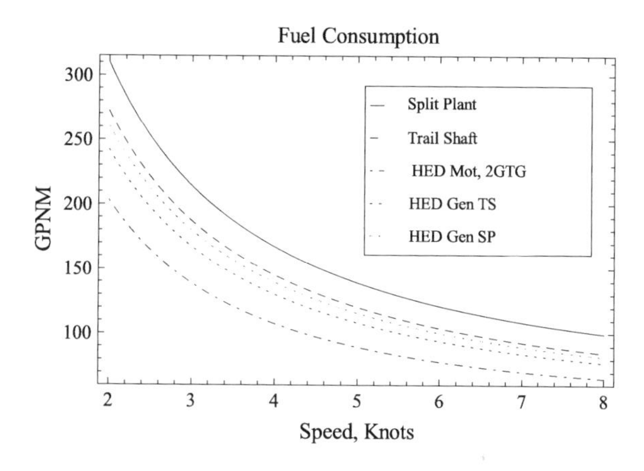

Fig. 8.11 Low-speed fuel consumption as a function of speed in various configurations. Electrical load fixed at 3,000 KW

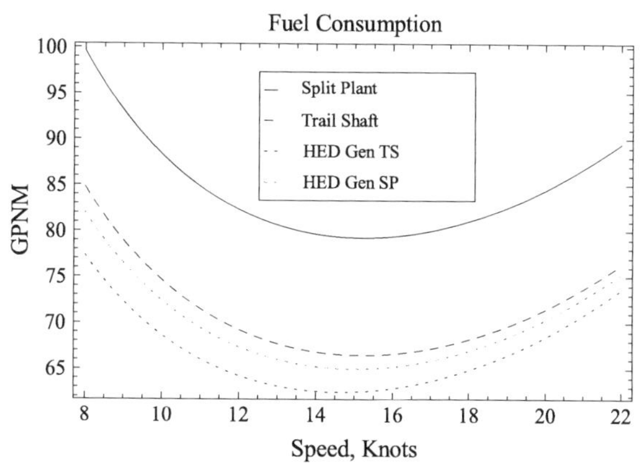

Fig. 8.12 High-speed fuel consumption as a function of speed in various configurations. Electrical load fixed at 3,000 KW

from the high-speed considerations in Figure 8.12. In the high-speed range, trail shaft HED generation is the most fuel-efficient operating configuration. Also, note that the optimal speed is in the range of 14-15 knots.

8.7.2 *Optimal Response to Contingencies*

From Section 8.7.1, it is clear that without consideration of supply reliability, the most efficient operational configuration at low speed is trail shaft HED motoring, and at high-speed operation, it is trail shaft HED generation. The question now turns to how QOS constraints alter this picture. In accordance with Definition 8.3, to answer this, it is necessary to evaluate the candidate configuration with respect to all contingency events in $\mathcal{R}$. This requires delineating the admissible corrective actions to each contingency and then evaluating the corresponding response in terms of continuity of supply variables $\mathcal{Y}$.

Example 8.4 Low-Speed Operation: Loss of Generator. As an example, consider operation of the system described in Appendix A at 7 knots, so the trail shaft HED motoring is the most fuel-efficient configuration. Suppose one of the specified contingencies is loss of one of the two GTGs. Figure 8.13 illustrates the situation in terms of a state diagram. The normal operating state q_1 consists of two GTGs, each producing 2250 KW. The system operating in state q_1 experiences an external event e_1 corresponding to a GTG failure inducing a transition to state q_2. From the failed state, it is desired to restore the system back to the HED motoring state with two GTGs and to do so without violating the QOS requirements. To accomplish this, the controller should react with a sequence of corrective actions. In this example, the actions to be taken include the following:

1. Start up the spare GTG (it takes 6 minutes to get from shutdown to full power).
2. Temporarily drop nonvital load (1000 KW),
3. Supply power, temporarily from the emergency storage module (ESM)
4. Use the generator crisis capacity (4500 KW for up to 5 minutes).

The discrete states $q_i, i = 2, \dots, 6$ are illustrated along with admissible controllable transitions $s_i, i = 1, \dots, 9$. The contingency triggering event causes the system to transition from q_1 to q_2. There are four controlled events leading to transition from

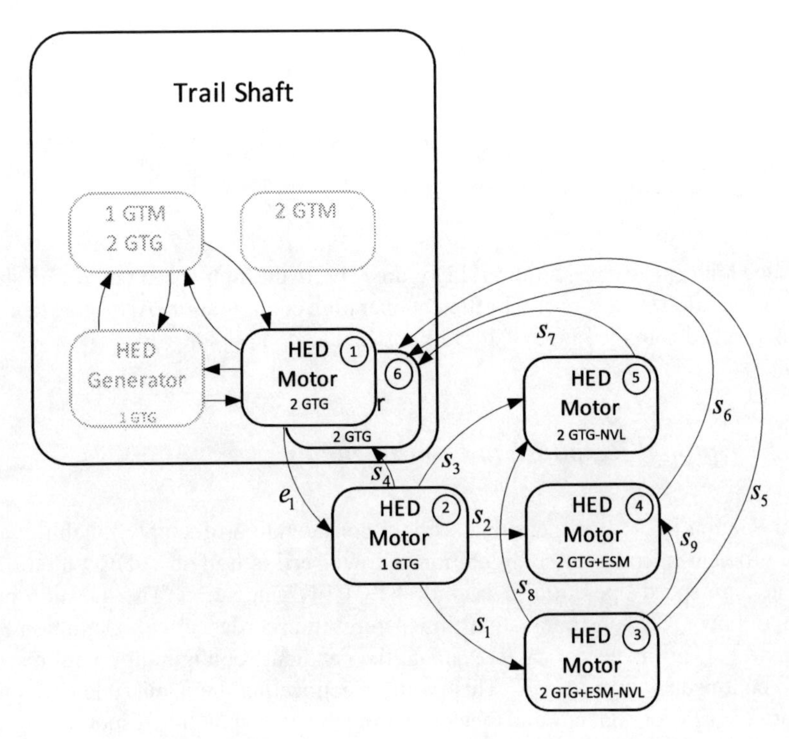

Fig. 8.13 Possible remedial strategies following loss of GTG from trail shaft motoring configuration

q_2. Any departure from state q_2 initiates start-up of GTG 3. Now, it is proposed to select the *best* sequence of controlled transitions aimed at satisfying the QOS constraints. If the best does indeed satisfy the constraints as specified in Definition 8.3, then the same process can be followed for the other contingencies until one fails the test. If all contingencies have an adequate response sequence, the mode is accepted as a valid operating configuration.

In earlier publications [118, 120], the authors introduced an approach that uses a nonlinear DAE model to describe the continuous state dynamics. In [108], new concepts were introduced for improving the efficiency of the dynamic programming computations. Logical specifications are used to define the admissible transition behavior of the discrete system, to incorporate saturation of the continuous control, and to characterize the algebraic constraints of the DAE model and in the definition of the cost function. Conversion of the logical specifications to integer formulas using symbolic computation enables the use of mixed-integer dynamic programming to derive an optimal feedback control.

8.7.2.1 Modeling

The system operates in one of the m modes denoted as $q_1, \ldots, q_m$. $Q = \{q_1, \ldots, q_m\}$ is the discrete state space. The continuous time differential-algebraic equation (DAE) describing operation in mode q_i is

$$\begin{aligned} \dot{x} &= f_i(x, y, u) \\ 0 &= g_i(x, y) \end{aligned} \qquad i = 1, \ldots, m \tag{8.24}$$

where $x \in X \subseteq R^n$ is the system continuous state, $y \in Y \subseteq R^p$ is the vector of algebraic variables, and $u \in U \subseteq R^l$ is the continuous control. Transitions can occur only between certain modes. The set of admissible transitions is $\mathcal{E} \subseteq Q \times Q$. It is convenient to view the mode transition system as a graph with elements of the set Q being the nodes and the elements of $\mathcal{E}$ being the edges. We assume that transitions are instantaneous. So, if a system transitions from mode q_1 to q_2 at time t, we would write $q(t) = q_1, q(t^+) = q_2$. We allow resets. State trajectories are assumed continuous through events, i.e., $x(t) = x(t^+)$, unless a reset is specified.

Transitions are triggered by external *events* and *guards*. Events are designated s and belong to a set Σ. A guard is a subset of the continuous state space X that enables/disables a transition. A transition enabled by a guard might represent a protection device. Not all transitions have guards, and some transitions might require simultaneous satisfaction of a guard and the occurrence of an event.

Each discrete state label, $q \in Q$, and each event label, $s \in \mathcal{E}$, are considered to be a logical variable that takes the value true or false. Guards also are specified as logical conditions. In this way, the transition system can be defined by a logical specification (formula) $\mathcal{L}$.

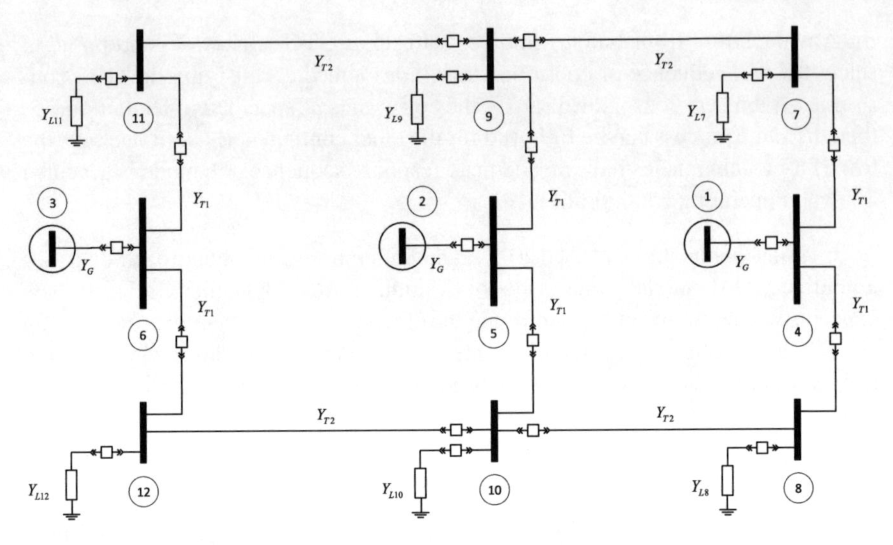

Fig. 8.14 The distribution network 12-bus configuration includes the generator internal buses

For computational purposes, it is useful to associate with each logical variable, say α, a binary variable or indicator function, δ_α, such that δ_α assumes the values 1 or 0 corresponding, respectively, to α being true or false. It is convenient to define the discrete state vector $\delta_q = \left[\delta_{q_1}, \ldots, \delta_{q_m}\right]$. Precisely, one of the elements of δ_q will be unity and all others will be zero.

With the introduction of the binary variables, the set of dynamical equations (8.24) can be replaced with the single DAE:

$$\begin{aligned} \dot{x} &= f\left(x, y, \delta_q, u\right) = \textstyle\sum_{i=1}^{m} \delta_{q_i} f_{q_i}(x, y, u) \\ 0 &= g\left(x, y, \delta_q\right) = \textstyle\sum_{i=1}^{m} \delta_{q_i} g_{q_i}(x, y) \end{aligned} \tag{8.25}$$

Remark 8.5 (*Power System DAE Models*) Power systems are typically modeled by sets of semiexplicit DAEs as given by (8.24). In any mode q_i, the flow defined by (8.24) is constrained to the set $M_i \subset X \times Y$ defined by $0 = g_i(x, y)$. Ordinarily, it is assumed that M_i is a regular submanifold of $X \times Y$.

Example 8.6 Loss of Generator, Continued. The dynamical behavior in each of the six discrete states shown in Figure 8.13 will be modeled with reference to the network illustrated in Figure 8.14. Note that the initial state involves two generators corresponding to buses 1 and 2. The spare generator corresponds to bus 3. It is assumed that the bus 2 generator fails. The difference between the initial state q_1 and the final state q_6 in Figure 8.13 is that the replacement generator is on a different bus. In summary, the reduced bus network models for the 6 states are as follows:

State q_1: Generator buses 1 and 2, PQ buses 4,5,6, full load.
State q_2: Generator bus 1, PQ bus 4, full load,
State q_3: Generator buses 1 and 3, PQ buses 4,6, vital load,
State q_4: Generator buses 1 and 3, PQ buses 4,6, ESM, full load,
State q_5: Generator buses 1 and 3, PQ bus 4,6, ESM, vital load
State q_6: Generator buses 1 and 3, PQ bus 4,6, full load.

8.7.2.2 The Control Problem

The system is observed in operation over some finite time horizon T that is divided into N discrete time intervals of equal length. A control policy is a sequence of functions

$$\pi = \left\{\mu_0\left(x_0, \delta_{q0}\right), \ldots, \mu_{N-1}\left(x_{N-1}, \delta_{q(N-1)}\right)\right\}$$

such that $[u_k, \delta_{sk}] = \mu_k\left(x_k, \delta_{q_k}\right)$. Thus, μ_k generates the continuous control u_k and the discrete control δ_{sk} that are to be applied at time k, based on the state $\left(x_k, \delta_{qk}\right)$ observed at time k.

Consider the set of m-tuples $\{0, 1\}^m$. Let Δ_m denote the subset of elements $\delta \in \{0, 1\}^m$ that satisfy $\delta_1 + \cdots + \delta_m = 1$. Denote by Π the set of sequences of functions $\mu_k : X \times \Delta_m \to U \times \{0, 1\}^{m_s}$ that are piecewise continuous on X.

The *optimal feedback control problem* is defined as follows. For each $x_0 \in X$, $\delta_{q0} \in \Delta_m$ determine the control policy $\pi^* \in \Pi$ that minimizes the cost

$$\begin{aligned} J_\pi\left(x_0, \delta_{q0}\right) = \\ g_N\left(x_N, \delta_{qN}\right) + \\ \sum_{k=0}^{N-1} g_k\left(x_k, \delta_{qk}, \mu_k\left(x_k, \delta_{qk}\right)\right) \end{aligned} \tag{8.26}$$

subject to the constraints (8.24) and the logical specification, i.e.,

$$J_{\pi^*}\left(x_0, \delta_{q0}\right) \le J_\pi\left(x_0, \delta_{q0}\right) \quad \forall \pi \in \Pi \tag{8.27}$$

8.7.3 *Example*

Consider, again, the loss of generator 2. This event causes the transition from state q_1 to q_2 as indicated in Figure 8.13. The goal now is to determine an optimal response strategy for this contingency. Departure from q_2 to any of the states $q_3, \ldots, q_6$ initiates start-up of the spare generator (GTG3). It is assumed that the generator power increases at a conservative rate of 250 KW/minute. In units of pu per sec,

$$\dot{P}_3 = 1/1200 \tag{8.28}$$

The goal is to steer the system from the initial state $P_3 = 0, q = q_2$ to the terminal state $P_3 = 0.45, q = q_6$. This will take 9 minutes since P_3 must reach 0.45 pu from 0 pu. The fast electrical dynamics will be neglected so that the only dynamics are associated with equation (8.28). Each mode is described by (8.28) and a set of algebraic equations describing the network.

The nine-minute interval is divided into nine one-minute segments, and (8.28) is replaced by the discrete time equation

$$P_{3,i+1} = P_{3,i} + 60/1200 \tag{8.29}$$

The goal is to find a sequence of state transitions that steer the system from the initial state $\{0, q_2\}$ to the final state $\{0.45, q_r\}$ such that QOS constraints are met. To do this, an optimal control is sought that minimizes a cost defined to reflect the QOS objectives. In this example, the cost J is

$$J = \sum_{i=4}^{12} |V_i - 1| + \max[0, P_1 - 0.5] + 0.3\,\delta_{ESM} + 0.15\,\delta_{NVL}$$

where δ_{ESM} and δ_{NVL} are binary variables that take the values 0 or 1. $\delta_{ESM} = 1$ denotes that the ESM is active and $\delta_{NVL} = 1$ denotes that the nonvital load is dropped, whereas in each case, the value zero denotes the opposite. Dynamic programming is used to obtain the switching strategy illustrated in Figure 8.15. The weights assigned to $\delta_{ESM} = 1, \delta_{ESM} = 1$ are selected to reflect a judgment of the relative cost of employing these actions.

Notice that following the failure, the controller immediately switches to configuration q_3, which means that the nonvital load is dropped and the ESM turned on providing 1000 KW of supporting power. It is worth noting that the power provided by GTG1 is $P_1 = 0.494$ pu, which is still below the unit's normal rating of 0.5 pu. If no action is taken, GTM1 would provide 0.786 pu power, which is just below

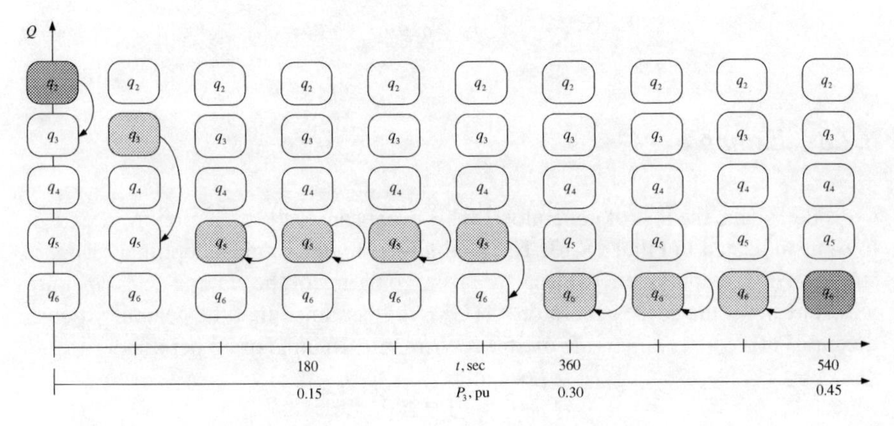

Fig. 8.15 The optimal strategy is shown in terms of the time period and GTG3 power level

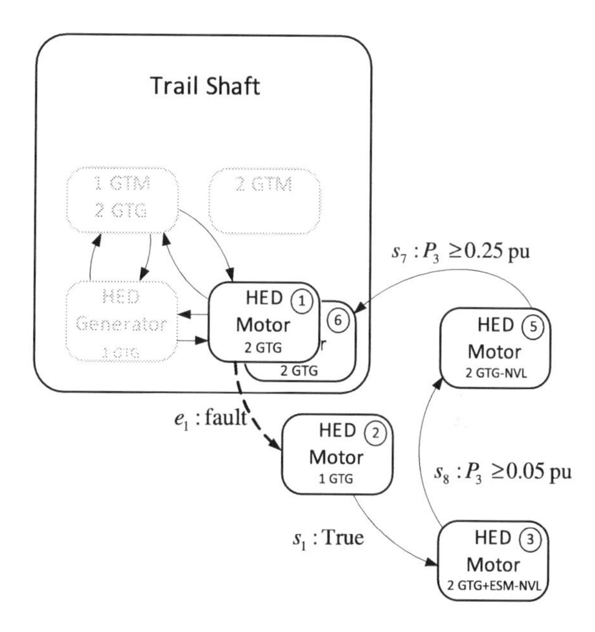

Fig. 8.16 The optimal strategy is shown as a discrete state transition diagram

the unit's five-minute crisis capability (0.9 pu). However, the voltage levels are also unacceptably low. After one minute, the optimal strategy switches to q_5, in which the ESM is turned off, but the nonvital loads remain disconnected. The GTG1 power output increases to 0.642 pu. The system remains in this state for four minutes by which time the GTG1 power output has dropped below its normal rating to 0.444 pu. At this point, the configuration is switched to q_6, the nonvital load is picked up, and the GTG1 power output increases to 0.640 pu. The system remains in this configuration and reaches the target state in four minutes as the GTG1 power output reduces linearly to its target value. Throughout this trajectory, the bus voltages remain within acceptable limits. The optimal recovery sequence is illustrated in Figure 8.16.

In summary, using the engine fuel consumption data, a set of possible operational configurations, and mission-specific electric load and ship speed requirements, it is a straightforward matter to compute the most fuel-efficient operating configuration. However, when QOS constraints are imposed, the problem is more complicated. In this case, it is necessary to delineate all credible contingencies and eliminate any configuration which violates the QOS constraints for any one of the contingent events. The occurrence of a contingency should trigger a remedial action designed to prevent violation of the QOS constraints. The goal is to design an optimal sequence of available remedial actions. The cost function is constructed from penalties associated with QOS violations which are balanced against costs associated with using the available remedial actions. With a remediation strategy defined, the response to a contingency can be evaluated to determine whether a QOS constraint is violated.

Appendix A
Ship Hybrid Electric Propulsion System

Figure A.1 illustrates a ship electric power and propulsion system that is used to demonstrate the concepts discussed throughout the text. The system is loosely based on a notional DDG-51 class naval ship with a hybrid electric drive, example [40, 144, 146]. The system includes three gas turbine-driven electric generators (GTGs) and four propulsion gas turbines (GTM), two on each of two propulsion shafts. Two bidirectional permanent magnet synchronous machine (PMSM), one geared to each propulsion shaft, can be used as motors to drive the shaft or as generators to provide electric power to ship power network.

The system can be operated in six configurations defined as follows.

1. *Full Power* - 2 turbines (GTM)per shaft driving 2 shafts; 1 or more GTGs supply the electric loads.
2. *Split Plant* - 1 turbine (GTM)per shaft driving 2 shafts; 1 or more GTGs supply the electric loads.
3. *Trail Shaft* - 1 turbine (GTM) driving 1 shaft, other shaft free; 1 or more GTGs supply the electric loads.
4. *HED Motoring* - 1 PMSM driving 1 shaft; 2 or more GTGs supply the electric power.
5. *HED generation Split Plant* - 1 turbine (GTM) per shaft driving 2 shafts with PSMS generators supporting 1 or more GTGs in supplying electrical loads.
6. *HED Generation Trail Shaft* - 1 turbine (GTM) driving 1 shafts with PSMS generator supporting 1 or more GTGs in supplying electrical loads.

A.1 12-Bus Network

The electric power distribution network is shown as a single- line diagram in Figure A.2. The system is designed to accommodate 5.0 MW of load. All numbers used below are on a per unit basis. The power basis is 5 MVA, and the voltage basis is 450 V. The admittance parameter generator (Y_G) and transmission lines (Y_{T1}, Y_{T2}) are assumed fixed with the following values:

H.G. Kwatny and K. Miu-Miller, *Power System Dynamics and Control*, Control Engineering, DOI 10.1007/978-0-8176-4674-5

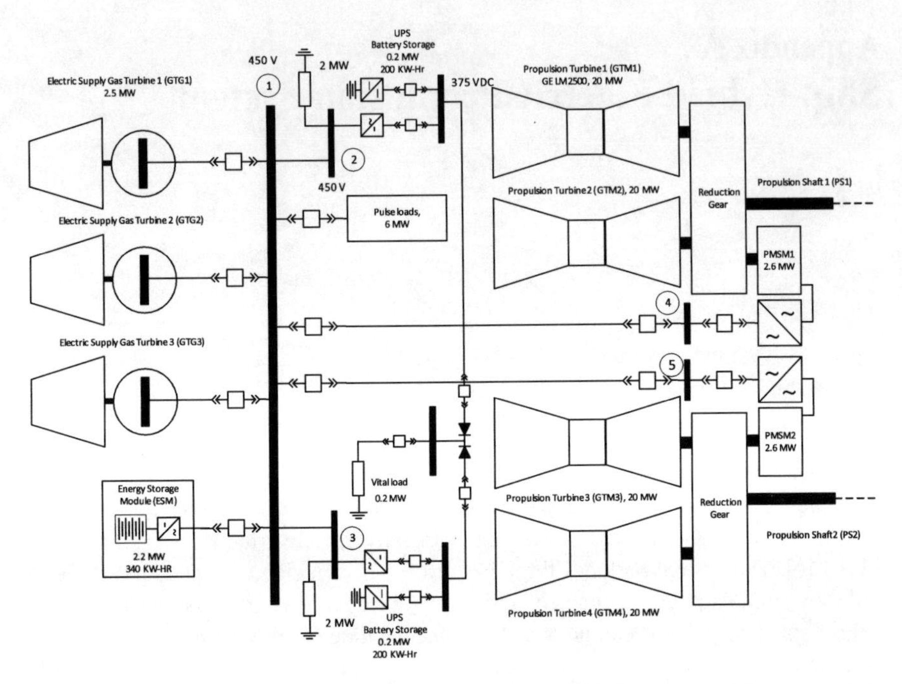

Fig. A.1 Notional hybrid electric power system based on a DDG 51 class naval ship with hybrid electric propulsion drive

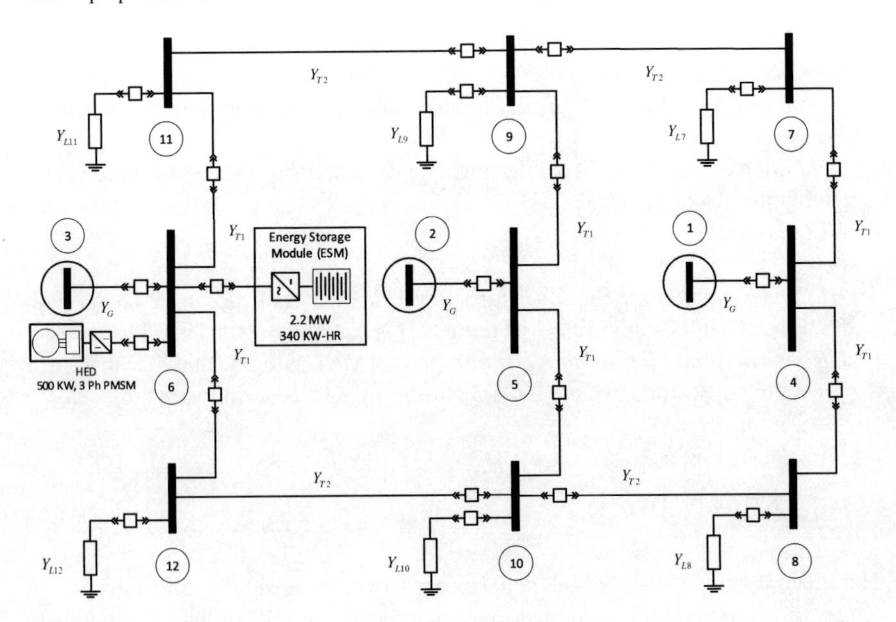

Fig. A.2 The distribution network 12-bus configuration includes the generator internal buses

$$Y_G = -j3.636,\ Y_{T1} = 0.40672 - j40.672,\ Y_{T2} = 0.2032 - j20.32$$

Notice that loads are shown on the six buses, numbers 7 through 12. It is assumed, for now, that the loads are equal and they are purely resistive, with $Y_{Li} = 1/R$, for $i = 7, \ldots, 12$, and R variable.

The no load admittance matrix is

$$B = \begin{pmatrix} -3.636 & 0 & 0 & 3.636 & 0 & 0 & 0 & 0 & 0 & 0 & 0 & 0 \\ 0 & -3.636 & 0 & 0 & 3.636 & 0 & 0 & 0 & 0 & 0 & 0 & 0 \\ 0 & 0 & -3.636 & 0 & 0 & 3.636 & 0 & 0 & 0 & 0 & 0 & 0 \\ 3.636 & 0 & 0 & -84.98 & 0 & 0 & 40.672 & 40.672 & 0 & 0 & 0 & 0 \\ 0 & 3.636 & 0 & 0 & -84.98 & 0 & 0 & 0 & 40.672 & 40.672 & 0 & 0 \\ 0 & 0 & 3.636 & 0 & 0 & -84.98 & 0 & 0 & 0 & 0 & 40.672 & 40.672 \\ 0 & 0 & 0 & 40.672 & 0 & 0 & -60.992 & 0 & 20.32 & 0 & 0 & 0 \\ 0 & 0 & 0 & 40.672 & 0 & 0 & 0 & -60.992 & 0 & 20.32 & 0 & 0 \\ 0 & 0 & 0 & 0 & 40.672 & 0 & 20.32 & 0 & -81.312 & 0 & 20.32 & 0 \\ 0 & 0 & 0 & 0 & 40.672 & 0 & 0 & 20.32 & 0 & -81.312 & 0 & 20.32 \\ 0 & 0 & 0 & 0 & 0 & 40.672 & 0 & 0 & 20.32 & 0 & -60.992 & 0 \\ 0 & 0 & 0 & 0 & 0 & 40.672 & 0 & 0 & 0 & 20.32 & 0 & -60.992 \end{pmatrix} \quad (A.1)$$

$$G = \begin{pmatrix} 0\ 0\ 0 & 0 & 0 & 0 & 0 & 0 & 0 & 0 & 0 & 0 \\ 0\ 0\ 0 & 0 & 0 & 0 & 0 & 0 & 0 & 0 & 0 & 0 \\ 0\ 0\ 0 & 0 & 0 & 0 & 0 & 0 & 0 & 0 & 0 & 0 \\ 0\ 0\ 0 & -0.81344 & 0 & 0 & 0.40672 & 0.40672 & 0 & 0 & 0 & 0 \\ 0\ 0\ 0 & 0 & -0.81344 & 0 & 0 & 0 & 0.40672 & 0.40672 & 0 & 0 \\ 0\ 0\ 0 & 0 & 0 & -0.81344 & 0 & 0 & 0 & 0 & 0.40672 & 0.40672 \\ 0\ 0\ 0 & 0.40672 & 0 & 0 & -0.60992 & 0 & 0.2032 & 0 & 0 & 0 \\ 0\ 0\ 0 & 0.40672 & 0 & 0 & 0 & -0.60992 & 0 & 0.2032 & 0 & 0 \\ 0\ 0\ 0 & 0 & 0.40672 & 0 & 0.2032 & 0 & -0.81312 & 0 & 0.2032 & 0 \\ 0\ 0\ 0 & 0 & 0.40672 & 0 & 0 & 0.2032 & 0 & -0.81312 & 0 & 0.2032 \\ 0\ 0\ 0 & 0 & 0 & 0.40672 & 0 & 0 & 0.2032 & 0 & -0.60992 & 0 \\ 0\ 0\ 0 & 0 & 0 & 0.40672 & 0 & 0 & 0 & 0.2032 & 0 & -0.60992 \end{pmatrix} \quad (A.2)$$

To accommodate the resistive loads simply add to G as follows

$$G \to G + \operatorname{diag}\left(0, 0, 0, 0, 0, 0, \frac{1}{R}, \frac{1}{R}, \frac{1}{R}, \frac{1}{R}, \frac{1}{R}, \frac{1}{R}\right) \quad (A.3)$$

Notice that full load corresponds $R = 1/6$.

A.2 The 5-Bus Network

In this section and the next, reduced order models are derived that are used in various examples. The system ordinarily operates with two of the three generators. Assuming the generators in use are generators 1 and 2, bus number 3 (the internal bus of generator 3) can be eliminated by dropping the third row and column of the 12-bus admittance matrix. This leaves an 11-bus system. Load buses 7 through 12 have only constant admittance loads. There are no current injections. Consequently, they can be eliminated. The result is a five-bus reduced network in which the retained buses are 1, 2, 4, 5, and 6. With $R = 0.10$, i.e., the total resistive load is 0.60 per unit, the reduced admittance matrix is

$$B_{5bus} = \begin{pmatrix} -3.636 & 0 & 3.636 & 0 & 0 \\ 0 & -3.636 & 0 & 3.636 & 0 \\ 3.636 & 0 & -25.3183 & 16.2635 & 5.41829 \\ 0 & 3.636 & 16.2635 & -36.1635 & 16.2635 \\ 0 & 0 & 5.41829 & 16.2635 & -25.3183 \end{pmatrix}$$

$$G_{5bus} = \begin{pmatrix} 0. & 0 & 0. & 0 & 0 \\ 0 & 0. & 0 & 0. & 0 \\ 0. & 0 & -0.100345 & 0.218627 & 0.0817217 \\ 0 & 0. & 0.218627 & -0.237251 & 0.218627 \\ 0 & 0 & 0.0817217 & 0.218627 & -0.100345 \end{pmatrix}$$

A.3 The 2-Bus Network

When the generator internal bus is included in the network, as is done here, and the generator terminal buses have no other current injections, it is possible to exclude those as well. Normally, they would be retained because the terminal bus is usually voltage controlled. In the 5-bus model, buses 4 and 5 were retained. Bus 6 was also retained in the event that the motor load or battery supply was to be used. It is possible to eliminate these as well leaving a simple two-bus network. Of course, such a model would have much more limited value in application, but is useful for illustration. The admittance matrix for the two-bus network is

$$B_{2bus} = \begin{pmatrix} -2.17074 & 1.12413 \\ 1.12413 & -2.25972 \end{pmatrix}, \ G_{2bus} = \begin{pmatrix} 0.0648096 & 0.0671797 \\ 0.0671797 & 0.0649872 \end{pmatrix}$$

Appendix B
Computational Tools

This appendix provides information about computing tools and links to Mathematica notebooks and Simulink models of examples used through out the book.

B.1 ProPac - A Mathematica Toolbox

ProPac is a toolbox originally developed to support symbolic computations required for modeling complex mechanical and electrical systems as linear and nonlinear control system design. An early version was included with the book [114]. Functions are provided for the manipulation of linear control systems in state space or frequency domain forms and for the conversion of one form to the other. The software also contains functions required to apply modern geometric methods of control system design to nonlinear systems. These include tools for the design of adaptive as well as variable structure control systems. Functions are provided that assist model building, simulation in MATLAB/Simulink, control design, and implementation. Functions are available to assemble optimized C-code that compiles as a Simulink S-function, thereby enabling system models and controllers to be easily included in Simulink simulations. The stand-alone control modules can be directly converted to real-time code in a target DSP board. An updated version of ProPac is available at http://www.pages.drexel.edu/~hgk22/. Mathematica notebooks that illustrate the application of these tools can be found at http://www.pages.drexel.edu/~hgk22/notebooks/. These notebooks include generator modeling examples and MEX function construction as well as network modeling and MEX function construction. Simulink examples of a complete simulation are also available.

H.G. Kwatny and K. Miu-Miller, *Power System Dynamics and Control*,
Control Engineering, DOI 10.1007/978-0-8176-4674-5

B.2 Power System Modeling

Electric power system dynamics can typically be represented by DAE models of the semi-implicit form

$$\begin{aligned} \dot{x} &= f(x, y, u, \eta) \\ 0 &= g(x, y, u, \eta) \end{aligned} \tag{B.1}$$

where the variables include the state vector $x \in R^n$, a vector of auxiliary variables $y \in R^p$, a vector of input variables $u \in R^m$, and a set of parameters denoted by the vector $\eta \in R^k$. Ordinarily, the dynamical equations are associated with generation and loads and the algebraic portions with the network. Some explanatory remarks are given in the following paragraphs, and illustrative mathematica notebooks can be found at http://www.pages.drexel.edu/~hgk22/notebooks/notebooks.htm.

B.2.1 Machine Simulation Models

Three phase machine models are considered in Section 4.3.2. The governing equations for synchronous, permanent magnet synchronous, and induction motors and generators are derived. Balanced operation allows simplification of these models. The simplified synchronous machine model is developed in Section 4.4.2. The governing equations are given in Equations (4.99) – (4.107). The interconnection of the simplified model with the network is shown in Figure 4.23.

B.2.2 Power Network Simulation

Electric power system dynamics can typically be represented by DAE models in the semi-implicit form of (B.1). The ability to run fast simulations of power networks is essential to the design and implementation of intelligent, high-performance power management systems. However, fast simulation is impeded by the structure of power system DAE models. Simulink, for example, one of the most important model building and simulation tools for control system designers, does not contain reliable computational tools for large sets of DAEs. The approach described here provides a method of addressing modeling and simulation of power networks within Simulink.

In the MATLAB-based voltage stability toolbox, [17], the fourth- and fifth-order Runge–Kutta–Fehlberg method (one of MATLAB's standard tools for solving ordinary differential equations) was adapted to suit semi-implicit DAE models of power systems. At each time step, a standard Newton-Raphson procedure was used to solve the algebraic equations for y. This is feasible so long as the trajectory does not pass through a "noncausal" point (see Chapter 6 and, in particular, Example 6.3 for a discussion of noncausal points and the related "singularity-induced" bifurcation).

This approach has been extended so that it can be used in the Simulink environment. This allows engineers to use Simulink graphical interface to build and share models in a modular framework and to exploit Simulink computational capabilities and its various toolboxes. While toolboxes exist for power system modeling within Simulink, they solve the DAE computational issue by including enough parasitic dynamics in the models to break all of the many algebraic loops, thus producing a set of stiff differential equation instead of a DAE system of equations. The resulting simulations are painfully slow and inadequate for many applications.

An electric power system is a collection of components including generators, motors, storage devices, static and dynamic loads, and others connected together by a network of transmission lines. Building a mathematical model of the system can be facilitated by a graphical interface in which models of individual components and the network can be connected together in the desired configuration. Simulink provides such an interface where the system model can be assembled, modified, and simulated. In the approach used here, the network is isolated thereby segregating all of the algebraic loops in a single block that implements a specified number of Newton iterations as a discrete time dynamical element. The simulation can be organized so that one or more Newton iterations are performed during each continuous time step. The result is a very fast simulation because there is no need to include artificial parasitic dynamics.

This approach views the power system in terms of the "network" which embodies the critical algebraic equations to which generators and loads of various types and their controllers are attached. The network algebraic equations are implemented in a single discrete time module. The assembly of the network model block is accomplished symbolically resulting in an optimized C-code program that compiles as a Simulink S-function. The following paragraphs define the overall concept and provides examples of its application.

The general structure of the network is defined in Figure 4.25. It is viewed as a multi-port block. Each port (or bus) interacts with the external world through four variables P, Q, V, θ. Two of these are inputs, and two are outputs. The most common input pairs are P, Q, P, V,and V, θ. Note that because of the translational symmetry in the network equations, at least one-bus angle, θ, must be an input variable. This would most likely be a reference generator, corresponding to a V, θ bus.

B.3 Power System Management

Computational tools are also available for mixed logical-dynamical optimization problems associate with power system control problems. A tutorial notebook that implements tools for converting logic statements to mixed-integer programming formulas is available at http://www.pages.drexel.edu/~hgk22/, as is a notebook that implements some elementary constructions for dynamic programming. Examples are also included.

B.4 Voltage Stability Toolbox

The Voltage Stability Toolbox (VST) is a stand-alone software environment that implements computational methods for investigating power system voltage stability. It allows the user to specify the system configuration from which the toolbox assembles the relevant equations and computational algorithms. These techniques, which include convergent load flow procedures that work near collapse points and can locate the precise point of collapse (POC), involve computations that are more complex than conventional load flow methods [17]. Compared to traditional methods, they require additional information (such as 2nd derivatives), which necessitates far more elaborate code for their implementation and makes them computationally intensive. For these reasons, key elements of the VST are automatic model assembly, code generation, and code optimization. The current system integrates numerical and symbolic computing. It is built in MATLAB and exploits its symbolic toolbox (Maple) and GUI tools. VST is available at http://www.pages.drexel.edu/~hgk22/ or at the Web site of the Center for Electric Power Engineering (CEPE) http://power.ece.drexel.edu/VST/.

References

1. S. Abe, N. Hamada, A. Isono, K. Okuda, Load flow convergence in the vicinity of a voltage instability limit. IEEE Trans. Power Appar. Syst. **PAS-97**(6), 1983–1993 (1978)
2. E. Abed, P. Varaiya, Nonlinear oscillations in power systems. Int. J. Electr. Power Energy Syst. **6**, 37–43 (1984)
3. V. Ajjarapu, Identification of steadystate voltage stability in power systems. Int. J. Energy Syst. **11**(1), 43–46 (1991)
4. J.C. Alexander, Oscillatory solutions of a model system of nonlinear swing equations. Int. J. Electr. Power Energy Syst. **8**, 130–136 (1986)
5. F.L. Alvarado, T.H. Jung, Direct detection of voltage collapse conditions, in *Bulk Power Systems Voltage Phenomena Voltage Stability and Security*, vol. EL6183 (EPRI, 1988), pp. 5.23–5.38
6. M. Anghel, F. Milano, A. Papachristodoulou, Algorithmic construction of lyapunov functions for power system stability analysis. IEEE Trans. Circ. Syst. **60**(92533–2546) (2013)
7. P.J. Antsaklis, A.N. Michel, *Linear Systems* (McGraw-Hill, New York, 1997)
8. V.I. Arnold, *Geometrical Methods in the Theory of Ordinary Differential Equations* (Springer, New York, 1983)
9. V.I. Arnold, *Mathematical Methods of Analytical Mechanics*, 2nd edn. (Springer, New York, 1989)
10. V.I. Arnold, V.V. Kozlov, A.I. Neishtadt, *Mathematical Aspects of Classical and Celestial Mechanics*, vol. 3, Encyclopedia of Mathematical Sciences (Springer, Heidelberg, 1988)
11. G.V. Aronovich, N.A. Kartvelisvili, Application of stability theory to static and dynamic stability problems of power systems, in *Second All-Union Conference on Theoretical and Applied Mechanics*, ed. by L.I. Sedov. Moscow, 1965. English Trans. NASA TT F-503, Part 1, pp. 15–26, 1968
12. D.K. Arrowsmith, C.M. Place, *An Introduction to Dynamical Systems* (Cambridge University Press, Cambridge, 1990)
13. M. Athans, P.L. Falb, *Optimal Control: An Introduction to the Theory and Its Applications* (McGraw-Hill, New York, 1966)
14. T. Athay, R. Podmore, S. Virmani, A practical method for direct analysis of power system stability. IEEE Trans. Power Appar. Syst. **PAS-98**, 573–584 (1979)
15. S. Ayasun, C.O. Nwankpa, H.G. Kwatny, Computation of singular and singularity-induced bifurcation points of differential-algebraic power systems models. IEEE Trans. Circuts Syst. **51**(8), 1525–1538 (2004)

H.G. Kwatny and K. Miu-Miller, *Power System Dynamics and Control*, Control Engineering, DOI 10.1007/978-0-8176-4674-5

16. S. Ayasun, C.O. Nwankpa, H.G. Kwatny, An efficient method to compute singularity induced bifurcations of decoupled parameter-dependent differential-algebraic power system model. Appl. Math. Comput. **167**(1), 435–453 (2005)
17. S. Ayasun, C.O. Nwankpa, H.G. Kwatny, Voltage stability toolbox (vst) for power system education and research. IEEE Trans. Educ. **49**, 432–442 (2006)
18. L.Y. Bahar, H.G. Kwatny, Some remarks on the derivability of linear nonconservative systems from a lagrangian. Hadron. J. **3**, 1264–1280 (1980)
19. L.Y. Bahar, H.G. Kwatny, Extension of noethers theorem to constrained nonconservative dynamical systems. Int. J. Nonlinear Mech. **22**(2), 125–138 (1987)
20. J. Baillieul, C.I. Byrnes, Geometric critical point analysis of lossless power system models. IEEE Trans. Circuits Syst. **CAS-29**(11), 724–737 (1982)
21. L. Bao, X. Duan, T.Y. He. Analysis of voltage collapse mechanisms in state space. IEE Proc. Gener. Transm. Distrib. **147**(6), 395–400 (2000)
22. T.E. Bechert, H.G. Kwatny, On the optimal dynamic dispatch of real power. IEEE Trans. Power Appar. Syst. **91**(3), 889–898 (1972)
23. R.E. Bellman, *Dynamic Programming* (Princeton University Press, Princeton, 1957)
24. Alberto Bemporad, Manfred Morari, Control of systems integrating logic, dynamics, and constraints. Automatica **35**(3), 407–427 (1999)
25. J. Berg, H.G. Kwatny, A canonical parameterization of the kronecker form of a matrix pencil. Automatica **31**(5), 669–680 (1995)
26. A.R. Bergen, D.J. Hill, A structure preserving model for power system stability analysis. IEEE Trans. Power Appar. Syst. **PAS-100**, 25–35 (1981)
27. A.R. Bergen, *PowerSystems Analysis*. Prentice Hall Series in Electrical and computer Engineering (Prentice–Hall, Inc., Englewood Cliffs, 1986)
28. M.W. Boothby, *An Introduction to Differentiable Manifolds and Riemannian Geometry* (Academic Press, San Diego, 1986)
29. F. Borrelli, M. Baotic, A. Bemporad, M. Morari, Dynamic programming for constrained optimal control of discrete-time linear hybrid systems. Automatica **41**, 1709–1721 (2005)
30. M.S. Branicky, V.S. Borkar, S.K. Mitter, A unified framework for hybrid control: model and optimal control theory. IEEE Trans. Autom. Control **43**(1), 31–45 (1998)
31. R.K. Brayton, J.K. Moser, A theory of nonlinear networks, parts 1, 2. Quart. Appl. Math. 12(1 and 2), 1–33, 81–104 (1964)
32. K.E. Brenan, S.L. Cambell, L.R. Petzold, *Numerical Solution of Initial-Value Problems in Differential-Algebraic Equations* (Elsevier Science Publishing Co., New York, 1989)
33. N.G. Bretas, L.F.C. Alberto, Lyapunov function for power systems with transfer conductances: extension of the invariance principle. IEEE Trans. Power Syst. **18**(2), 769–777 (2003)
34. P.R. Bryant, The order of complexity of electrical networks. *The Institution of Electrical Engineers, Monograph No. 335 E*, pp. 174–188, 1959
35. C.A. Cañizares, F.L. Alvarado, Computational experience with the point of collapse method on very large ac/dc systems, in *Bulk Power Systems Voltage II Phenomena–Voltage Stability and Security* (Deep Creek Lake, 1991), pp. 103–112
36. C.A. Cañizares, F.L. Alvarado, Point of collapse and continuation methods for large ac/dc systems. IEEE Trans. Power Syst. **8**(1), 1–7 (1993)
37. C.A. Cañizares, On bifurcations, voltage collapse and load modeling. IEEE Trans. Power Syst. **10**(1), 512–518 (1995)
38. M.S. Calovic, Linear regulator design for a load and frequency control. IEEE Trans. Power Appar. Syst. **PAS-91**(6), 2271–2285 (1972)
39. E.G. Carpaneto, G. Chicco, R. Napoli, F. Piglione, A newtonraphson method for steadystate voltage stability assessment, in *Bulk Power Systems Voltage II Phenomena Voltage Stability and Security* (Deep Creek Lake, 1991) pp. 341–345
40. G. Castles, G. Reed, A. Bendre, R. Pitsch, Economic benefits of hybrid drive propulsion for naval ships, in *Electric Ship Technologies Symposium*, pp. 515–520 (2009)
41. C.T. Chen, *Linear Systems Theory and Design* (Oxford University Press, New York, 1999)

42. R.L. Chen, P. Varaiya, Degenerate hopf bifurcation in power systems. IEEE Trans. Circuits Syst. **CAS-35**(7), 818–824 (1988)
43. W.-K. Chen, R.-S. Liang, A general n-port network reciprocity theorem. IEEE Trans. Educ. **33**(4), 360–362 (1990)
44. E.C. Cherry, Some general theorems for nonlinear systems possessing reactance. Philos. Mag. **42**(333), 1161–1177 (1951)
45. N.G. Chetaev, On the equations of poincar. PMM (Appl. Math. Mech.) **5**, 253–262 (1941)
46. N.G. Chetaev, *Theoretical Mechanics* (Springer, New York, 1989)
47. H.-D. Chiang, C.-C. Chu, Theoretical foundation of the bcu method for direct stability analysis of network-reduction power system models with small transfer conductances. IEEE Trans. Circuits Syst. **42**(5), 252–265 (1995)
48. H.-D. Chiang, A.J. Fluek, K.S. Shah, N. Balu, Cpflow: a practical tool for tracing power system steady-state stationary behavior due to load and generation variations. IEEE Trans. Power Syst. **10**(2), 623–630 (1995)
49. H.D. Chiang, C.W. Liu, P.P. Varaiya, F.F. Wu, M.G. Lauby, Chaos in a simple power system. IEEE Trans. Power Syst. **8**(4), 1407–1414 (1993)
50. H.-D. Chiang, F.F. Wu, P.P. Varaiya, Foundations of the potential energy boundary surface method for power system transient stability analysis. IEEE Trans. Power Circuits Syst. **35**(6), 712–726 (1988)
51. S.-N. Chow, J.K. Hale, *Methods of Bifurcation Theory* (Springer-Verlag, New York, 1982)
52. B.H. Chowdhury, S. Rahman, A review of recent advances in economic dispatch. IEEE Trans. Power Syst. **5**(4), 1248–1257 (1990)
53. L. Chua, Memristor—the missing circuit element. IEEE Trans. Circuit Theory **CT-18**(5), 507–519 (1971)
54. L.O. Chua, J.D. McPherson, Explicit topological formulation of lagrangian and hamiltonian equations for nonlinear circuits. IEEE Trans. Circuits Syst. **CAS-21**(March), 277–285 (1974)
55. N. Cohn, Some aspects of tie-line bias control of interconnected power systems. *Transactions American Institute of Electrical Engineers*, pp. 1415–1436, 1957
56. N. Cohn, *Control of Generationa and Power Flow of Interconnected Systems* (Wiley, New York, 1966)
57. S.H. Crandall, D.C. Karnopp, E.F. Kurtz Jr, D.C. PridmoreBrown, *Dynamics of Mechanical and Electromechanical Systems* (McGrawHill, New York, 1968)
58. R.J. Davy, I.A. Hiskins, Lyapunov functions for multimachine power systems with dynamic loads. IEEE Trans. Circuits Syst. 1: Fundam. Theory Appl. **44**(9), 796–812 (1997)
59. S.E.M. de Oliveira, Synchronizing and damping torque coefficients and power system steady state stability as affected by static var compensators. IEEE Trans. Power Syst. **9**(1), 109–119 (1994)
60. C.L. DeMarco, A.R. Bergen, Applications of singular perturbation techniques to power system transient stability analysis, in *IEEE International Symposium on Circuits and Systems*, pp 597–601 (1984)
61. F.P. DeMello, R.J. Mills, W.F. B'Rells, Automatic genera tion control, part i–process modelling. IEEE Trans. Power Appar. Syst. **PAS-92**(2), 710–715 (1973)
62. U. DiCaprio, Use of lyapunov and energy methods for stability analysis of multimachine power systems, in *International Symposium on Circuitsand Systems*, volume IEEE Publicaion CH16B1-6/82, Rome, Italy, 1982. IEEE
63. U. DiCaprio, Accounting for transfer conductance effects in lyapunov transient stability analysis of a multimachine power system. Electr. Power Energy Syst. **8**(1), 27–41 (1986)
64. I. Dobson, Observations on geometry of saddle node bifurcation and voltage collapse in electric power systems. IEEE Trans. Circuits Syst. Part I **39**(3), 240–243 (1992)
65. I. Dobson, H.-D. Chiang, Towards a theory of voltage collapse in electric power systems. Syst. Control Lett. **13**, 253–262 (1989)
66. I. Dobson, L. Lu, Using an iterative method to compute a closest saddle node bifurcation in the load parameter space of an electric power system, in *Bulk Power Systems Voltage II PhenomenaVoltage Stability and Security* (Deep Creek Lake, 1991), pp. 157–161

67. I. Dobson, L. Lu, Computing an optimal direction in control space to avoid saddle node bifurcation and voltage collapse in electric power systems. IEEE Trans. Autom. Control **37**(10), 1616–1620 (1992)
68. I. Dobson, L. Lu, New methods for computing a closest saddlenode bifurcation and worst case load power margin for voltage collapse. IEEE Trans. Power Syst. **8**(3), 905–913 (1993)
69. N.H. Doerry, J.V. Amy Jr. Implementing quality of service in shipboard power system design. in *IEEE Electric Ship Technology Symposium* (Alexandria, 2011)
70. N.H. Doerry, Designing electric power systems for survivability and quality of service. ASNE Naval Eng. J. **119**(2), 25–34 (2007)
71. R.D. Dunlop, D.N. Ewart, System requirements for dynamic response of generating units. IEEE Trans. Power Appar. Syst. **94**(3), 838–849 (1973)
72. O.I. Elgerd, C.E. Fosha, Optimum megawatt-frequency control of multiarea electric energy systems. IEEE Trans. Power Appar. Syst. **PAS-89**(4), 556–563 (1970)
73. C.E. Fosha, O. I. Elgerd. The megawatt-frequency control problem: a new approach via optimal control theory. IEEE Trans. Power Appar. Syst. **PAS-89**(4), 563–577 (1970)
74. Bruce A. Francis, The linear multivariable regulator problem. SIAM J. Control Optim. **15**(3), 486–505 (1977)
75. F. Gantmacher, *Lectures in Analytical Mechanics* (Mir, Moscow, english translation edition, 1975)
76. Felix R. Gantmacher, *The Theory of Matrices*, vol. 1 (Chelsea Publishing Co., New York, 1959)
77. T. Geyer, M. Larsson, M. Morari, Hybrid emergency voltage control in power systems, in *European Control Conference* (Cambridge, 2003)
78. R. Gilmore, *Catastrophe Theory for Scientists and Engineers* (Wiley, New York, 1981)
79. H. Glavitsch, Where developments in power system stability should be directed, in *International Symposium on Power System Stability* (Ames, 1985), pp. 61–68
80. M. Golubitsky, V. Guilleman, *Stable Mappings and their Singularities* (Springer, New York, 1973)
81. M. Golubitsky, D.G. Schaeffer, *Singularities and Groups in Bifurcation Theory*, vol. 1 (Springer, New York, 1984)
82. J.J. Grainger, W.D. Stevenson, *Power System Analysis* (McGraw-Hill, Electrical engineering series, 1994)
83. J. Guckenheimer, P. Holmes, *Nonlinear Oscillations, Dynamical Systems, and Bifurcation of Vector Fields* (Springer, New York, 1983)
84. J.K. Hale, *Ordinary Differential Equations* (Wiley, New York, 1969)
85. W.R. Hamilton. On a general method in dynamics. Philos. Trans. Roy. Soc. **1834**(Part II), 247–308 (1834)
86. S. Hedlund, A. Rantzner, Optimal control of hybrid systems, in *Conference on Decision and Control* (Pheonix, 1999), pp. 3972–3977
87. D.J. Hill, Special issue on nonlinear phenomena in power systems. Proc. IEEE **83**(11), 1437–1596 (1995)
88. M.W. Hirsch, *Differential Topology* (Springer, New York, 1976)
89. M.W. Hirsch, S. Smale, *Differential Equations, Dynamical Systems, and Linear Algebra* (Academic Press, New York, 1974)
90. I. Hiskins, D.J. Hill, Energy functions, transient stability and voltage behavior in power systems with nonlinear loads. IEEE Trans. Power Syst. **4**(4), 1525–1533 (1989)
91. I.A. Hiskins, D.J. Hill, Failure modes of a collapsing power system, in *Bulk Power System Voltage Phenomena II: Voltage Stability and security*, ed. by L.H. Fink (Deep Creek Lake, 1991), pp. 53–63
92. J. Hooker, *Logic–Based Methods for Optimization: Combining Optimization and Constraint Satisfaction* (Wiley–Interscience, 2000)
93. K. Huseyin, *Vibrations and Stability of Multiple Parameter Sytems* (Sijthoff and Noordhoff, Leiden, 1978)

94. K. Huseyin, P. Hagedorn, The effect of damping on the stability of pseudo-conservative systems. *ZAMM (Z. fur angew. Math. und Mech.)***58**, 147–148 (1978)
95. K. Iba, H. Suzuki, M. Egawa, T. Watanabe, Calculation of critical loading with nose curve using homotopy continuation method. IEEE Trans. Power Syst. **6**(2), 584–590 (1991)
96. IEEE. Voltage stability of power systems: concepts, tools and industry experience. System Dynamic Performance Subcommittee Report 90TH0358-2-PWR (IEEE, 1990)
97. IEEE Std 1709-2010. IEEE recommended practice for 1 kv to 35 kv medium voltage dc power systems on ships, 2010
98. D.J. Inman, Dynamics of assymetric nonconservative systems. ASME J. Appl. Mech. **50**(1), 199–203 (1983)
99. M.R. Iravani, A. Semlyen, Hopf bifurcations in torsional dynamics. IEEE Trans. Power Syst. **7**(1), 28–35 (1992)
100. A. Isidori, *Nonlinear Control Systems*, 3rd edn. (Springer, London, 1995)
101. N. Jaleeli, L.S. VanSlyck, D.N. Ewart, L.H. Fink, A.G. Hoffmann, Understanding automatic generation control. IEEE Trans. Power Syst. **7**(3), 1106–1122 (1992)
102. J. Jarjis, F.D. Galiana, Quantitative analysis of steady state stability in power networks. IEEE Trans. Power Appar. Syst. **PAS-100**(1), 318–326 (1981)
103. C.D. Johnson, Accomodation of external disturbances in linear regulator and servomechanism problems. IEEE Trans. Autom. Control **16**, 635–644 (1971)
104. D.L. Jones, F.J. Evans, A classification of physical variables and its applications in variational methods. J. Franklin Inst. **29**(6), 449–467 (1971)
105. Y. Kataoka, An approach for the regularization od a power flow solution around the maximum loading point. IEEE Trans. Power Syst. **7**(3), 1068–1077 (1992)
106. H.K. Khalil, *Nonlinear Systems* (MacMillan, New York, 1992)
107. M. Kubicek, Dependence of systems of nonlinear equations on a parameter. ACM Trans. Math. Softw. **2**(1), 98–107 (1976)
108. H. Kwatny, G. Bajpai, M. Yasar, K. Miu, Fuel optimal control with service reliability constraints for ship power systems, in *IFAC World Conference*, vol 19 (Capetown, South Africa, 2014), pp. 6386–6391
109. H.G. Kwatny, Stability enhancement via secondary voltage regulation, in *Bulk Power System Voltage Phenomena II: Voltage Stability and Security*, ed. by L.H. Fink (ECC Inc, Deep Creek Lake, 1991), pp. 147–155
110. H.G. Kwatny, T.A. Athay. Coordination of economic dispatch and load-frequency control in electric power systems. in *IEEE Conference on Decision and Control* (IEEE, Fort Lauderdale, 1979), pp. 793–714
111. H.G. Kwatny, L.Y. Bahar, F.M. Massimo, Some remarks on the nonderivability of linear nonconservative systems from a lagrangian. Hadronic J. **3**(2), 1159–1177 (1979)
112. H.G. Kwatny, L.Y. Bahar, A.K. Pasrija, Energylike lyapunov functions for power system stability analysis. IEEE Trans. Circuits Syst. **CAS-32**(11), 1140–1149 (1985)
113. H.G. Kwatny, T.E. Bechert, On the structure of optimal area controls in electric power networks. IEEE Trans. Autom. Control **18**(2), 167–174 (1973)
114. H.G. Kwatny, G.L. Blankenship, *Nonlinear Control and Analytical Mechanics: A Computational Approach* (Control Engineering. Birkhauser, Boston, 2000)
115. H.G. Kwatny, R.F. Fischl, C. Nwankpa, Local bifurcations in power systems: theory, computation and application. Proc. IEEE **83**(11), 1456–1483 (1995)
116. H.G. Kwatny, K.C. Kalnitsky, A.D. Bhatt, An optimal regulator approach to loadfrequency control. IEEE Trans. Power Appar. Syst. **PAS-94**(5), 16351643 (1975)
117. H.G. Kwatny, F.M. Massimo, L.Y. Bahar, The generalized lagrange equations for nonlinear rlc networks. IEEE Trans. Circuits Syst. **CAS-29**(4), 220–233 (1982)
118. H.G. Kwatny, E. Mensah, D. Niebur, G. Bajpai, C. Teolis, Logic based design of optimal reconfiguration strategies for ship power systems. in *7th IFAC Symposium on Nonlinear Control Systems* (Pretoria, 2007)
119. H.G. Kwatny, E. Mensah, D. Niebur, C. Teolis, Optimal shipboard power system management via mixed integer dynamic programming. in *2005 IEEE Electric Ship Technologies Symposium (IEEE Cat. No. 05EX1110)* (IEEE, Philadelphia, 2005), pp 55–62

120. H.G. Kwatny, E. Mensah, D. Niebur, C. Teolis, Optimal power system management via mixed integer dynamic programming. in *IFAC Symposium on Power plants and Systems* (Kananaskis, Canada, 2006)
121. H.G. Kwatny, A.K. Pasrija, L.Y. Bahar, Static bifurcations in electric power networks: loss of steady state stability and voltage collapse. in *IEEE Transactions on Circuits and Systems*, **CAS-33**(10), 981–991 (1986)
122. H.G. Kwatny, G.E. Piper, Frequency domain analysis of hopf bifurcations in electric power networks. IEEE Trans. Circuits Syst. **37**(10), 1317–1321 (1990)
123. H.G. Kwatny, Yu. XiaoMing, Energy analysis of loadinduced flutter instability in classical models of electric power networks. IEEE Trans. Circuits Syst. **CAS-36**(12), 1544–1557 (1989)
124. Kwatny, H.G., Kalnitsky, K.C.: On alternative methodologies for the design of robust linear multivariable regulators. IEEE Trans. Autom. Control. **AC-23**(5), 930–933 (1978)
125. S. Lahiri, K. Miu, H.G. Kwatny, G. Bajpai, A. Beyton, J. Patel, Fuel optimization under quality of service constraints for shipboard hybrid electric drive. in *4th International Symposium on Resilient Control Systems* (IEEE, Boise, 2011)
126. C. Lanczos, *The Variational Principles of Mechanics*, 4th edn. (Dover Publications, New York, 1970)
127. J. LaSalle, S. Lefschetz, *Stability by Liapunovs Direct Method* (Academic Press, New York, 1961)
128. H.H.E. Leipholz, On the sufficiency of the energy criterion for the stability of of certain non-conservative systems of the follower type. J. Appl. Mech. **39**, 712–722 (1972)
129. Q. Li, Y. Guo, T. Ida, Transformation of logical specification into ip–formulas. in *3rd International Mathematica Symposium (IMS '99)* (Computational Mechanics Publications, WIT Press, Hagenburg, 1999)
130. Q. Li, Y. Guo, T. Ida, Modelling integer programming with logic: language and implementation. in *IEICE Transactions on Fundamentals of Electronics, Communications and Computer Sciences*, **E83-A**(8), 1673–1680 (2000)
131. T.E. Dy Liacco, The adaptive reliability control system. IEEE Trans. Power Appar. Syst. **PAS-86**(5), 517–528 (1967)
132. T.E. Dy Liacco, *Control of Power Systems via the Multi-Level Concept*. Ph.D. thesis, Case Western Reserve, 1968
133. T.E. Dy Liacco, Processing by logic programming of circuit-breaker and protective-relaying information. IEEE Trans. Power Appar. Syst. **PAS-82**(2), 171–175 (1969)
134. B. Lincoln, A. Rantzer, Relaxing dynamic programming (2003)
135. T. Lindl, K. Fox, A. Ellis, R. Broferick, Integrated distribution planning concept paper: a proactive approach for accommodating high penetrations of distributed generation resources (2013)
136. K.A. Loparo, G.L. Blankenship, A probabilistic mechanism for small disturbance instabilities in electric power systems. IEEE Trans. Circuits Syst. **CAS-32**(2), 177–185 (1985)
137. J. Lu, C.W. Liu, J.S. Thorp, New methods for computing a saddle-node bifurcation point for voltage stability analysis. IEEE Trans. Power Syst. **10**(2), 978–985 (1995)
138. A.G.J. MacFarlane, Formulation of the state-space equations for nonlinear networks. Int. J. Control **5**(2), 145–161 (1967)
139. A.G.J. MacFarlane, An integral invariant formulation of a canonical equation set for non-linear electrical networks. Int. J. Control **11**(3), 449–470 (1970)
140. P.C. Magnusson, Transient energy method of calculating stability. Trans. Am. Inst. Electr. Eng. **66**, 747–755 (1947)
141. J.E. Marsden, M. McCracken, *The Hopf Bifurcation and its Applications* (Springer, New York, 1976)
142. J.E. Marsden, *Introduction to Mechanics and Symmetry*, 2nd edn., Texts in Applied Mathematics (Springer, New York, 1998)
143. J.C. Maxwell, On governors. Proc. Roy. Soc. (100) (1868)
144. T. McCoy, J. Zgliczynski, N.W. Johanson, F.A. Puhn, T.W. Martin, Hybrid electric drive for ddg-51 class destroyers. Nav. Eng. J. **2**, 83–91 (2007)

145. K. McKinnon, H. Williams, Constructing integer programming models by the predicate calculus. Ann. Oper. Res. **21**, 227–246 (1989)
146. D. McMullen, T. Dalton, Hybrid electric drive—enhancing energy security. in *Maritime Systems and Technologies (MAST) Americas Conference* (Washington, 2011)
147. A.I. Mees, L. Chua, The hopf bifurcation theorem and its applications to nonlinear oscillations in circuits and systems. IEEE Trans. Circuits Syst. **26**, 235254 (1979)
148. L. Meirovitch, *Methods of Analytical Dynamics* (McGrawHill Inc, New York, 1970)
149. A.N. Michel, A.A. Fouad, V. Vittal, Power system transient stability using individual machine energy functions. IEEE Trans. Circuits Syst. **CAS-30**(5), 266–276 (1983)
150. M.M. Milic, L.A. Novak, Formulation of equations in terms of scalar functions for lumped non-linear networks. Circuit Theory Appl. **9**, 15–32 (1981)
151. W. Millar, Some general theorems for non-linear systems possessing resistance. Philos. Mag. **42**(333), 1150–1160 (1951)
152. J. Moiola, G. Chen, Computations of limit cycles via higher order harmonic balance approximations. IEEE Trans. Autom. Control **38**(5), 782790 (1993)
153. J. Momoh, *Smart Grid: Fundamentals of Design and Analysis* (Wiley, Hobocken, 2012)
154. N. Narasimhamurthi, On the existence of energy functions for power systems with transmission losses. IEEE Trans. Circuits Syst. **CAS-31**(2), 199–203 (1984)
155. J.I. Neimark, N.A. Fufaev, *Dynamics of Nonholonomic Systems*, volume 33 of *Translations of Mathematical Monographs* (American Mathematical Society, Providence, 1972)
156. N.S. Nice, *Control Systems Engineering. Control Systems Technology*, 7th edn. (Wiley, New York, 2015)
157. B. Noble, M.J. Sewell, On dual exremum principles in applied mathematics. J. Inst. Math. Appl. **9**, 123–193 (1972)
158. E. Noether, Invariant variation problems. Transp. Theory Stat. Phys. **1**(3), 186–207 (1971). Translater: Mort Tavel
159. J.T. Oden, J.N. Reddy, *Variational Methods in Theoretical Mechanics* (Universitext. Springer, Berlin, 2012)
160. Presidents Council of Economic Advisors, U.S. Department of Energys Office of Electricity Delivery, Energy Reliability, and White House Office of Science and Technology. Economic benefits of iincreasing electric grid resilence to weather outages, 2013
161. H. Ohtsuki, A. Yokoyama, Y. Sekine, Reverse action on-load tap changer in association with voltage collapse. IEEE Trans. Power Syst. **6**(1), 300–306 (1991)
162. L. Page, *Introduction to Theoretical Physics* (D. Van Nostrand Company, Princeton, 1952)
163. M.A. Pai, *Power System Stability Analysis by Direct Method of Lyapunov* (North Holland, New York, 1981)
164. M.K. Pal, Voltage stability: analysis needs, modelling requirement, and modelling adequacy. IEE Proc. C **140**(4), 279–286 (1993)
165. H.M. Paynter, *Analysis and Design of Engineering Systems* (MIT Press, Cambridge, 1961)
166. C. Rajagopalan, P.W. Sauer, M.A. Pai, An integrated approach to dynamic and static voltage stability. in *American Control Conference*, pp 1231–1236 (1989)
167. S. Ramo, J.R. Whinnery, *Fields and Waves in Modern Radio* (Wiley, New York, 1964)
168. R. Riaza, *Differential-Algebraic Systems: Analytical Aspects and Circuit Applications* (World Scientific Publishing Co., Singapore, 2008)
169. D. Roose, An algorithm for computing hopf bifurcations in comparison with other methods. J. Comput. Appl. Math. **12-13**, 517–529 (1985)
170. D. Roose, V. Hlavacek, A direct method for the computation of hopf bifurcation points. SIAM J. Appl. Math. **45**(6), 879–894 (1985)
171. C.W. Ross, Error adaptive control computer for interconnected power systems. IEEE Trans. Power Appar. Syst. **PAS-85**, 742–749 (1966)
172. D.W. Ross, S. Kim, Dynamic economic dispatch of generation. IEEE Trans. Power Appar. Syst. **99**(6), 2060–2068 (1980)
173. R.M. Santilli, *Foundations of Theoretical Mechanics I: The Inverse Problem in Newtonian Mechanics* (Springer, New York, 1978)

174. A.J. Van der Schaft, H. Schumacher, *An Introduction to Hybrid Dynamical Systems*, Lecture Notes in Control and Information Sciences (Springer, London, 2000)
175. National Science and Technology Council. A policy framework for the 21st century grid: Enabling our secure energy future, 2011
176. R. Seydel, Numerical computation of branch points in nonlinear equations. Numer. Math. **33**, 339–352 (1979)
177. K.S. Sibirsky, *Introduction to Topological Dynamics* (Noordhoff International Publishing, Leyden, 1975)
178. F.H.J.R. Silva, L.F.C. Albert, J.B.A. London, N.G. Bretas, Smooth perturbation on a classical energy function for lossy power system stability analysis. IEEE Trans. Circuits Syst. **52**(1), 222–229 (2005)
179. D.B. Strukov, G.S. Snider, D.R. Stewart, R.S. Williams, The missing memristor found. Nature **453**(May), 80–83 (2008)
180. A. Szatkowski, Remarks on 'explicit topological formulation of lagrangian and hamiltonian equations for nonlinear networks'. IEEE Trans. Circuits Syst. **CAS-26**(May), 350–369 (1979)
181. Y. Tamura, H. Mori, S. Iwamoto, Relationship between voltage instability and multiple load flow solutions in electric power systems. IEEE Trans. Power Appar. Syst. **PAS-102**(5), 1115–1125 (1983)
182. O. Taussky, Positive definite matrices and their role in the study of the characteristic roots of general matrices. Adv. Math. **2**, 175–186 (1963)
183. C.J. Tavora, O.J.M. Smith, Equilibrium analysis of power systems. IEEE Trans. Power Appl. Syst. **PAS-91**, 1131–1137 (1972)
184. C.J. Tavora, O.J.M. Smith, Stability analysis of power systems. IEEE Trans. Power Appar. Syst. **PAS-91**, 1138–1144 (1972)
185. J.M.T. Thomson, G.W. Hunt, *Elastic Instability Phenomena* (Wiley, New York, 1984)
186. J.A. Thorpe. *Elementary Topics in Differential Geometry. Undergraduate Texts in Mathematics* (Springer, New York, 1979)
187. F.D. Torrisi, A. Bemporad, Hysdel-a tool for generating computational hybrid models for analysis and synthesis problems. IEEE Trans. Control Syst. Technol. **12**(2), 235–249 (2004)
188. N.A. Tsolas, A. Arapostathis, P.P. Varaiya, A structure preserving energy function for power system stability analysis. IEEE Trans. Circuits Syst. **CAS-32**(10), 1041–1049 (1985)
189. M.L. Tyler, M. Morari, Propositional logic in control and monitoring problems. Automatica **35**(4), 565–582 (1999)
190. J.E. van Ness, F.M. Brash Jr, G.L. Landgren, S.I. Neuman, Analytic investigation of dynamic instability occuring at powerton station. IEEE Trans. Power Appar. Syst. **PAS-99**, 1386–1395 (1980)
191. P.P. Varaiya, F.F. Wu, R.-L. Chen, Direct methods for transient stability analysis of power systems: recent results. Proc. IEEE **73** (1985)
192. V.A. Venikov, V.A. Stroev, V.I. Idelchick, V.I. Tarasov, Estimation of power system steadystate stability in load flow calculations. IEEE Trans. Power Appar. Syst. **PAS-94**(3), 1034–1038 (1975)
193. V. Venkatasubramanian, H. Schattler, J. Zaborsky, A taxonomy of the dynamics of the large power system with emphasis on its voltage stability, in *Bulk Power System Voltage Phenomena II: Voltage Stability and Security*, ed. by L.H. Fink (Deep Creek Lake, 1991), pp. 9–44
194. V. Venkatasubramanian, H. Schattler, J. Zaborsky, Voltage dynamics: study of a generator with voltage control, transmission, and matched mw load. IEEE Trans. Autom. Control **37**(11), 1717–1733 (1992)
195. B.D. Vujanovic, Conservation laws of dynamical systems via d'alembert's principle. Int. J. Nonlinear Mech. **13**, 185–197 (1978)
196. J.A. Walker, W.E. Schmitendorf, A simple test for asymptotic stability in partially dissipative symmetric systems. ASME J. Appl. Mech. **40**(4), 1120–1121 (1973)
197. H.O. Wang, E.H. Abed, A.M.A. Hamden, Bifurcation, chaos, and crisis in voltage collapse of a model power system. IEEE Trans. Circuits Syst. I: Fundam. Theory Appl. **41**(3), 294–302 (1994)

198. J.L. Willems, A partial stability approach to the problem of power system stability. Int. J. Control **19**, 1–4 (1974)
199. H.P. Williams. *Model Building in Mathematical Programming* (Wiley, 1993)
200. W.M. Wonham, *Linear Multivariable Control: A Geometric Approach*, 3rd edn. (Springer, New York, 1985)
201. J.L. Wyatt, L.O. Chua, J.W. Gannett, I.C. Goknar, D.N. Green, Energy concepts in state-space theory of nonlinear n-ports: Part i—passivity. IEEE Trans. Circuits Syst. **CAS-31**(1), 48–61 (1981)
202. X. Xia, A.M. Elaiw, Optimal dynamic economic dispatch: a review. Electr. Power Syst. Res. **80**, 975–986 (2010)
203. X.M. Yu. *Stability and Bifurcation of Equilibria in Electric Power Networks*. Ph.d. thesis, Drexel University, 1991

Index

H.G. Kwatny and K. Miu-Miller, *Power System Dynamics and Control*,
Control Engineering, DOI 10.1007/978-0-8176-4674-5

Printed in the United States
By Bookmasters